TRANSFORM CIRCUIT ANALYSIS FOR ENGINEERING AND TECHNOLOGY

SECOND EDITION

William D. Stanley

Eminent Professor
College of Engineering and Technology
Old Dominion University
Norfolk, Virginia

PRENTICE HALL, *Englewood Cliffs, New Jersey 07632*

Library of Congress Cataloging-in-Publication Data

Stanley, William D.
    Transform circuit analysis for engineering and technology /
William D. Stanley. —2nd ed.
        p.   cm.
    Includes index.
    ISBN 0-13-928896-1
    1. Electric circuit analysis.   2. Transformations (Mathematics)
I. Title.
TK454.S7 1989
621.319′2—dc19   88-12433
                                                       CIP

Editorial/production supervision
and interior design: Anne Kenney
Cover design: Wanda Lubelska
Manufacturing buyer: Robert Anderson

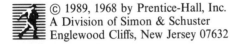 © 1989, 1968 by Prentice-Hall, Inc.
A Division of Simon & Schuster
Englewood Cliffs, New Jersey 07632

Printed in the United States of America

10   9   8   7   6   5   4   3   2   1

ISBN   0-13-928896-1

PRENTICE-HALL INTERNATIONAL (UK) LIMITED, *London*
PRENTICE-HALL OF AUSTRALIA PTY. LIMITED, *Sydney*
PRENTICE-HALL CANADA INC., *Toronto*
PRENTICE-HALL HISPANOAMERICANA, S.A., *Mexico*
PRENTICE-HALL OF INDIA PRIVATE LIMITED, *New Delhi*
PRENTICE-HALL OF JAPAN, INC., *Tokyo*
SIMON & SCHUSTER ASIA PTE. LTD., *Singapore*
EDITORA PRENTICE-HALL DO BRASIL, LTDA., *Rio de Janeiro*

To
Mary Lou

CONTENTS

CHAPTER 3 CIRCUIT PARAMETERS   63

CHAPTER 4 THE BASIC TIME-DOMAIN CIRCUIT   98

PREFACE

Approximately twenty years have passed between the publication of the original first edition and the publication of this second edition. Yet the first edition was still being used at the time that preparation of the second edition was completed. The long-term usage of the first edition must indicate that the presentation was sufficiently basic and general to avoid rapid obsolescence. That philosophy has continued in the preparation of the second edition.

Virtually all the successful features of the first edition have been retained. A large number of sections in which the material has not become dated, and in which the writing style was judged to be appropriate, have been retained without change. The titles of the original eight chapters have also been retained without change. However, some important changes and additions have been made, of which some of the most significant are the following:

1. A new chapter on Fourier analysis (Chapter 9) has been added.
2. A new chapter on discrete-time systems (Chapter 10) has been added.
3. Most of the original problems have either been replaced by new ones or revised to be more appropriate.
4. Most chapters include computer problems, in which students can be assigned specific programs to prepare with a computer or programmable calculator.
5. Problems are divided into three categories: General Problems, Derivation Problems, and Computer Problems.
6. Each chapter begins with an overview and a detailed list of chapter objectives.
7. Answers to most odd-numbered General Problems are included at the back of the book.
8. A section on convolution (Chapter 7) has been added.

9. A section on block diagram algebra (Chapter 7) has been added.

10. Dependent sources are introduced (Chapter 1).

As in the case of the first edition, the second edition was designed for an advanced undergraduate circuit analysis course in an applied engineering curriculum or in an upper-division engineering technology curriculum. It is also expected that the book could serve as a self-study reference for engineers and technologists.

The reader should be familiar with the fundamentals of differential and integral calculus and with basic dc circuit analysis techniques. It is anticipated that most readers will also be familiar with steady-state ac circuit theory. However, the latter condition is not a necessity since the major portion of the book may be mastered without a background in ac circuits.

The first four chapters of the book are devoted to time-domain considerations. Chapter 1 is an introduction to the general philosophy of the book. The fundamentals of waveform analysis are presented in Chapter 2. The reader should find that his or her knowledge of differentiation and integration will be strengthened after mastering this chapter, particularly in regard to graphical techniques. The voltage–current relationships for each of the basic circuit elements are explained in Chapter 3 and developed fully in Chapter 4.

The next four chapters are devoted to transform-domain considerations. Following a detailed development of the Laplace transform and inverse transform in Chapter 5, the use of transform techniques in obtaining complete circuit responses is presented in Chapter 6. In Chapter 7, the emphasis shifts to the system concepts of circuit theory. Among the topics considered are transfer functions, impedance functions, convolution, and stability. In Chapter 8, sinusoidal steady-state techniques are developed and compared with Laplace transform techniques. The frequency response concept is developed, and the use of pole–zero methods for obtaining frequency response plots is explored.

Chapter 9 deals with Fourier analysis and the concept of a spectrum. Both the Fourier series and the Fourier transform are covered.

Chapter 10 provides an introduction to discrete-time systems. Topics covered include sampled signals, the sampling theorem, difference equations, and the z-transform. This chapter represents a modern and timely supplement to the continuous-time material predominant through the remainder of the book.

Among the topics presented in the appendices are complex algebra, normalization, and several proofs of transform theorems. These topics may be considered at the discretion of the instructor whenever desirable.

Depending on the nature of the course and the background of the students, it is felt that there is reasonable latitude available on the depth and rigor at which the material in this book can be presented. On one hand, the derivation and formulation of the principles involved can be emphasized. On the other hand, the use of principles as "tools" for solving and interpreting practical problems can be emphasized with only casual consideration of the mathematical "fine points." Any suitable compromise between these limits should be possible.

The completion of this second edition brings back special memories for me. The original edition was my first book, and it was mostly written when I was in my late twenties. Many changes have occurred in my career since that time, including the completion of six more books. However, this book was the one that represented the start of my book-writing activities, and it will always hold a special place in my heart. I owe much appreciation to a Prentice-Hall representative, Harry Gaines, who encouraged me to prepare the original manuscript and who provided much needed support in the early stages of that all-important first draft.

As always, I owe much appreciation to Mrs. Estelle B. Walker, who typed all the new portions of the second edition.

William D. Stanley
Norfolk, Virginia

1

INTRODUCTORY CONSIDERATIONS

OVERVIEW
Chapter 1 is by far the shortest chapter in the book. Unlike all other chapters, it has no example problems or problems for the reader. The purpose of this brief chapter is to introduce a few definitions and conventions that are fundamental in establishing the approach for all subsequent chapters. In addition, a list of the concepts for which the reader should be familiar for further study will be given.

OBJECTIVES
After completing this chapter, the reader should be able to

1. State the three types of circuit parameters and show their schematic symbols.
2. State the two forms of ideal electrical energy sources and show their schematic symbols.
3. Show the two forms for representing an actual source with internal resistance.
4. Discuss the terms *excitation* and *response*.
5. Discuss the difference between *circuit analysis* and *circuit synthesis*.
6. Define the four types of *dependent* sources and show their schematic forms.

1-1 CIRCUIT ELEMENTS

A linear electric consists of some combination of *three* types of passive circuit *parameters* and *two* types of *energy sources*. The three types of circuit parameters are (a) *resistance*, (b) *inductance*, and (c) *capacitance*. The schematic representations of the three circuit parameter components are shown in Fig. 1-1. An extensive treatment of these parameters is presented in Chapter 3.

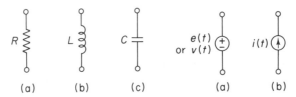

Figure 1-1 Circuit parameters:
(a) resistance; (b) inductance;
(c) capacitance.

Figure 1-2 Ideal energy
sources: (a) voltage
(b) current.

The two ideal types of energy sources are (a) the ideal *voltage source* and (b) the ideal *current source*. The schematic representations of the ideal sources are shown in Fig. 1-2. The hypothetical ideal voltage source is assumed to maintain the same voltage across its terminals regardless of the load. The hypothetical ideal current source is assumed to deliver the same current regardless of the load.

No actual electrical energy source quite fits the ideal models, although many are sufficiently close that they may be approximated by ideal sources under many conditions. An actual energy source will neither maintain a constant voltage across its terminals nor deliver a constant current from its terminals for all values of load. The actual behavior of such sources can usually be described by a model composed of a hypothetical ideal source and one or more passive circuit parameters. The effect of the passive elements is to control the voltage or current available to the remainder of the circuit in essentially the same manner as the terminals of the actual sources.

In a wide variety of cases, a single resistance in conjunction with an ideal energy source is sufficient to characterize the external behavior of the actual source. The two possible forms for this case are shown in Fig. 1-3. These forms are: (a) an ideal voltage source in series with a resistance and (b) an ideal current source in parallel with a resistance. In general, a given source may be represented by either form.

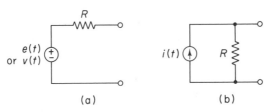

Figure 1-3 Two models for real energy
sources.

1-2 EXCITATION AND RESPONSE

Assume that a given circuit is excited by one or more voltage and/or current sources. Throughout the circuit, currents will begin to flow through passive elements, and voltages will appear across them. In general, all voltages and currents will vary with *time* in some fashion. Letting t represent time, we can assert that, in general, all voltages and currents in the circuit are functions of the independent variable t. In most cases in this text, we will use the symbol e to represent a voltage source and the symbol v to represent a voltage drop across a passive element. The symbol i will be used to represent a current without regard to whether it is a source or the current through a passive element. For special reasons there are a few exceptions to this notation, so it should not be interpreted as ironclad.

To emphasize that voltages and currents are functions of the variable t, we will often write these quantities in the forms $e(t)$, $v(t)$, and $i(t)$. The quantity in the parentheses is called the *argument* of the voltage or current which, of course, is the variable t for the quantities just stated. In an algebraic expression involving several operations, one must be careful not to confuse this notation with multiplication which is also expressed in some cases by parentheses.

The voltage and/or current sources that excite a circuit are called the *excitations*, and the resulting voltages and currents associated with passive components are called the *responses*. In many applications, the excitations are regarded as the *inputs*, and the responses are regarded as the *outputs*.

1-3 ANALYSIS AND SYNTHESIS

The study of circuit theory can be divided into two separate areas: (a) *circuit analysis* and (b) *circuit synthesis*. The main topics in this text will be devoted to circuit analysis. The basic theme of circuit analysis is as follows: If the voltage and current source excitations and the network are known, determine the voltage and current responses within the network. Frequently, the analysis theme is slightly modified in that perhaps a source or a component must be determined, but a basic underlying property of analysis problems is that the circuit configuration is known.

The basic theme of circuit synthesis is as follows: If we are given the response and the excitation, or a relationship between the response and the excitation, determine the circuit. In essence, the principles of circuit synthesis permit one to determine a circuit that will provide a prescribed processing on a given signal. Practical circuit design differs from the science of network synthesis in the sense that actual circuit design frequently utilizes the concepts of both analysis and synthesis (along with engineering economics) to accomplish its goal.

1-4 DEPENDENT SOURCES

The voltage and current source models introduced in Section 1-1 represent the common *independent* source models. Independent sources are those whose values are independent of the levels of any voltages or currents in the circuit. The values of all voltage and current sources encountered thus far have either been independent constant values (corresponding to dc sources) or independent time-varying quantities.

Dependent (or controlled) source models are a special class in which the voltage or current is dependent on some other voltage or current within the circuit. Such sources arise from complex physical interactions in many electronic and electrical devices. For example, the base current of a bipolar junction transistor causes a much larger collector current to flow, and this process is modeled by a dependent source.

A dependent source, like an independent source, may either be a voltage source or a current source. However, it may also be controlled by either a voltage or a

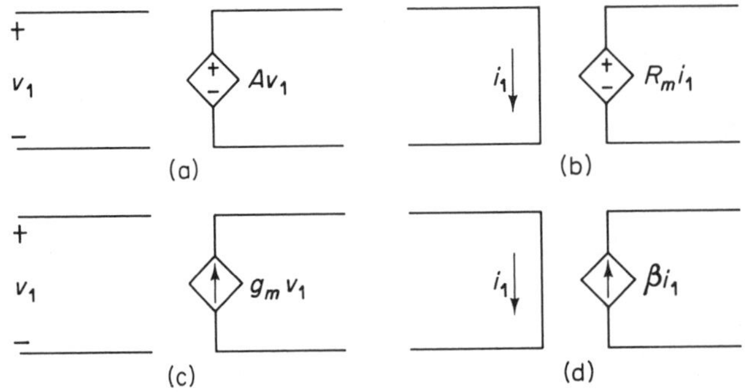

Figure 1-4 Four possible models of ideal dependent (or controlled) sources.

current. This results in four different combinations for dependent source models, and these are illustrated in Fig. 1-4. The symbol used for dependent source models is in accordance with an increasing portion of the literature, although earlier texts employ the same symbol (circle) as for independent source models. Each of the four forms will now be discussed briefly.

Voltage-Controlled Voltage Source

The model for the *voltage-controlled voltage source* (*VCVS*) is shown in Fig. 1-4a. An independent control voltage v_1 is assumed to exist across certain control terminals. This voltage controls a dependent voltage whose value is Av_1. The quantity A is dimensionless and corresponds to a voltage gain when the model represents a voltage amplifier.

Current-Controlled Voltage Source

The model for the *current-controlled voltage source* (*ICVS*) is shown in Fig. 1-4b. An independent control current i_1 is assumed to be flowing in certain control terminals. This current controls a dependent voltage whose value is $R_m i_1$. The quantity R_m, relating the dependent voltage to the controlling current, has the dimensions of ohms and is called the *transresistance*.

Voltage-Controlled Current Source

The model for the *voltage-controlled current source* (*VCIS*) is shown in Fig. 1-4c. An independent control voltage v_1 is assumed to exist across certain control terminals. This voltage controls a dependent current whose value is $g_m v_1$. The quantity g_m, relating the dependent current to the controlling voltage, has the dimensions of siemens and is called the *transconductance*.

Current-Controlled Current Source

The model for the *current-controlled current source* (*ICIS*) is shown in Fig. 1-4d. An independent control current i_1 is assumed to be flowing in certain control terminals. This current controls a dependent current whose value is βi_1. The quantity β is dimensionless and corresponds to a current gain when the model represents a current amplifier.

1-5 OUTLINE OF CIRCUIT ANALYSIS

The study of basic linear circuit analysis may be arbitrarily classified in the following four categories:

1. Steady-state dc resistive circuit analysis
2. Steady-state sinusoidal ac circuit analysis
3. General linear circuit analysis from the classical differential equation approach
4. General linear circuit analysis from the Laplace transform approach

The steady-state responses of a linear circuit excited by dc sources are all constants, independent of time, and the main parameter of interest is usually resistance since the primary effects of inductances and capacitances disappear in this case. The study of electric circuits usually begins with a treatment of dc resistive circuits.

If a linear circuit is excited only by sinusoidal excitations, all steady-state responses in the circuit will also be sinusoidal. An elegant body of theory has been built around this concept, and a number of texts are devoted solely to this topic. In a sense, category 1 may be regarded as a special case of category 2, in which the frequency of dc is considered to be zero.

Categories 3 and 4 actually form the most general approaches for linear circuit analysis since they apply to a circuit with arbitrary excitations, and the responses obtained by using these methods are complete solutions containing both *transient* and *steady-state* responses. (The concepts of transient and steady state will be explained in Chapter 4.) Actually, categories 3 and 4 both accomplish the same general task. However, the Laplace transform approach is very popular for dealing with electric circuits, due primarily to the following reasons: (a) one can learn to apply transform methods to solve circuit problems using only algebraic manipulations in conjunction with tables; (b) the procedure governing transform methods may be readily developed as an extension of the methods of steady-state analysis; and (c) the use of transform methods permits the engineering analyst to simplify, analyze, and interpret problems of greater complexity than with the differential equation approach.

From the title of this book it is evident that our primary goal is the development of transform analysis techniques. However, it would be a serious mistake to study this method without first developing a suitable background to help the reader understand the overall perspective.

First, it is assumed that the reader has a reasonable background in dc resistive circuit analysis. It is desirable that the reader have a working knowledge of at least the following basic concepts for the dc case:

1. Ohm's law
2. Kirchhoff's voltage law (including mesh current equations)

3. Kirchhoff's current law (including node voltage equations)

4. Thévenin's theorem

5. Norton's theorem

6. Determination of the equivalent resistance of a passive network

In addition, it is felt that most readers of this book will also have a background in steady-state ac circuit theory using complex algebra; such a background will enhance the understanding of the transform method. However, the latter desirability is not an absolute necessity since Appendix A is devoted to complex algebra, and a basic treatment of steady-state ac concepts is developed along with the transform method within the text.

As far as the classical differential equation approach is concerned, enough is included to provide the reader with a basic understanding of this approach. In particular, the reader will be taught to express and interpret the differential equation relationships for circuit components and networks, as such concepts are employed in the development of the transform approach.

2

WAVEFORM ANALYSIS

OVERVIEW

Before developing a detailed treatment of transient circuit analysis, it is necessary to understand the mathematical forms of the waveforms that are used in describing the phenomena. Virtually all voltage and current waveforms that occur in network analysis can be described in terms of a few basic mathematical functions. These functions are investigated in detail in this chapter and will be used to describe voltage and current waveforms throughout the remainder of the book.

OBJECTIVES

After completing this chapter, the reader should be able to

1. State the units and abbreviations for the most common circuit quantities.
2. Sketch a *step function* and express its mathematical form.
3. Sketch the form of a *switched function* and express its mathematical form.
4. Sketch a *ramp function* and express its mathematical form.
5. State the derivative and integral relationships between *step*, *ramp*, and *parabolic functions*.
6. Sketch the *exponential function* and express its mathematical form.
7. Define the *damping constant* and the *time constant* of an exponential function, and state the relationship between them.
8. Sketch the *sinusoidal function* and express its mathematical form.
9. Define *angular frequency*, *repetition frequency*, *phase*, *angle*, and *amplitude* of a sine wave.
10. State the relationship between the *frequency* and *period* of a sine wave.
11. Draw a diagram showing the relative phase shifts between the different forms of sine and cosine functions.

12. **Express sine or cosine function with any phase angle in terms of the different forms of sine and cosine functions.**

13. **Add any number of sinusoids of the same frequency, but with arbitrary phase angles, to obtain a single sinusoid at the same frequency.**

14. **Sketch the *damped sinusoid function* and express its mathematical form.**

15. **Sketch the form of a *shifted function* and express its mathematical form.**

16. **Express various complex waveforms in terms of simple waveforms such as steps, ramps, and sinusoids.**

17. **Define the concept of an *impulse function* and express its mathematical form.**

18. **State the mathematical relationships between step and impulse functions.**

19. **Define the condition for a function to be *periodic* and sketch a representative periodic function.**

20. **Determine the *average* or *dc* value of a periodic waveform and discuss its significance.**

21. **Determine the *rms* or *effective* value of a periodic waveform and discuss its significance.**

22. **Determine the average power dissipated in a resistor by a periodic voltage or current.**

2-1 NOTATION AND UNITS

Most of the mathematical functions presented in this chapter are considered to be functions of the independent variable time. Whenever the framework is that of a general waveform without regard to either voltage or current, general symbols such as $f(t)$, $g(t)$, and $h(t)$ will be employed. As stated in Chapter 1, the argument (t) identifies such waveforms as functions of time. Whenever a waveform is to be used to represent a voltage or current, specific symbols like $v(t)$, $e(t)$, or $i(t)$ will be employed. As a general rule, lowercase letters will be used to specify time-dependent quantities. Occasionally, the argument (t) will be omitted, but the presence of the lowercase letter will still identify the quantity as a possible time-dependent function.

The use of uppercase letters for voltages and currents will be reserved for fixed quantities such as peak and average values, and for steady-state phasors and transforms. More will be said about phasors and transforms later.

The primary quantities of interest in this text, along with their symbols, basic units, and abbreviations for units, are summarized in Table 2-1. Prefixes and their abbreviations are summarized in Table 2-2.

For simplicity in notation and expression, we will frequently omit units from the expressions for complicated waveform functions and circuit diagrams. *The absence of specific units in a given problem will be understood to mean that all quantities involved are expressed in their basic units given in Table 2-1.* However, when a quantity requires a prefix from Table 2-2, it will be necessary to explicitly specify the required unit. A general waveform such as $f(t)$ will usually not require units to be specified.

TABLE 2-1   Most common quantities used in electrical circuit analysis

Quantity	Symbols	Unit	Abbreviation of unit
Time	t	second	s
Energy	w, W	joule	J
Power	p, P	watt	W
Charge	q, Q	coulomb	C
Current	i, I	ampere	A
Voltage	v, V, e, E	volt	V
Resistance	R	ohm	Ω
Conductance	G	siemens	S
Inductance	L	henry	H
Capacitance	C	farad	F
Impedance	$\bar{Z}$	ohm	Ω
Reactance	X	ohm	Ω
Admittance	$\bar{Y}$	siemens	S
Susceptance	B	siemens	S
Frequency (cyclic)	f	hertz	Hz
Frequency (radian)	ω	radians/second	rad/s

TABLE 2-2   Most common prefixes used in electrical systems

Value	Prefix	Abbreviation
10^{-18}	atto	a
10^{-15}	femto	f
10^{-12}	pico	p
10^{-9}	nano	n
10^{-6}	micro	μ
10^{-3}	milli	m
10^{3}	kilo	k
10^{6}	mega	M
10^{9}	giga	G
10^{12}	tera	T

When in doubt about handling prefixed units, the reader should always convert such units to their most basic form.

2-2 STEP FUNCTION

In the solution of general transient problems, it is often necessary to specify a particular value of the independent variable time at which the excitation is applied to the circuit. Since this is usually arbitrary, the time $t = 0$ is often chosen for convenience. Suppose then that an excitation is a simple dc voltage or current that is assumed to be switched into a circuit at $t = 0$. The switching process can be described mathematically by defining a special function called the *step function*.

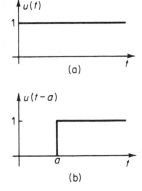

Figure 2-1 Step functions.

The basic *unit step function* u(t) can be defined as follows:

$$u(t) = \begin{cases} 0 & \text{for } t < 0 \\ 1 & \text{for } t > 0 \end{cases} \qquad (2\text{-}1)$$

This function is shown in Fig. 2-1a. If the step function is used to represent a voltage of E volts or a current of I amperes, the step function is multiplied by the appropriate constant. Thus the function $Eu(t)$ represents a dc voltage of value E switched on at $t = 0$, and $Iu(t)$ represents a dc current of value I switched on at $t = 0$.

In many complex transient problems, a source may be switched on at some initial time, while at some later time, a delayed source may be switched into the circuit. A delayed dc function may be represented by means of the *shifted* or *delayed unit step function*. The shifted unit step function is denoted by $u(t - a)$, and it is defined by

$$u(t - a) = \begin{cases} 0 & \text{for } t < a \\ 1 & \text{for } t > a \end{cases} \qquad (2\text{-}2)$$

This function is shown in Fig. 2-1b.

Occasionally, the argument of the step function may be more complex than just t or $t - a$. In general, let x represent an arbitrary function of time. Then $u(x)$ is defined by

$$u(x) = \begin{cases} 0 & \text{for } x < 0 \\ 1 & \text{for } x > 0 \end{cases} \qquad (2\text{-}3)$$

Thus Eqs. (2–1) and (2–2) may be considered to be special cases of Eq. (2–3).

Example 2-1

A dc current of 25 mA is switched on at $t = 5$ ms. Write an expression for the current $i(t)$.

        Solution   Working with the basic units of amperes and seconds, we have

$$i(t) = 0.025u(t - 0.005) \qquad (2\text{-}4)$$

2-3 ELEMENTARY SWITCHING FUNCTION

We saw in Section 2-2 how the unit step function concept allows us to mathematically describe the process of switching on a dc waveform at any desired time. We now wish to consider the problem of mathematically describing the switching operation for more complex waveforms.

In this section we consider only mathematical functions switched on at $t = 0$. The extension to general delayed functions will be made in a later section after further study of the basic circuit waveforms.

Figure 2-2a illustrates some general waveform $f(t)$ that is defined for all time. We wish to describe mathematically the process of forcing the portion of $f(t)$, for $t < 0$, to be zero while maintaining $f(t)$, for $t > 0$, to be unchanged. We will let $f_s(t)$ represent the switched waveform. One obvious way to describe the desired function, which is shown in Fig. 2-2c, is to write the two separate equations:

$$f_s(t) = \begin{cases} 0 & \text{for } t < 0 \\ f(t) & \text{for } t > 0 \end{cases} \tag{2-5}$$

However, we can observe that if the unit step function of Fig. 2-2b is multiplied by $f(t)$ in Fig. 2-2a, the same resulting function is obtained. The important point is: *Multiplication of a continuous function of time, $f(t)$, by the unit step function, $u(t)$, forces the resulting function to be zero for $t < 0$ and produces the original $f(t)$ for $t > 0$.* In this case, we can write

$$f_s(t) = f(t)u(t) \tag{2-6}$$

In many problems, all the excitations will be applied to the circuit at the same instant of time. Usually, this reference time is chosen as $t = 0$. In such cases, it is usually obvious that we are not interested in the behavior of the resulting mathematical expressions for negative time. Consequently, it is not always necessary to carry along the $u(t)$ factor as it amounts to "extra baggage." *In problems where all*

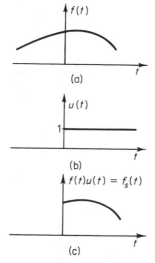

(a)

(b)

(c)

Figure 2-2 Switching function concept.

functions in the solution begin at t = 0, we will often omit the u(t) factor as it is inter-preted to be understood in these cases.

The reason for introducing the switching function concept at this point is to acquaint the reader with the basic idea for proper interpretation in subsequent sections. In later sections we develop the concept of the general switching function for use in problems where several functions are switched into a circuit at various times. In such cases it is necessary to keep track of the times at which various responses begin, and thus the step function factors will be necessary.

Example 2-2

A certain voltage source is expressed by the equation

$$e(t) = 2t + 1 \qquad (2\text{-}7)$$

The voltage is switched into a circuit at $t = 0$. Write an expression for the switched voltage.

Solution   The function is shown in Fig. 2-3a as it would appear for both positive and negative time. However, as far as the circuit is concerned, the portion of the function for $t < 0$ does not exist. The actual waveform seen by the circuit is shown in Fig. 2-3b. Letting $e_s(t)$ represent this function, we have

$$e_s(t) = e(t)u(t) = (2t + 1)u(t) \qquad (2\text{-}8)$$

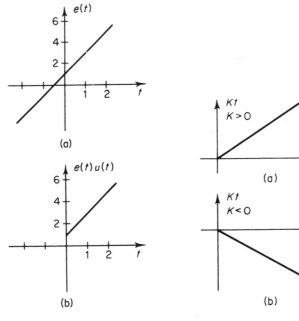

Figure 2-3   Voltage of Example 2-2.   **Figure 2-4**   Ramp functions.

2-4 RAMP FUNCTION

The basic *ramp function* is a straight line beginning at the origin and increasing (or decreasing) linearly with time as shown in Fig. 2-4. No special symbol is necessary for the ramp, but for reference in this section, we will simply use $f(t)$ to refer to the ramp. The ramp can then be defined by

$$f(t) = \begin{cases} 0 & \text{for } t < 0 \\ Kt & \text{for } t > 0 \end{cases} \tag{2-9}$$

If desired, we could of course write Eq. (2-9) in the form

$$f(t) = Ktu(t) \tag{2-10}$$

The constant K represents the *slope* of the ramp. The slope of a voltage ramp is measured in volts/second = V/s, while the slope of a current ramp is measured in amperes/second = A/s. If K is positive, the slope is upward as shown in Fig. 2-4a, whereas if K is negative, the slope is downward as shown in Fig. 2-4b. If $K = 1$, the ramp is called a *unit ramp*.

The reason that no special symbol will be assigned to the ramp is the fact that the expression $Ktu(t)$ adequately describes the function. As we develop further results, we shall see that most other basic waveforms can be described by ordinary mathematical equations in combination with the step function. The impulse function to be introduced later will be an exception to this rule.

Example 2-3

A ramp function voltage $e(t)$ with a slope of -3 V/s is switched on at $t = 0$. Write an equation for the function and sketch it.

      Solution   We may write

$$e(t) = -3t \tag{2-11}$$

with the idea understood that we are not interested in the region $t < 0$. We may also write

$$e(t) = -3tu(t) \tag{2-12}$$

The function is shown in Fig. 2-5.

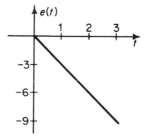

Figure 2-5   Voltage of Example 2-3.

2-5 GENERAL t^n FUNCTIONS

A general set of possible functions, which includes the step and ramp functions as special cases, is given by

$$f_n(t) = \begin{cases} 0 & \text{for } t < 0 \\ \dfrac{Kt^n}{n!} & \text{for } t > 0 \end{cases} \tag{2-13}$$

or equivalently by

$$f_n(t) = \frac{Kt^n}{n!} u(t) \tag{2-14}$$

where n is a positive integer representing the degree of a given function. The quantity $n!$ has been included to provide a convenient correlation between the derivatives and integrals of the general form. Curves for $n = 0$, 1, and 2 are shown in Fig. 2-6 with $K = 1$ in all cases. These three cases will be considered individually.

($n = 0$): *step function.* If $n = 0$, Eq. (2-14) reduces to

$$f_0(t) = Ku(t) \tag{2-15}$$

which is immediately recognized as a step function with magnitude K. (Recall that $0! = 1$.)

($n = 1$): *ramp function.* If $n = 1$, Eq. (2-14) becomes

$$f_1(t) = Ktu(t) \tag{2-16}$$

which is recognized as a ramp function with slope K.

($n = 2$): *parabolic function.* If $n = 2$, Eq. (2-14) becomes

$$f_2(t) = \frac{Kt^2}{2} u(t) \tag{2-17}$$

This function is called the *parabolic function.* It occurs occasionally in circuit solutions, but not nearly as frequently as the step and ramp functions. The higher-order functions for $n > 2$ occur only rarely in circuit solutions and hence require no further consideration at this time.

We now wish to illustrate how the derivatives and integrals of various members of this set are related. First of all we recognize that the $u(t)$ factor poses a new problem in differentiation and integration since we have not defined the derivative

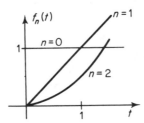

Figure 2-6   Several t^n functions.

and integral of $u(t)$. Later, we will define some precise relationships concerning the derivative or integral of a step function. However, at this time we may accomplish the same purpose by using the form of Eq. (2-13) with the step function factor omitted. Observing Fig. 2-6 and Eq. (2-13), we note that, with the exception of the case $n = 0$ (the basic step function itself), all members of the set are continuous at $t = 0$ and can thus be differentiated by ordinary calculus.

Suppose we start with the parabolic function

$$f_2(t) = \frac{Kt^2}{2} \qquad \text{for } t > 0 \tag{2-18}$$

Now let us differentiate the parabolic function. We have

$$f_2'(t) = Kt \qquad \text{for } t > 0 \tag{2-19}$$

This is immediately recognized to be the ramp function. Hence,

$$f_2'(t) = f_1(t) \tag{2-20}$$

Thus, the slope of the parabolic function at all points is a ramp function. Equivalently, *the derivative of a parabolic function is a ramp function.*

Next, let us differentiate the ramp function. We have

$$f_1'(t) = K \qquad \text{for } t > 0 \tag{2-21}$$

This is immediately recognized to be the step function. Hence

$$f_1'(t) = f_0(t) \tag{2-22}$$

Thus, *the derivative of a ramp function is a step function.*

It can be readily deduced that, in general,

$$f_n'(t) = f_{n-1}(t) \qquad \text{for } n > 0 \tag{2-23}$$

Thus, the derivative of any function belonging to the class of Eq. (2-13), whose degree is higher than the step function, is another function of the same class corresponding to the next lower degree. Notice that Eq. (2-23) excludes the possibility of differentiating the step function, a procedure to be considered in a later section.

Now let us observe the integral relationships existing between the various members of this set. Consider first the step function expressed in the form

$$f_0(t) = K \qquad \text{for } t > 0 \tag{2-24}$$

Now let us integrate Eq. (2-24) from 0 to any arbitrary time t. We have

$$\int_0^t f_0(t)\, dt = \int_0^t K\, dt = Kt \qquad \text{for } t > 0 \tag{2-25}$$

This result is recognized to be the ramp function. Hence

$$\int_0^t f_0(t)\, dt = f_1(t) \tag{2-26}$$

Thus, the area under the step function curve from the origin to time t represents the value of the ramp function at time t. Equivalently, *the integral of a step function is a ramp function.* We will leave as an exercise for the reader to demonstrate that *the integral of the ramp function is a parabolic function.*

It can be deduced that, in general,

$$\int_0^t f_n(t) \, dt = f_{n+1}(t) \tag{2-27}$$

Thus the integral of any function belonging to the class of Eq. (2-13) is another function of the same class corresponding to the next higher degree.

Example 2-4

The block diagram of Fig. 2-7 represents two integrator circuits connected in cascade. The output of the first integrator is given by

$$e_2(t) = \int_0^t e_1(t) \, dt \tag{2-28}$$

The output of the second integrator is given by

$$e_3(t) = \int_0^t e_2(t) \, dt \tag{2-29}$$

A step function, $e_1(t) = 10u(t)$, is applied at the input. Calculate and plot $e_2(t)$ and $e_3(t)$.

       Solution   The input $e_1(t)$ is shown in Fig. 2-8a. The area under this curve from 0 to t represents the value of $e_2(t)$ for any value of t. We have

$$e_2(t) = \int_0^t e_1(t) \, dt = \int_0^t 10 \, dt = 10t \qquad \text{for } t > 0 \tag{2-30}$$

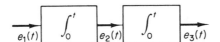

Figure 2-7   Block diagram of integrators for Example 2-4.

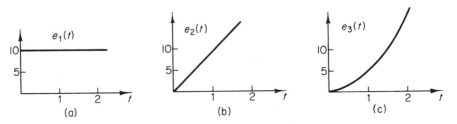

Figure 2-8   Waveforms of Example 2-4.

or equivalently,

$$e_2(t) = 10tu(t) \tag{2-31}$$

which is shown in Fig. 2-8b. For $e_3(t)$ we have

$$e_3(t) = \int_0^t e_2(t)\, dt = \int_0^t 10t\, dt = 5t^2 \qquad \text{for } t > 0 \tag{2-32}$$

or

$$e_3(t) = 5t^2u(t) \tag{2-33}$$

which is shown in Fig. 2-8c.

2-6 EXPONENTIAL FUNCTION

A differential equation that occurs frequently in mathematics and engineering is the equation

$$\frac{dy}{dx} = y \tag{2-34}$$

The basic solution of this equation is called the *exponential function*, and it is usually expressed in the form

$$y = \epsilon^x \tag{2-35}$$

The quantity ϵ is the base of the system of natural logarithms. The value of ϵ to four significant figures is 2.718.

The general shape of the exponential function for both positive and negative x is shown in Fig. 2-9. If desired, we can "rotate" the entire curve around by replacing x by $-x$. The function obtained is

$$y = \epsilon^{-x} \tag{2-36}$$

which is shown in Fig. 2-10.

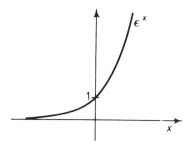

Figure 2-9   Exponential function ϵ^x.

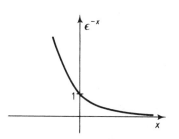

Figure 2-10   Exponential function ϵ^{-x}.

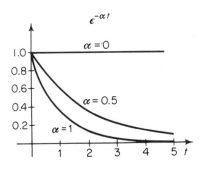

Figure 2-11   Decaying exponential functions.

In most passive network applications the form of Eq. (2-36) and Fig. 2-10 is of primary interest. Let us introduce a positive *damping constant* α and replace x in Eq. (2-36) by αt. Furthermore, we are interested only in the region for $t > 0$. We then define the *decaying exponential* function as

$$f(t) = \begin{cases} 0 & \text{for } t < 0 \\ \epsilon^{-\alpha t} & \text{for } t > 0 \end{cases} \qquad (2\text{-}37)$$

which is equivalent to

$$f(t) = \epsilon^{-\alpha t} u(t) \qquad (2\text{-}38)$$

The function expressed by Eqs. (2-37) and (2-38) is shown in Fig. 2-11 for several possible values of α. Notice that for $\alpha = 0$, there is no damping or decay at all, and the resulting function is simply $f(t) = u(t)$, the unit step function. As α increases, the function diminishes more rapidly. Observe that the decaying exponential function always begins at unity. Furthermore, it always approaches zero for very large time, except for the trivial case when $\alpha = 0$.

Another important form widely used in expressing the decaying exponential function is the expression

$$f(t) = \epsilon^{-t/\tau} \qquad \text{for } t > 0 \qquad (2\text{-}39)$$

This equation is equivalent to Eq. (2-37) if

$$\alpha = \frac{1}{\tau} \qquad (2\text{-}40)$$

The quantity τ is called the *time constant*. In the remainder of the text, we will encounter both forms quite frequently.

Values of the exponential function are well-tabulated in many mathematical handbooks, and virtually all scientific calculators provide this function. In tabular form the decaying exponential form is simply presented in the form ϵ^{-x}. Suppose then that we need to tabulate an exponential whose form is $\epsilon^{-\alpha t}$ or $\epsilon^{-t/\tau}$. To accomplish this goal we let $x = \alpha t$ or t/τ. Knowing α or τ, we first calculate x for different values of t. Then the values of x are used with a calculator. Thus a single exponential curve can be considered to be universal in the sense that it can be used for any damping constant or time constant.

As an example of the numbers involved, let $\alpha = 1$ and $t = 3$. Then $x = 3$ and $\epsilon^{-3} = 0.0498$. Suppose now that we increase the damping by letting $\alpha = 2$. Now

$x = 3$ when $t = 1.5$, producing again $\epsilon^{-3} = 0.0498$. However, the same point on the curve now occurs at an earlier time, indicating that the second function is approaching zero faster.

Except for the trivial case $\alpha = 0$, the decaying exponential function approaches zero as t approaches infinity (∞). Practically speaking, the concept of $t = \infty$ needs some clarification since we cannot wait forever as the pure definition would imply. The engineering concept of $t = \infty$ is simply that it is a sufficiently large time such that some final condition is reached. In the case of the exponential function, this limiting value is zero.

An inspection of the exponential function indicates that for $x = 5$, the decaying exponential function is down to approximately 0.7% of its initial value. Using the time constant form $t/\tau = x = 5$, we note that this point corresponds to $t = 5\tau$. Therefore, a somewhat arbitrary but very useful rule of thumb is that *in a length of time equal to approximately $5\tau = 5/\alpha$, the decaying exponential function has, for most practical purposes, reached its limiting value of zero.*

We now turn our attention to the derivative of the exponential function. As in a previous section we will omit the $u(t)$ factor when differentiating. Consider then

$$f(t) = \epsilon^{-\alpha t} = \epsilon^{-t/\tau} \qquad \text{for } t > 0 \qquad (2\text{-}41)$$

as shown in Fig 2-12a. The possible discontinuity at $t = 0$ poses a problem since the derivative of a discontinuity has not been defined. This subject will be explored in due time, but for now let us "close our eyes" to what might happen at $t = 0$ and consider only the region $t > 0$. In this region, the exponential is "well-behaved" and can be differentiated.

A basic property of the exponential is that the derivative is also an exponential. The derivative of Eq. (2-41) is

$$f'(t) = -\alpha\epsilon^{-\alpha t} = \frac{-1}{\tau}\epsilon^{-t/\tau} \qquad \text{for } t > 0 \qquad (2\text{-}42)$$

This function is shown in Fig. 2-12b. Notice that the derivative is everywhere negative, a simple result of the fact that the slope of the decaying exponential is negative.

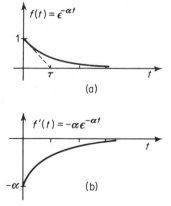

Figure 2-12   Exponential function and its derivative.

The value of $f'(t)$ at $t = 0$ is $f'(0) = -\alpha = -1/\tau$. If the exponential $f(t)$ continued to decrease at this initial rate, the function would reach zero at $t = \tau$ as shown by the dashed line in Fig. 2-12a. Actually, the exponential is down to 36.8% of its initial value at $t = \tau$.

The integral of Eq. (2-41) from 0 to t is

$$\int_0^t f(t)\, dt = \int_0^t \epsilon^{-\alpha t}\, dt = \frac{-1}{\alpha} \epsilon^{-\alpha t} \Big]_0^t$$

$$= \frac{1}{\alpha}(1 - \epsilon^{-\alpha t}) \qquad \text{for } t > 0 \tag{2-43}$$

Thus the definite integral of an exponential function is another exponential function plus a dc term.

So far we have considered only the decaying exponential function $\epsilon^{-\alpha t}$ due to its importance in circuit applications. However, we should also briefly look at the other possible form. Let us define the *growing exponential function* as

$$f(t) = \epsilon^{\alpha t} u(t) \tag{2-44}$$

where α is a positive constant. The form of this function is shown in Fig. 2-13. Notice that this function increases quite rapidly and grows without bound for large t. Quite often, the growing exponential occurs unintentionally, as in unstable feedback systems. A more detailed discussion of this topic will be presented in a later chapter.

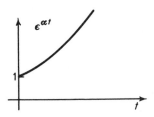

Figure 2-13　Growing exponential function.

Example 2-5

The voltage response in a certain circuit is given by

$$v(t) = 100\epsilon^{-50t} \qquad \text{for } t > 0 \tag{2-45}$$

Plot the response over the range of 5τ.

　　　Solution　The damping constant is $\alpha = 50$. The time constant is $\tau = \frac{1}{50} = 0.02\text{ s} = 20\text{ ms}$. Although it is not necessary to do so, for convenience we will redefine the exponential in terms of the time constant as

$$v(t) = 100\epsilon^{-t/20 \times 10^{-3}} \tag{2-46}$$

Since the ratio t/τ is dimensionless, we can drop the 10^{-3} factor if t is expressed in ms. We can now tabulate the values as shown in Table 2-3. Accuracy has been retained to three significant figures. The curve is shown in Fig. 2-14.

TABLE 2-3   Table of values for Example 2-5

t (ms)	$x = t/\tau$	ϵ^{-x}	$v(t)$ (V)
0	0	1.000	100
10	0.5	0.607	60.7
20	1.0	0.368	36.8
30	1.5	0.223	22.3
40	2.0	0.135	13.5
50	2.5	0.082	8.2
60	3.0	0.050	5.0
70	3.5	0.030	3.0
80	4.0	0.018	1.8
90	4.5	0.011	1.1
100	5.0	0.007	0.7

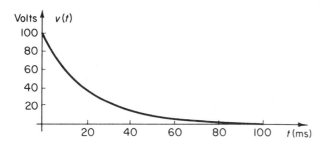

Figure 2-14   Function of Example 2-5.

2-7 SINUSOIDAL FUNCTION

The sinusoidal function is probably the most important single waveform that occurs in electric circuits. Not only is the sinusoid the almost universal waveform for commercial power generation and transmission, but throughout the entire domain of electronics and communications, sinusoidal waveforms and techniques are widely used.

Due to the importance of the sinusoidal function, a special branch of circuit analysis dealing with the steady-state solutions of circuit responses excited by sinusoidal sources evolved very early in electrical engineering history. This analysis, which depends quite heavily on complex algebra, is usually called *steady-state alternating current (ac) circuit analysis.* Later in this book, we will show that ac steady-state theory can be developed as a special case of transform circuit analysis.

Since this text is primarily concerned with developing the more generalized transform approach, we will not concern ourselves with a great deal of the details of steady-state ac analysis. Such treatments can be found in many general coverage circuit analysis texts. In the next few sections we consider some of the basic properties of the sinusoidal function. Since the manipulation of sinusoidal functions is facilitated by complex algebra, the reader not familiar with this analysis will wish to study Appendix A. We have chosen to include the treatment of complex algebra in an appendix since we anticipate that most readers will be familiar with steady-state ac circuit theory.

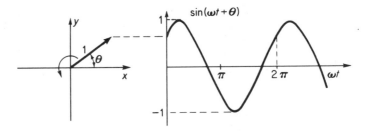

Figure 2-15   Generation of the sinusoidal function.

As an aid in understanding certain properties of the sinusoidal function, consider the rotating arm with unit length shown in Fig. 2-15. At $t = 0$, the arm is set at an angle θ with respect to the horizontal axis. If the arm is allowed to rotate at a constant angular speed around a circle, the vertical projection of the arm plotted versus time results in the sinusoidal function.

In one complete revolution of the arm, the sinusoidal function will complete one *cycle*. In one cycle, the angle increases by *2π radians* or *360°*. The time required for the sinusoid to complete one cycle is called the *period* and is denoted by T. The number of cycles completed per second is called the *repetition frequency* and is denoted by f. The hertz (Hz) has been adopted as the unit for frequency (1 Hz = 1 cycle/second). The period and frequency are simply related by

$$T = \frac{1}{f} \tag{2-47}$$

If the sinusoid completes f cycles/second, the change of the total angle is $2\pi f$ radians/second since one cycle corresponds to 2π radians. The number of radians/second is called the *angular velocity* in rotational mechanics. However, in electrical engineering, this quantity is usually called the *angular frequency*. The angular frequency is denoted by ω and is measured in radians/second = rad/s. The quantities ω and f are related by

$$\omega = 2\pi f \tag{2-48}$$

Due to the dual use of the word frequency, one must be careful about which is implied in a given situation. It is for this reason that we have defined the adjectives *repetition* and *angular*. These adjectives will be employed throughout the book except in cases where it is obvious which quantity is implied.

The basic sinusoidal function can be written as

$$f(t) = \sin(\omega t + \theta) \tag{2-49}$$

Two possible alternative forms are

$$f(t) = \sin(2\pi f t + \theta) = \sin\left(\frac{2\pi t}{T} + \theta\right) \tag{2-50}$$

The quantity θ represents an arbitrary initial *phase angle*.

Referring again to Fig. 2-15, we see that the basic sinusoid oscillates between the limits of $+1$ and -1. If we wish the sinusoid to oscillate between $+A$ and

$-A$, we write

$$f(t) = A \sin(\omega t + \theta) \tag{2-51}$$

The factor A is called the *amplitude, peak value,* or *crest value* of the sinusoid.
Let us now consider two special values for the phase angle θ.

1. For $\theta = 0°$, Eq. (2-49) reduces to

$$f(t) = \sin \omega t \tag{2-52}$$

Traditionally in mathematics this function is defined as the *sine* function, although we will refer to all possible cases as sinusoidal functions. A sketch of the sine function is shown in Fig. 2-16a.

2. For $\theta = 90°$, Eq. (2-49) reduces to

$$f(t) = \sin(\omega t + 90°) = \cos \omega t \tag{2-53}$$

This function is called the *cosine* function. A sketch of the cosine function is shown in Fig. 2-16b.

Both the sine and cosine functions are tabulated in most mathematical and engineering handbooks. Scientific calculators also have sine and cosine functions.

Next we will look at the derivatives and integrals of the sinusoidal functions. As in the case of the exponential function, we will avoid the possible discontinuity at $t = 0$, resulting from "switching on" the function, and consider only the region $t > 0$. Thus consider

$$f(t) = A \sin(\omega t + \theta) \qquad \text{for } t > 0 \tag{2-54}$$

The derivative of this function is

$$f'(t) = \omega A \cos(\omega t + \theta) \qquad \text{for } t > \theta \tag{2-55}$$

The *indefinite* integral of Eq. (2-54) is

$$\int f(t)\, dt = \frac{-A}{\omega} \cos(\omega t + \theta) + C \tag{2-56}$$

The presence of definite limits on the integral could result in a possible dc term for Eq. (2-56). Ignoring this possible dc term, Eqs. (2-54), (2-55), and (2-56) point out an

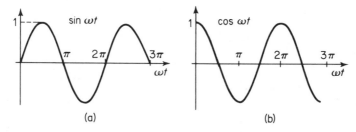

(a)   **(b)**

Figure 2-16   Sine and cosine functions.

important property of the sinusoidal function: *The derivative or integral of a sinuisoidal function is also a sinusoidal function of the same frequency, but with different amplitude and phase angle.* Thus, while the basic waveforms of most functions are changed by differentiation or integration, the sinusoid is unchanged in basic form.

Example 2-6

A certain current is given by

$$i(t) = 0.01 \sin (6283t + 30°) \tag{2-57}$$

Determine (a) amplitude, (b) angular frequency, (c) repetition frequency, and (d) period. Plot the current over the range of several cycles.

    Solution   (a) By inspection, the amplitude is 0.01 A = 10 mA. (b) The angular frequency is $\omega = 6283$ rad/s. (c) The repetition frequency is

$$f = \frac{6283}{2\pi} = 1000 \text{ Hz} = 1 \text{ kHz} \tag{2-58}$$

(d) The period is

$$T = \frac{1}{1000} = 1 \times 10^{-3} \text{ s} = 1 \text{ ms} \tag{2-59}$$

    To enhance plotting this function, a short tabulation of values is presented in Table 2-4. Note that for convenience, the quantity ωt is expressed in degrees rather than radians. Introduction of the artificial variable x is a convenient aid in determining the values of the sine function. The extension to several cycles can be done by adding 360° to the ωt values and repeating the values in the extreme right column. The function obtained is shown in Fig. 2-17.

TABLE 2-4   Table of values for Example 2-6

ωt (deg)	$x = \omega t + 30°$	$\sin x$	i (mA)
0°	30°	0.500	5.00
30°	60°	0.866	8.66
60°	90°	1.000	10.00
90°	120°	0.866	8.66
120°	150°	0.500	5.00
150°	180°	0.000	0.00
180°	210°	−0.500	−5.00
210°	240°	−0.866	−8.66
240°	270°	−1.000	−10.00
270°	300°	−0.866	−8.66
300°	330°	−0.500	−5.00
330°	360°	0.000	0.00
360°	390° or 30°	0.500	5.00

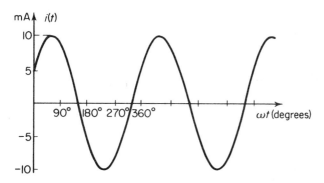

Figure 2-17   Current of Example 2-6.

2-8 RELATIVE PHASE OF SINUSOIDAL FUNCTIONS

Under the general heading of sinusoidal functions, it is possible to obtain functions of the forms $A \sin (\omega t + \theta)$ and $B \cos (\omega t + \phi)$. In circuit problems, we will frequently encounter several different sinusoidal functions of the same frequency that have different phase angles with respect to each other. Let us again refer to the rotating arm analogy. In Fig. 2-18, we have two arms that start, at $t = 0$, to rotate at the same speed, but with arm B at an angle $60°$ ahead of arm A. Arm A makes an angle $45°$ with respect to the x-axis. For illustrative purposes, both arms have been chosen to have lengths of 100 units. The corresponding sinusoidal functions are shown in Fig. 2-18b. Since sinusoid B crosses the ωt axis $60°$ sooner than A, then B is said to *lead* A by $60°$. By the same token, since A crosses the ωt axis $60°$ later than B, then A is said to *lag* B by $60°$. The phase difference between A and B is then said to be $60°$ with respect to each other. If the functions are assumed to be voltages, the equations for the two sinusoids are

$$e_a(t) = 100 \sin (\omega t + 45°) \tag{2-60}$$

and

$$\begin{aligned} e_b(t) &= 100 \sin (\omega t + 45° + 60°) \\ &= 100 \sin (\omega t + 105°) \end{aligned} \tag{2-61}$$

with the subscripts a and b representing sinusoids A and B, respectively.

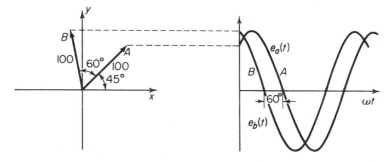

Figure 2-18   Illustration of phase difference between sinusoids.

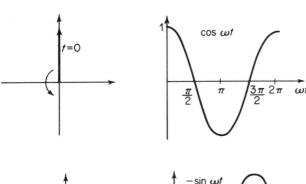

Figure 2-19   Generation of the cosine function.

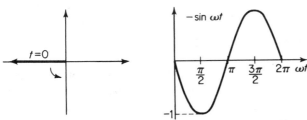

Figure 2-20   Generation of the negative sine function.

It should be observed that if the rotating arm starts at the vertical axis as shown in Fig. 2-19, it will generate the cosine function. Similarly, if the arm starts on the negative horizontal axis, as shown in Fig. 2-20, it will generate the negative sine function, which is equivalent to the sine function displaced by 180°. Finally, if the arm starts on the negative vertical axis, as shown in Fig. 2-21, it will generate the negative cosine function, which is the cosine function displaced by 180°. The following identities are thus developed:

$$\sin(\omega t \pm 180°) = -\sin \omega t \tag{2-62}$$

and

$$\cos(\omega t \pm 180°) = -\cos \omega t \tag{2-63}$$

In the case where there are a number of sinusoids at the same frequency, the speeds of all the corresponding generating arms are the same. Hence, the relative

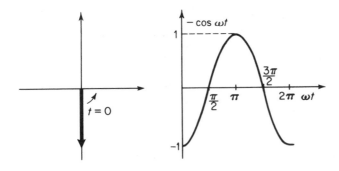

Figure 2-21   Generation of the negative cosine function.

speeds between the arms are zero, and thus the angles between the various arms will remain constant. Therefore, why allow the arms to rotate in looking at the relative phase? We can obtain enough information from looking at the relative positions of the arms.

We can thus "freeze" the rotating arms at $t = 0$ and characterize time-varying sinusoids by means of the arms which, if allowed to rotate, would generate the required functions. The representation for the case previously considered (Fig. 2-18) is shown in Fig. 2-22. The arms represent a special class of vectors since they have both magnitude and phase. In steady-state ac circuit theory, these vectors are usually called *phasors*.

The study of steady-state ac circuit theory is built on the foundation of phasors and complex algebra. It is rigorously justified in ac circuit analysis textbooks, and to some extent in Chapter 8 of this book, that these phasors may be represented by complex numbers and that they obey the laws of complex algebra. In this and the next section, we will discuss certain basic ideas that depend on the use of the phasor concept and, in some cases, complex algebra. The reader should refer to Appendix A when in doubt about any basic rules of complex algebra.

To summarize the preceding few paragraphs, a sinusoidal time function may be represented by a stationary phasor whose length represents the amplitude of the sinusoid and whose angle represents the initial phase angle of the sinusoid. The resulting phasor is considered to be a complex number. Notice in Fig. 2-22 that we have used the notation $\bar{E}_a$ and $\bar{E}_b$ on the phasor diagram to represent $e_a(t)$ and $e_b(t)$, respectively. The *capital* letters emphasize that the phasors are now considered to be fixed quantities rather than time-varying quantities, and the *bars* emphasize that the phasors are considered to be complex numbers.

The problem we wish to consider in the remainder of this section is that of expressing a sine function in terms of a cosine function, or vice versa, by means of the phasor diagram. From the earlier work of this section, we can display a relative phase diagram of the sine and cosine functions as shown in Fig. 2-23. Remembering that the positive direction of rotation is counterclockwise, we can deduce a few properties from this figure.

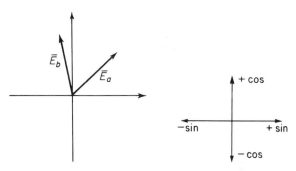

Figure 2-22   Phasor diagram of the sinusoids of Fig. 2-18.

Figure 2-23   Relative phase diagram for sinusoidal functions.

1. The positive cosine function leads the positive sine function by 90°.

$$\cos \omega t = \sin (\omega t + 90°) \qquad (2\text{-}64)$$

2. The negative sine function leads the positive cosine function by 90°.

$$-\sin \omega t = \cos (\omega t + 90°) \qquad (2\text{-}65)$$

Whenever an arbitrary phase angle is associated with a sine or cosine function of amplitude A, we sketch a phasor of length A at an angle of θ degrees with respect to the appropriate axis. We may then express the phasor in question in terms of any other unit phasor by measuring the angle with respect to the position of the new reference axis.

Example 2-7

Express $5 \cos (\omega t + 30°)$ in terms of the sine function.

   Solution   We wish to express the function in the form $5 \sin (\omega t + \theta)$, where θ is to be determined. We first sketch a phasor of length 5 at an angle 30° greater (more counterclockwise) than the positive cosine axis as shown in Fig. 2-24. The angle with respect to the positive sine axis is $30° + 90° = 120°$. Hence

$$5 \cos (\omega t + 30°) = 5 \sin (\omega t + 120°) \qquad (2\text{-}66)$$

An alternative choice is to write the function in terms of the negative sine axis. The angle is $-60°$ in this case, since the given function lags the negative sine axis by 60°. Hence

$$5 \cos (\omega t + 30°) = -5 \sin (\omega t - 60°) \qquad (2\text{-}67)$$

Example 2-8

A certain reference sinusoidal voltage is given by $e_1 = 5 \sin (\omega t - 30°)$. Another voltage e_2 is given by $e_2 = -10 \sin (\omega t - 40°)$.
   (a) Express e_2 in terms of the positive sine function.
   (b) Determine the phase of e_2 with respect to e_1.
   Solution   We first draw a phasor diagram as shown in Fig. 2-25 with the two phasors placed as shown. The angle of $\bar{E}_2$ with respect to the positive sine axis is 140° in a leading sense. Hence:

(a) $$e_2 = 10 \sin (\omega t + 140°). \qquad (2\text{-}68)$$

   (b) The angle between $\bar{E}_1$ and $\bar{E}_2$ is $140° + 30° = 170°$, with $\bar{E}_2$ leading. Thus e_2 *leads* e_1 by 170°; equivalently, e_1 *lags* e_2 by 170°.

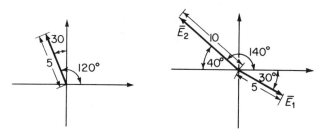

Figure 2-24 Phasor diagram for Example 2-7.

Figure 2-25 Phasor diagram for Example 2-8.

2-9 ADDITION OF SINUSOIDS OF THE SAME FREQUENCY

The phasor diagram serves as a basis for adding sinusoids of the same frequency but with different amplitudes and phases. Consider an expression of the form

$$f(t) = A_1 \sin (\omega t + \theta_1) + A_2 \sin (\omega t + \theta_2) + \cdots + A_n \sin (\omega t + \theta_n) \qquad (2\text{-}69)$$

A basic property of sinusoids is that *the sum of an arbitrary number of sinusoids of the same frequency is equivalent to a single sinusoid of the given frequency.* We emphasize the restriction that all sinusoids must be of the *same frequency.* Thus the sum represented by Eq. (2-69) can be expressed in the form

$$f(t) = A \sin (\omega t + \theta) \qquad (2\text{-}70)$$

where A and θ are to be determined.

This entire operation can be performed by means of a phasor diagram. Each of the sinusoids given in Eq. (2-69) is represented by a phasor of the appropriate magnitude and phase. Note that one or more of the sinusoids could be expressed as cosine functions as long as the appropriate axis is used in determining the position of a given phasor. Once the phasors are specified, the phasors are added. This operation can be done graphically or by means of complex algebra. It can then be shown that the magnitude of the resulting phasor is the amplitude of the equivalent sinusoid, and the phase angle of the resulting phasor with respect to a reference axis is the phase of the equivalent sinusoid. Let us illustrate with some examples.

Example 2-9

Determine a single sinusoid equivalent to the sum

$$e(t) = 100 \sin 50t + 80 \sin (50t + 60°) \qquad (2\text{-}71)$$

Solution   The two sinusoids are first represented by phasors as shown in Fig. 2-26. As shown on the figure, the phasors could be added by a graphical procedure if desired. However, we will choose to combine them by complex algebra. We have

$$\bar{E} = 100\underline{/0^\circ} + 80\underline{/60^\circ}$$
$$= 100 + 40 + j40\sqrt{3}$$
$$= 140 + j69.28 \tag{2-72}$$
$$= 156.2\underline{/26.33^\circ}$$

Thus, in terms of the positive sine function, $e(t)$ can be written as

$$e(t) = 156.2 \sin (50t + 26.33^\circ) \tag{2-73}$$

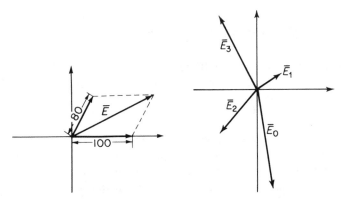

Figure 2-26   Phasor diagram for   **Figure 2-27**   Phasor diagram
Example 2-9.   for Example 2-10.

Example 2-10

A certain circuit combines three input signals to yield an output signal as follows:

$$e_0(t) = e_1(t) + e_2(t) - e_3(t) \tag{2-74}$$

Determine $e_0(t)$ for the following input signals:

$$e_1(t) = 10 \sin (500t + 30^\circ) \tag{2-75}$$

$$e_2(t) = -20 \cos (500t - 45^\circ) \tag{2-76}$$

$$e_3(t) = 30 \cos (500t + 30^\circ) \tag{2-77}$$

   Solution   We can first draw phasors representing the three signals as shown in Fig. 2-27. The complex representations for the phasors are

$$\bar{E}_1 = 10\underline{/30^\circ} = 8.66 + j5 \tag{2-78}$$

$$\bar{E}_2 = 20\underline{/-135^\circ} = -14.14 - j14.14 \tag{2-79}$$

$$\bar{E}_3 = 30\underline{/120^\circ} = -15 + j25.98 \tag{2-80}$$

The phasor combination is

$$\bar{E}_0 = \bar{E}_1 + \bar{E}_2 - \bar{E}_3$$

$$= (8.66 + j5) + (-14.14 - j14.14) - (-15 + j25.98)$$

$$= 8.66 - 14.14 + 15 + j(5 - 14.14 - 25.98) \tag{2-81}$$

$$= 9.52 - j35.12 = 36.39\underline{/-74.83°}$$

In terms of the positive sine function, we may express $e_0(t)$ as

$$e_0(t) = 36.39 \sin(500t - 74.83°) \tag{2-82}$$

Alternatively, we may express $e_0(t)$ as

$$e_0(t) = -36.39 \cos(500t + 15.17°) \tag{2-83}$$

2-10 DAMPED SINUSOIDAL FUNCTION

The *damped sinusoidal function* is defined by

$$f(t) = \epsilon^{-\alpha t} \sin(\omega t + \theta) \qquad \text{for } t > 0 \tag{2-84}$$

Thus, at any time t, the value of the damped sinusoidal function is the product of the decaying exponential function and the sinusoidal function, both evaluated at time t. This idea is illustrated in Fig. 2-28. The exponential function is shown in (a), and the sinusoidal function is shown in (b). The damped sinusoidal function is shown

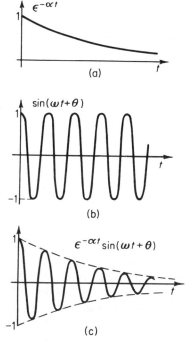

Figure 2-28 Damped sinusoidal function.

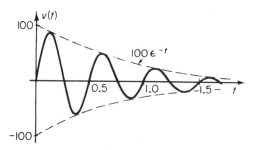

Figure 2-29   Voltage of Example 2-11.

in (c). Note that the envelope of the curve, shown by the dashed lines, is the exponential function, and the sinusoid simply oscillates between the bounds established by the exponential.

To determine the exact behavior of a given damped sinusoid, three parameters must be known. The frequency (or period) and the phase angle of the sinusoid must be known. In addition, the damping constant (or time constant) of the exponential must be known. As the damping constant is increased, the sinusoidal oscillations damp out more rapidly. Since the oscillations are confined within the bounds of the exponential, the time required for the function to approach zero, for practical purposes, can be determined by the "rule of thumb" for the exponential function.

Example 2-11

A certain voltage is given by

$$v(t) = 100\epsilon^{-t} \sin 4\pi t \qquad (2\text{-}85)$$

Sketch the function over the range of several cycles.

Solution   The damping constant α is 1 and the repetition frequency is 2 Hz. The envelope of the curve ($100\epsilon^{-t}$) is first drawn as shown in Fig. 2-29. The damped sinusoid is then sketched to meet the envelope at maxima and minima. If desired, an exact tabulation could be performed, but in many cases a simple scheme such as this will enable the most important properties of the function to be determined.

2-11 SHIFTED FUNCTIONS

We saw in Section 2-3 that multiplication of any function $f(t)$ by the unit step function $u(t)$ results in a mathematical switching action at $t = 0$. On the other hand, we have seen that whenever all the functions in a given problem start at $t = 0$, it is not always necessary to carry along the $u(t)$ factor so long as we understand the nature of the given function. In this section we turn our attention to functions that are "switched on" at times other than $t = 0$. Such waveforms appear in problems in which several apparent sources occur at different times. We will see that in such cases the $u(t)$ factor, along with other notations to be introduced, will prove to be essential in describing the behavior of the waveforms.

First let us observe the behavior of the function described by

$$g(t) = f(t)u(t - a) \tag{2-86}$$

The process of describing such a function is shown in Fig. 2-30. The original $f(t)$ is shown in (a), $u(t - a)$ is shown in (b), and $g(t)$ is shown in (c). From the curve it can be seen that *some of the positive time region of $f(t)$ is lost by this process.* We will call this form of a function an *unshifted delayed function.*

We now wish to turn our attention to another type of general switched function which is somewhat similar in nature to the previous case, but must be carefully distinguished from it. Consider first the ideal *delay line* shown in Fig. 2-31. This device has the property that it will hold a signal for a length of time a. The input signal to the delay line is assumed to be of the form $f(t)u(t)$. The delay line will store the signal a seconds and then release it in its original form. No output will appear before $t = a$, and so at $t = a$, the release of signal from the storage corresponds to a switching operation $u(t - a)$.

The question now is raised: Does $f(t)$ require alteration, and if so, what type of alteration does it require? The answer to the first part of the question is that some alteration is needed since the product $f(t)u(t - a)$ does not describe the operation as we have previously noted. We must start the original function at $t = a$ without losing any information.

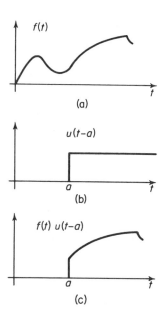

(a)

(b)

(c)

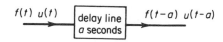

Figure 2-30 Development of the unshifted delayed function.

Figure 2-31 Delay line used to develop shifted function.

Let us observe the process of replacing t by $t - a$ as we did it shifting the step function. For $t = a$, $t - a = 0$, and hence, the argument of the function is zero. For any original $t = t_1$ the quantity $t - a$ is equal to t_1 for $t = t_1 + a$. Thus the entire function is shifted to the right by a units, and no information is lost. We shall call this function a *shifted delayed function*, or for simplicity, a *shifted* function. Mathematically, we represent it as

$$g(t) = f(t - a)u(t - a) \tag{2-87}$$

An illustration of the shifting operation is shown in Fig. 2-32.

To summarize the discussion, we note that:

1. Replacement of t by $t - a$ in the argument of $f(t)$ shifts the original function a units to the right.
2. Multiplication of $f(t - a)$ by $u(t - a)$ removes the portion of the function for $t < a$ and thus completes the process of shifting the original function without unwanted modification.
3. This operation is identical to passing $f(t)$ through an ideal delay line with a delay of a.

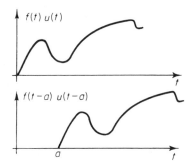

Figure 2-32 Shifted delayed function.

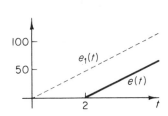

Figure 2-33 Ramp functions of Example 2-12.

Example 2-12

A ramp voltage with slope $+25$ V/s is to start at $t = 2$ s, as shown by the solid line in Fig. 2-33, and is to be zero for $t < 2$ s. Write an equation for the voltage.

Solution Using the shifting function concept, we first construct an unshifted ramp function with slope $+25$ V/s, as shown by the dashed line in Fig. 2-33. Calling this function $e_1(t)$, we have

$$e_1(t) = 25t \qquad \text{for } t > 0 \tag{2-88}$$

or

$$e_1(t) = 25tu(t) \tag{2-89}$$

if desired.

Let $e(t)$ represent the final translated function. We can generate $e(t)$ by replacing t in $e_1(t)$ by $t - 2$ and multiplying by $u(t - 2)$. Note that if the form $25tu(t)$ is first considered, we simply replace t by $t - a$ in both places, and the delayed step factor is automatically generated. Hence

$$e(t) = e_1(t - 2)$$
$$= 25(t - 2)u(t - 2) \qquad (2\text{-}90)$$

A common error is to write the translated function as $e(t) = (25t - 2) \cdot u(t - 2)$. We emphasize that the quantity $t - a$ is to be treated as an *exact* replacement for t.

We might point out an alternative method that the reader will recognize as an application of analytic geometry. Let us consider the desired ramp to extend for all time as shown in Fig. 2-34. The equation of the resulting straight line is

$$y = mt + b \qquad (2\text{-}91)$$

Evaluation of m and b results in $m = 25$ and $b = -50$. Hence

$$y = 25t - 50 \qquad (2\text{-}92)$$

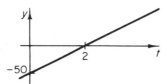

Figure 2-34   Modified function of Example 2-12.

In this case, we obtain $e(t)$ by simply multiplying y by $u(t - 2)$, as no shifting is needed. In other words, by writing the equation of the function directly and cancelling out the undesired portion, we have in effect made use of the unshifted delayed function concept. The result is

$$e(t) = (25t - 50)u(t - 2)$$
$$= 25(t - 2)u(t - 2) \qquad (2\text{-}93)$$

which agrees with the previous result. For most circuit waveforms, the shifting scheme is more convenient.

Example 2-13

A certain voltage is given by

$$e_1(t) = 10\epsilon^{-t}u(t) \qquad (2\text{-}94)$$

This voltage is applied to the input of an ideal delay line with 6 s delay. Determine the output voltage $e_2(t)$.

Solution   The effect of the delay line is to shift $e_1(t)$ to the right by 6 s. Since $e_1(t)$ has been expressed with the $u(t)$ factor, we merely replace t by $t - 6$ in both the exponential and the $u(t)$ factor. Thus $e_2(t)$ is

$$e_2(t) = 10\epsilon^{-(t-6)}u(t-6) \tag{2-95}$$

Both functions are shown in Fig. 2-35.

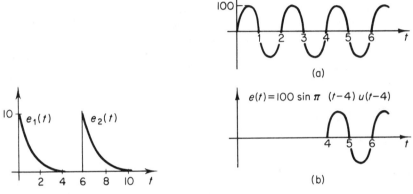

Figure 2-35   Waveforms of Example 2-13.   **Figure 2-36**   Waveforms of Example 2-14.

Example 2-14

A sinusoidal voltage with a period of 2 s and an amplitude of 100 V is to start at $t = 4$ as shown in Fig. 2-36b. Write the equation of the voltage.
      Solution   We first sketch an unshifted sinusoid as shown in Fig. 2-36a. Its equation is

$$e_1(t) = 100 \sin \pi t u(t) \tag{2-96}$$

The shifted function is

$$e(t) = 100 \sin \pi(t-4)u(t-4) \tag{2-97}$$

2-12 GENERAL WAVEFORMS

The material thus far has treated a number of the more basic waveforms that may represent the voltages or currents in electric circuits. We will see in this section how these basic waveforms can be combined by addition or subtraction to represent a wide variety of more general waveforms. Although a few general rules can be developed, a certain amount of intuition and practice is necessary to achieve proficiency in this task. For that reason, much of this section is devoted to practical examples. The reader should carefully study the examples that follow this section.

First, let us investigate a few basic operations:

Addition of step functions.   The sum of two or more step functions is the algebraic sum of the individual values, evaluated in each range where a new step function begins. Some possible results are shown in Figs. 2-37 and 2-38. The constants A and B are both assumed to be positive.

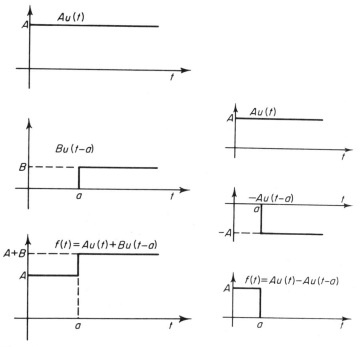

Figure 2-37   Addition of two positive step functions.

Figure 2-38   Addition of step functions producing pulse.

Addition of ramp functions.   The approach for adding ramp functions requires some discussion for proper interpretation. There are three helpful points to consider:

1. In each interval between times at which successive ramps begin, the resulting function is a straight line.

2. The slope of the resulting function in a given interval is the algebraic sum of the slopes of all ramps that have already been "switched on."

3. The starting point in a given interval is the value reached by the resulting function at the end of the preceding interval. In other words, the function is always continuous since it is the sum of continuous ramp functions.

Some possible results for two ramps are shown in Figs. 2-39, 2-40, and 2-41. The constants A and B are both assumed to be positive. Figure 2-41 presents a rather interesting result in that the net slope for $t > a$ is zero. Thus, the sum of the two ramps in this case produces an imperfect step function with nonzero rise time.

In the examples considered thus far, we have started with the basic components and developed the resulting function. The more general problem is to start with a desired waveform and to represent it as a combination of the basic waveforms. The following examples will help to clarify the approach to such problems.

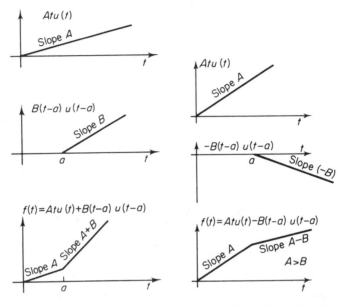

Figure 2-39   Addition of two positive ramps.

Figure 2-40   Addition of positive and negative ramps.

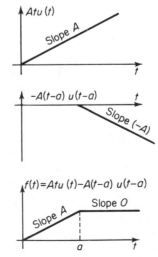

Figure 2-41   Addition of ramps producing imperfect step function.

Example 2-15

The repetitive square wave shown in Fig. 2-42a is given. Write an equation for the square wave in terms of simpler functions.

   Solution   Since a change occurs at intervals of 2 units, we must examine the transition point of each interval. We first observe that the basic unit step function $u(t)$ initiates the response in the first interval as shown in Fig. 2-42b. *Note that this particular component must be considered to continue for all time.* At $t = 2$, we bring the original function back to zero by adding the negative step $-u(t - 2)$ as shown in (c). Thus the function is zero for $2 < t < 4$. At $t = 4$, we simply add a positive step $u(t - 4)$ as shown in (d) and the process is repeated. Hence

$$f(t) = u(t) - u(t - 2) + u(t - 4) - u(t - 6) + \cdots$$

$$= \sum_{n=0}^{\infty} (-1)^n u(t - 2n) \tag{2-98}$$

Notice that an infinite number of terms are required to express the function for an infinite time range.

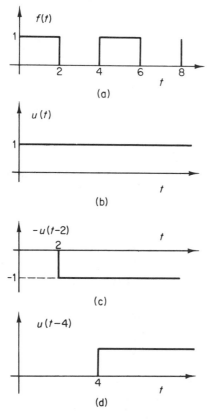

(a)

(b)

(c)

(d)

Figure 2-42   Square wave of Example 2-15.

Example 2-16

Write an equation for the single trapezoidal pulse shown in Fig. 2-43a.
  Solution   There are four intervals to be considered:

$$0 < t < 1, \quad 1 < t < 3, \quad 3 < t < 4, \quad \text{and} \quad t > 4$$

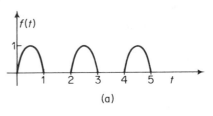

(a)

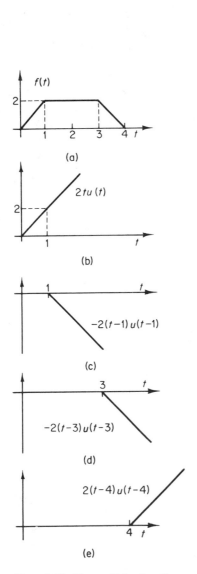

(a)

(b)

(c)

(d)

(e)

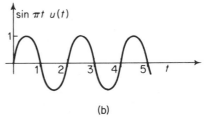

(b)

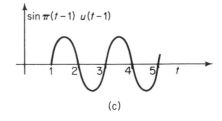

(c)

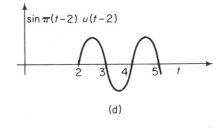

(d)

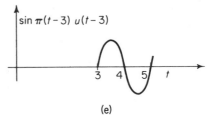

(e)

Figure 2-43 Trapezoidal pulse of
Example 2-16.

Figure 2-44 Waveforms of Example
2-17.

In the first interval, a single ramp of slope $+2$, as shown in Fig. 2-43b, initiates the linear rise. At $t = 1$, the effect of the positive ramp must be cancelled by a negative ramp of slope -2, as shown in (c). The composite function remains constant until $t = 3$, at which time a negative slope is required. The negative slope is achieved by the introduction of a negative ramp of slope -2 at $t = 3$, as shown in (d). Finally, the function must be completely eliminated for $t > 4$. To do this we introduce a positive ramp of slope $+2$ at $t = 4$, as shown in (e). The complete equation for the function is given by

$$f(t) = 2tu(t) - 2(t-1)u(t-1) - 2(t-3)u(t-3) + 2(t-4)u(t-4) \qquad (2\text{-}99)$$

Example 2-17

The waveform shown in Fig. 2-44a is the output waveform of a *half-wave rectifier*. The waveform represents a sinusoid with the negative half of each cycle clipped or removed. Write an equation for the waveform.

    Solution   Since the period is 2 s, the waveform is derived from the function $\sin \pi t$. We begin the function at $t = 0$ by introducing this function as shown in Fig. 2-44b. At $t = 1$, we can cancel the first negative half-cycle by adding the function $\sin \pi(t-1)u(t-1)$, as shown in (c). This completes the first cycle, and the process is repeated in subsequent cycles as shown in (d) and (e). The equation is thus

$$\begin{aligned} f(t) &= \sin \pi t u(t) + \sin \pi(t-1)u(t-1) + \sin \pi(t-2)u(t-2) + \cdots \\ &= \sum_{n=0}^{\infty} \sin \pi(t-n)u(t-n) \end{aligned} \qquad (2\text{-}100)$$

2-13 IMPULSE FUNCTION

We now wish to consider the concept of the *impulse function*, which is more of a mathematical limit than an actual physical reality. The impulse function will allow us to describe more precisely certain discontinuities that occur in analyzing waveforms.

    To develop the impulse function, let us first consider the narrow pulse $f(t)$ shown in Fig. 2-45a. The width of the pulse is t_1, and the height is $1/t_1$. Thus the area of the pulse is unity. As t_1 is allowed to decrease, the pulse width becomes smaller, the height becomes larger, but the area remains constant at unity. In the limit, we have a fictitious quantity whose height is infinite, whose width is zero, and whose area is unity! The limiting quantity appearing at the origin, as shown in Fig. 2-45b, is called a *unit impulse function* and is designated by the symbol $\delta(t)$. Mathematically, $\delta(t)$ is thus defined in terms of the pulse $f(t)$ as

$$\delta(t) = \lim_{t_1 \to 0} f(t) \qquad (2\text{-}101)$$

    As in the case of the step function, we can delay an impulse function by replacing the argument t by $t - a$. The delayed impulse function $\delta(t - a)$ is shown in Fig.

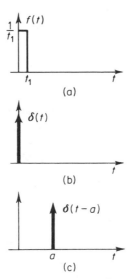

Figure 2-45   Development of the impulse function.

2-45c. In general, if x is any arbitrary function of t, the impulse function $\delta(x)$ is defined to be zero for $x \neq 0$, and to be infinite for $x = 0$.

To specify an impulse function other than the unit impulse, the area must be known. If the impulse is a voltage, the area is measured in volts × seconds = V · s, whereas the area of a current impulse is measured in amperes × seconds = A · s. For example, a voltage or current of the form $K\delta(t)$ represents an impulse with area K V·s or K A·s, depending on whether the impulse is a voltage or current. Since the function $K\delta(t)$ must have the units of volts or amperes, it is implied that the basic impulse function $\delta(t)$ has the dimension of seconds^{-1} = s^{-1}. With this assumption, the units of $K\delta(t)$ are $(V \cdot s) \cdot s^{-1} = V$ (volts) or $(A \cdot s) \cdot s^{-1} = A$ (amperes) as required.

The reader is probably wondering what possible benefit can result from such an abstract concept. At this point we can answer this question by four general statements:

1. Very often, a circuit excitation or response is a pulse whose width is short compared with the time constants of the system (a concept to be studied later). Under such conditions, we can approximate the actual response of the circuit by the response to an impulse having the same area and location in time.

2. Initial energy storage in electric circuits can often be conveniently represented by hypothetical impulse sources.

3. In our later studies of transform analysis, we will discover that the response of a network to an impulse will be one of the easiest types of responses to analyze. Therefore, the response of a circuit due to an impulse can be used as a property to characterize a system.

4. In utilizing circuit transformations, an equivalent impulse source often appears as a result of the mathematical manipulations, even though the actual source is not an impulse. Hence it is necessary to learn to manipulate circuits containing the impulse source.

Example 2-18

The voltage pulse of Fig. 2-46 is to be used in a circuit whose time constants are long compared with the width of the pulse. Define an impulse that will approximate the given pulse.

Solution   The area of the pulse is

$$K = 100 \text{ V} \times 0.5 \times 10^{-3} \text{ s} = 0.05 \text{ V} \cdot \text{s} \qquad (2\text{-}102)$$

The impulse is thus

$$v(t) = 0.05\delta(t - 25 \times 10^{-3}) \qquad (2\text{-}103)$$

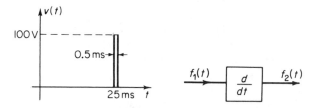

Figure 2-46 Pulse of Example 2-18.

Figure 2-47 Block diagram of differentiator.

2-14 RELATIONSHIP BETWEEN STEP AND IMPULSE FUNCTIONS

In considering the members of the t^n class in Section 2-5, we observed that the case for $n = 0$ corresponded to the step function $u(t)$. We avoided attempting to differentiate this function due to its discontinuous nature at $t = 0$. We now wish to show that by employing the impulse function, we can define a derivative for the step function.

Consider the differentiator circuit shown in block diagram form in Fig. 2-47. The output is given by

$$f_2(t) = \frac{df_1(t)}{dt} = f'_1(t)' \qquad (2\text{-}104)$$

First let us assume that the input is a unit dc excitation that has been present for some time prior to $t = 0$, as shown in Fig. 2-48a. Observation begins at $t = 0$, and we have

$$f_1(t) = 1 \qquad (2\text{-}105)$$

The output is

$$f_2(t) = \frac{d(1)}{dt} = 0 \qquad (2\text{-}106)$$

as shown in Fig. 2-48b. Thus we see no output signal since the dc input signal was already present at the point of observation, and the derivative of a constant is always zero.

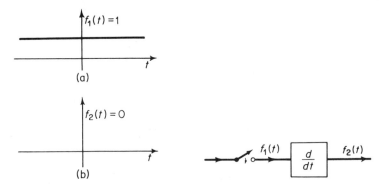

Figure 2-48 Constant dc waveform and its derivative.

Figure 2-49 Modified differentiator.

Now let us change the problem by introducing an ideal switch in series with the input as shown in Fig. 2-49. If the switch is closed at $t = 0$, the input "seen" by the differentiator will now suddenly rise from zero to 1 unit in a very short time and remain there until the switch is opened. Physically, it is impossible for any real variable to change values instantaneously as there will always be a small amount of *rise time*. Let us assume a rise time t_1 and an initial linear change of $f_1(t)$ as shown in Fig. 2-50a. (The rise time has been exaggerated for convenience.)

We are no longer interested in the region $t < 0$, and the output for $t > t_1$ is the same as before, namely zero. The interesting region in this case is $0 < t < t_1$. The slope of $f_1(t)$ in this region is $1/t_1$, and $f_2(t)$ is therefore a narrow pulse with amplitude $1/t_1$ and width t_1 as shown in Fig. 2-50b. Thus the pulse has unit area.

Now let t_1 become even smaller, as shown in Fig. 2-50c. The output pulse shown in (d) is now narrower and larger in amplitude, but the area remains constant at unity. Finally, if the rise time approaches zero as shown in (e), the output approaches a pulse with zero width, infinite height, and unit area. Such a function is represented by the impulse function as shown in (f).

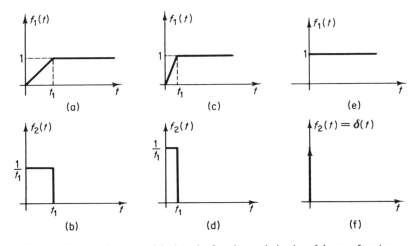

Figure 2-50 Development of the impulse function as derivative of the step function.

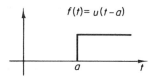

$$f(t) = u(t-a)$$

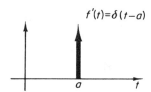

$$f'(t) = \delta(t-a)$$

Figure 2-51 Delayed step function and its derivative.

Thus, in a limiting sense, the derivative of a step function is

$$\frac{du(t)}{dt} = \delta(t) \tag{2-107}$$

If the step function is delayed, a similar development yields

$$\frac{d}{dt}\left[u(t-a)\right] = \delta(t-a) \tag{2-108}$$

as illustrated in Fig. 2-51.

In effect, Eqs. (2-107) and (2-108) say that the derivative of a step function is zero except at the instant of switching, and at this point, the derivative is very large if the rise time is very small. In a limiting sense, then, *the derivative of a step function is an impulse function*. Of course, it is impossible to have zero rise time, so the theoretical impulse never quite occurs; but the concept is quite useful in approximating many actual situations.

We wish to emphasize that we are not violating the reader's knowledge of differential calculus, which clearly states that the derivative of a constant is zero. The derivative is nonzero only for the case where the constant is "switched on" at some time, and then it is nonzero only at the point of switching.

It can now be seen that the dimensions of the unit impulse function are seconds^{-1}. This is true since the unit step function is dimensionless, and the derivative with respect to time of any function has the dimensions of seconds^{-1} times the dimensions of the function that is differentiated. The inverse relationships to Eqs. (2-107) and (2-108) are

$$\int_0^t \delta(t)\,dt = u(t) \tag{2-109}$$

$$\int_0^t \delta(t-a)\,dt = u(t-a) \tag{2-110}$$

Thus *the integral of an impulse function is a step function*. The step function begins at the same point that the impulse function occurs. This idea is intuitive since integration of a function represents the net area under the curve in moving along the direction of integration. As we pass the unit impulse function, we suddenly gain unit area which is maintained for all time thereafter.

Example 2-19

The imperfect step voltage $e_1(t)$ shown in Fig. 2-52a is applied to the input of a differentiator circuit whose output $e_2(t)$ is given by

$$e_2(t) = \frac{de_1(t)}{dt} \tag{2-111}$$

(a) Determine $e_2(t)$ for $e_1(t)$ as given.
(b) Determine $e_2(t)$ if the rise time of $e_1(t)$ were reduced to zero.

    Solution (a) Since the final value is 10 V and the rise time is 50 ms, the slope during the rise time is

$$\text{slope} = \frac{10 \text{ V}}{0.05 \text{ s}} = 200 \text{ V/s} \tag{2-112}$$

The output voltage $e_2(t)$ is equal to this slope as shown in Fig. 2-52b. (Note that although the slope has the dimensions of volts/second, the output has the dimensions of volts since the differentiation cancels out the seconds factor in the denominator.)

    (b) For the case of zero rise time, we may write $e_1(t)$ as

$$e_1(t) = 10u(t) \tag{2-113}$$

as shown in Fig. 2-52c. The output voltage is now

$$e_2(t) = \frac{de_1(t)}{dt} = \frac{d}{dt}[10u(t)] = 10\delta(t) \tag{2-114}$$

as illustrated in Fig. 2-52d. Note that the area of the impulse function is equal to the magnitude of the step function being differentiated.

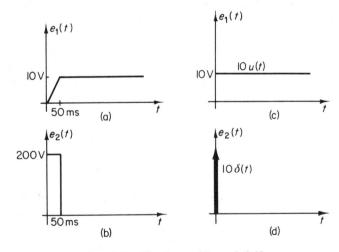

Figure 2-52 Waveforms of Example 2-19.

2-15 DIFFERENTIATION AND INTEGRATION OF WAVEFORMS

In previous sections we have considered the derivatives and integrals of several basic functions. In this section we wish to consider the derivatives and integrals of the more complicated waveforms involving the sums and products of several basic functions.

In general, differentiation and integration of complex waveforms can be performed by either of two separate approaches: (a) analytical calculus and (b) graphical calculus. Although in theory any problem can be handled by either approach, often one particular approach will stand out as the best procedure in a given problem. In other cases, it is a matter of practice and experience to determine the optimum procedure.

To employ analytical calculus, we write a specific equation for the waveform in question and differentiate or integrate each of the component parts. If desired, the resulting function can be sketched. In general, analytical techniques are best suited to waveforms that are easily described mathematically but are somewhat "curved" in form, such as the exponential and sinusoid.

Graphical techniques are best suited to functions that are composed of straight-line sections. Such functions are called *piecewise linear functions*. To differentiate piecewise linear functions, we calculate the slope of each segment. To integrate, we calculate the area under the curve for each segment and note the manner in which the integral will vary.

In differentiating by either analytical or graphical techniques, we will occasionally encounter discontinuities such as, for example, the step function. Prior to Section 2-14, we simply "closed our eyes" and avoided looking at such discontinuities. However, in Section 2-14 we introduced the impulse function as the derivative of a step function. We will now show how this concept may be extended to any function possessing a possible discontinuity.

Consider the function $g(t)$ defined by

$$g(t) = f(t)u(t) \tag{2-115}$$

Let us differentiate this function and retain the $u(t)$ factor instead of omitting it as we have done in the past. Recalling the formula for the derivative of a product, we have

$$\begin{aligned} g'(t) &= f'(t)u(t) + f(t)u'(t) \\ &= f'(t)u(t) + f(t)\delta(t) \end{aligned} \tag{2-116}$$

in which $u'(t) = \delta(t)$ from Eq. (2-107). Now what is the meaning of $f(t)\delta(t)$? Since $\delta(t)$ is zero everywhere but at $t = 0$, we need only consider $f(t)$ evaluated at $t = 0$. To emphasize the fact that the value of $f(t)$ immediately to the right of the discontinuity is of concern, we will use the notation $t = 0^+$. Thus the second term of Eq. (2-116) can be written as $f(0^+)\delta(t)$. The derivative of $f(t)u(t)$ can then be expressed as

$$\frac{d}{dt}\left[f(t)u(t)\right] = f'(t)u(t) + f(0^+)\delta(t) \tag{2-117}$$

If the function is continuous at $t = 0$, $f(0^+) = 0$. In this case the impulse function does not appear, and hence

$$\frac{d}{dt}\left[f(t)u(t)\right] = f'(t)u(t) \qquad \text{if } f(0^+) = 0 \tag{2-118}$$

For the case of a shifted function, a similar development yields

$$\frac{d}{dt} \left[f(t - a)u(t - a) \right] = f'(t - a)u(t - a) + f(a^+ - a)\delta(t - a) \qquad (2\text{-}119)$$

where $f(a^+ - a)$ is the value of $f(t - a)$ evaluated immediately to the right of $t = a$. If the function is continuous at $t = a$, Eq. (2-119) reduces to

$$\frac{d}{dt} \left[f(t - a)u(t - a) \right] = f'(t - a)u(t - a) \qquad \text{if } f(a^+ - a) = 0 \qquad (2\text{-}120)$$

The preceding four equations imply that the derivative of a switched function is also a switched function. If there is no discontinuity at the point of switching, we simply differentiate $f(t)$ and "switch it on" by multiplying by the step function. In addition, if there is a discontinuity, we add an impulse term whose area is equal to the value of the discontinuity. The impulse occurs at the point of discontinuity.

Now let us consider the process of integration. To integrate a function multiplied by the step function, we use the step function factor to define the lower limits of integration, but it is not necessary (or desirable) to include the step function factor in the actual integration process. After integration, the proper step function factor is again introduced. These results can be stated mathematically.

$$\int_0^t f(t)u(t) \, dt = \left[\int_0^t f(t) \, dt \right] u(t) \qquad (2\text{-}121)$$

$$\int_0^t f(t - a)u(t - a) \, dt = \left[\int_a^t f(t - a) \, dt \right] u(t - a) \qquad (2\text{-}122)$$

Observe in Eq. (2-122) that although the basic limits of integration given are from 0 to t, the *actual* integration is done from a to t since the function is zero for $t < a$.

Example 2-20

Determine the derivative and integral of the square wave function of Example 2-15. Do this (a) graphically and (b) analytically.

     Solution   (a) The function is shown again in Fig. 2-53a. The slope is zero everywhere except at the points of discontinuities. At these discontinuities, the derivative is a sequence of impulse functions as shown in (b). Note that the negative "jumps" produce negative impulse functions. All impulses in this problem have unit area. (The value 1 side of each impulse represents the area.)

     Since the area is positive, the integral is always increasing or constant as shown in Fig. 2-53c. Each pulse has an area of 2 units, so the integral increases linearly for a range of 2 units during a given pulse width.

     (b) We saw in Example 2-15 that this $f(t)$ could be expressed as

$$f(t) = u(t) - u(t - 2) + u(t - 4) - \cdots$$
$$= \sum_0^\infty (-1)^n u(t - 2n) \qquad (2\text{-}123)$$

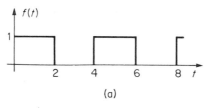

(a)

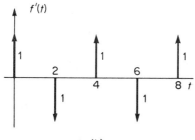

(b)

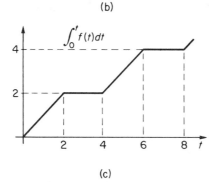

(c)

Figure 2-53 Waveforms of Example 2-20.

Analytically, the derivative is

$$f'(t) = \delta(t) - \delta(t - 2) + \delta(t - 4) - \cdots$$

$$= \sum_0^\infty (-1)^n \delta(t - 2n) \qquad (2\text{-}124)$$

and the integral of $f(t)$ is

$$\int_0^t f(t)\, dt = \int_0^t u(t)\, dt - \int_0^t u(t - 2)\, dt + \int_0^t u(t - 4)\, dt - \cdots \qquad (2\text{-}125)$$

By means of Eq. (2-122), this expression may be written as

$$\int_0^t f(t)\, dt = \left[\int_0^t (1)\, dt \right] u(t) - \left[\int_2^t (1)\, dt \right] u(t - 2)$$

$$+ \left[\int_4^t (1)\, dt \right] u(t - 4) - \cdots \qquad (2\text{-}126a)$$

$$= tu(t) - (t - 2)u(t - 2) + (t - 4)u(t - 4) - \cdots \qquad (2\text{-}126b)$$

The analytical results can be seen to correspond to the graphical results. For this example, the graphical procedure is probably simpler.

Example 2-21

Determine the integral of the pulse of Example 2-16 graphically.

      Solution   The function is shown again in Fig. 2-54a. We first note that since $f(t)$ is positive over the range $0 < t < 4$, its integral must be an increasing function of time over this range as shown in (b). In the interval $0 < t < 1$, $f(t)$ increases linearly with t, and its integral (area) must therefore follow a second-degree variation. The area of the first triangle is $(2 \times 1)/2 = 1$ unit, and thus the integral increases from 0 to 1 unit.

      In the interval $1 < t < 3$, $f(t)$ is constant, and its area must increase linearly. The area gained during this interval is $2 \times 2 = 4$ units, thus bringing the total value of the integral to 5 units. Finally, during the interval $3 < t < 4$, the area continues to increase, but since $f(t)$ is decreasing, the slope of the integral function decreases to zero. The area gained is 1 unit, thus bringing the integral to 6 units. Theoretically, the integral now remains at 6 units for all time thereafter.

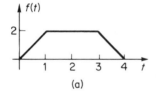

(a)

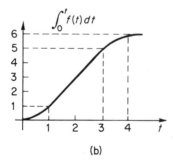

(b)

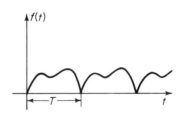

Figure 2-54 Arbitrary periodic function.

Figure 2-55 Waveforms of Example 2-21.

2-16 PERIODIC FUNCTIONS–AVERAGE AND EFFECTIVE VALUES

Among the waveforms that we have considered thus far have been a number of so-called periodic waveforms. *A function $f(t)$ is said to be periodic if $f(t + T) = f(t)$ for all t.* The quantity T is then the *period* of the function. Obviously, the sinusoidal function is one of the most common of periodic functions. A more general periodic function is shown in Fig. 2-55.

In this section we wish to consider two important characteristics of periodic functions: (a) the average or dc value and (b) the effective or rms value. These characteristics will be considered separately.

Average or DC Value

The average or dc value, F_{dc}, of a periodic function $f(t)$ is defined by the relationship

$$F_{dc} = \frac{\text{area under curve in 1 cycle}}{T}$$

$$= \frac{1}{T} \int_0^T f(t)\, dt \tag{2-127}$$

Thus $f(t)$ is integrated over a period and divided by the period. This may involve representing the integral as several integrals whenever the function is defined differently over different parts of the cycle. Whenever the function is composed of simple step or ramp combinations, the area may often be determined graphically.

The average value has a most interesting property whenever it represents a voltage or current waveform. *It can be shown that a true dc voltmeter or ammeter reads the average value of a periodic waveform.* For this reason, we prefer to use the subscript (dc) in referring to the average value of a waveform.

Effective or Root-Mean-Square Value

The effective or root-mean-square (rms) value of a periodic waveform is a term that has been defined for convenience in calculating the average power. To develop this concept, let us calculate the average power in a 1-Ω resistor due to a periodic waveform $f(t)$. The choice of the 1-Ω resistor allows us to consider $f(t)$ as being either a voltage or a current. A resistance other than 1 Ω would merely introduce a constant multiplier.

The instantaneous power $p(t)$ dissipated in the 1-Ω resistor is

$$p(t) = f^2(t) \tag{2-128}$$

Thus, the instantaneous power is, in general, a time-varying quantity that is always positive. A very important quantity is the average power P_{av}. The average power in terms of the instantaneous power is

$$P_{av} = \frac{1}{T} \int_0^T p(t)\, dt \tag{2-129}$$

The *effective* or *rms* value of the waveform F_{rms} is defined for the 1-Ω resistor case by

$$F_{rms}^2 = P_{av} \tag{2-130}$$

Comparison of Eqs. (2-130), (2-129), and (2-128) yields

$$F_{rms} = \sqrt{\frac{1}{T} \int_0^T f^2(t)\, dt} \tag{2-131}$$

Equation (2-131) is the basic defining relationship for the rms value of a periodic waveform $f(t)$. Note that the term *root mean square* is descriptive since we take the *root* of the *mean* of the *square*. ("Mean" is synonymous with "average.")

Considering now the case of a resistor R, the average power due to a periodic current $i(t)$ is

$$P_{av} = I_{rms}^2 R \qquad (2\text{-}132)$$

where I_{rms} is the effective value of $i(t)$. The average power due to a periodic voltage $e(t)$ is

$$P_{av} = \frac{E_{rms}^2}{R} \qquad (2\text{-}133)$$

where E_{rms} is the effective value of $e(t)$. Notice that Eqs. (2-132) and (2-133) are essentially the same type of relationships as are used in calculating power in dc circuits. In other words, the effective value of a periodic voltage or current is a value that is treated like a dc voltage or current in determining average power. Thus the periodic waveform is equivalent, insofar as average power is concerned, to a dc function whose value is the effective value.

Example 2-22

Determine (a) average and (b) rms values for a sinusoidal voltage of amplitude E.
    Solution (a) Since the positive area of a sinusoid over a period is equal in magnitude to the negative area, the net area is zero. Thus

$$E_{dc} = 0 \qquad (2\text{-}134)$$

    (b) To obtain the rms value, we will choose a phase angle of $0°$ for convenience. Thus let

$$e(t) = E \sin \omega t \qquad (2\text{-}135)$$

The mean of the square is

$$E_{rms}^2 = \frac{1}{T} \int_0^T E^2 \sin^2 \omega t \, dt \qquad (2\text{-}136)$$

and since

$$\sin^2 \theta = \tfrac{1}{2}(1 - \cos 2\theta) \qquad (2\text{-}137)$$

we have

$$E_{rms}^2 = \frac{E^2}{2T} \int_0^T (1 - \cos 2\omega t) \, dt$$

$$= \frac{E^2}{2T} \times T = \frac{E^2}{2} \qquad (2\text{-}138)$$

The rms value is

$$E_{rms} = \frac{E}{\sqrt{2}} = 0.707E \qquad (2\text{-}139)$$

which is a well-known result.

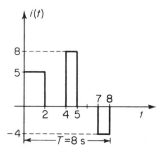

Figure 2-56 Waveform of Example 2-23.

Example 2-23

Determine (a) average and (b) rms values for the periodic current waveform of Fig. 2-56. (Only one cycle has been shown.)

      Solution (a) The area under the curve is readily determined by inspection. Thus

$$\text{area} = 5 \times 2 + 8 \times 1 - 4 \times 1 = 14 \tag{2-140}$$

The dc value is

$$I_{dc} = \frac{14}{8} = 1.75 \text{ A} \tag{2-141}$$

    (b) Since the square of a constant value is also a constant, we may also determine the area under the voltage-squared curve by inspection. Referring to Fig. 2-56, we have

$$\int_0^T i^2(t)\,dt = 25 \times 2 + 64 \times 1 + 16 \times 1 = 130 \text{ A}^2 \cdot \text{s} \tag{2-142}$$

The rms value is

$$I_{rms} = \sqrt{\frac{130}{8}} = 4.03 \text{ A} \tag{2-143}$$

GENERAL PROBLEMS

2-1. A dc voltage of 30 V is switched into a circuit at $t = 0$. Write an expression for the voltage $e(t)$ and sketch it.

2-2. A dc current of 200 mA is switched into a circuit at $t = 0$. Write an expression for the circuit $i(t)$ and sketch it.

2-3. A dc voltage of 12 V is switched on at $t = 5$ s. Write an expression for the voltage $e(t)$ and sketch it.

2-4. A dc current of 80 mA is switched on at $t = 200$ ms. Write an expression for the current $i(t)$ and sketch it.

2-5. A certain time is used as a reference for $t = 0$ in a problem. A certain dc voltage of 60 V is switched on 8 s prior to the reference time. Write an expression for the voltage $e(t)$ and sketch it.

2-6. Sketch the voltage waveform given by

$$e(t) = 20u(3t - 15)$$

2-7. A voltage $v = 5t + 8t^2$ is switched into a circuit at $t = 0$. Write an equation for the switched voltaged v_s.

2-8. A circuit $i = 5t + 3t^2$ is switched into a circuit at $t = 0$. Write an equation for the switched current i_s.

2-9. A certain voltage is described by the equations

$$e = \begin{cases} 0 & \text{for } t < 0 \\ 2t + 6 & \text{for } t > 0 \end{cases}$$

Write the voltage as a single equation and sketch it.

2-10. A certain current is described by the equations

$$i = \begin{cases} 0 & \text{for } t < 0 \\ 0.01t - 0.02 & \text{for } t > 0 \end{cases}$$

Write the current as a single equation and sketch it.

2-11. A ramp voltage with a slope of 8 V/s is switched on at $t = 0$. Write an expression for the function v and sketch it.

2-12. A current that changes at a rate of -2 A/s is switched into a circuit at $t = 0$. If the function starts at a level of zero, write an expression for the current i and sketch it.

2-13. A certain voltage is given by

$$v = -0.5 \times 10^6 tu(t)$$

Sketch the voltage and determine the slope.

2-14. A certain current is given by

$$i = 60tu(t)$$

Sketch the current and determine the slope.

2-15. The output voltage $e_2(t)$ of a certain differentiator is related to the input voltage $e_1(t)$ by the relationship

$$e_2(t) = \frac{de_1(t)}{dt}$$

Determine the output when the input is

$$e_1(t) = (6t + 4t^2)u(t)$$

2-16. For the differentiator of Problem 2-15, determine the output when the input is

$$e_1(t) = (7t^2 + 8t^3)u(t)$$

2-17. The output voltage $e_2(t)$ of a certain integrator is related to the input voltage $e_1(t)$ by the relationship

$$e_2(t) = \int_0^t e_1(t)\, dt$$

Determine the output when the input is

$$e_1(t) = (4 + 6t)u(t)$$

2-18. For the integrator of Problem 2-17, determine the output when the input is

$$e_1(t) = (8t + 9t^2)u(t)$$

2-19. Determine the damping constant and time constant for each of the following functions.
(a) $v(t) = 8\epsilon^{-10^6 t}$ (b) $i(t) = 4\epsilon^{-t/5}$
(c) $v(t) = 6\epsilon^{-0.02t}$ (d) $i(t) = 9\epsilon^{-t/0.04}$

2-20. Determine the damping constant and time constant for each of the following functions.
(a) $e(t) = 20\epsilon^{-5000t}$ (b) $i(t) = 6\epsilon^{-t/40}$
(c) $f(t) = 15\epsilon^{-0.04t}$ (d) $v(t) = 25\epsilon^{-t/0.02}$

2-21. A certain voltage $v(t)$ decays exponentially from an initial level of 80 V with a time constant of 2 ms.
(a) Write an expression for the voltage.
(b) Calculate the voltage for the following values of t: 1 ms, 2 ms, 4 ms, 10 ms.

2-22. A certain current $i(t)$ decays exponentially from an initial level of 20 A with a time constant of 50 ms.
(a) Write an expression for the current.
(b) Calculate the current for the following values of t: 25 ms, 50 ms, 100 ms, 250 ms.

2-23. A certain sinusoidal voltage is given by

$$v(t) = 50 \sin (2000t + 60°)$$

Determine the following properties of the sinusoid.
(a) Amplitude   (b) Angular frequency
(c) Repetition frequency   (d) Period
Sketch the form of the function over the range of cycle.

2-24. Repeat the analysis of Problem 2-23 for the sinusoidal current

$$i(t) = 4 \sin (2\pi \times 10^6 t - 30°)$$

2-25. A certain sinusoidal voltage $e(t)$ has an amplitude of 50 V and a period of 2 ms. At $t = 0$, $e(0) = 0$ and the slope is positive. Write an equation for the voltage and sketch it.

2-26. A certain sinusoidal current $i(t)$ has an amplitude of 8 A and a repetition frequency of 200 Hz. At $t = 0$, the current has its maximum value. Write an equation for the current and sketch it.

2-27. The input voltage to an ideal differentiator is

$$e_1(t) = 4 \sin 3t \qquad \text{for } t > 0$$

Determine the output voltage $e_2(t)$ and sketch the two functions.

2-28. The input voltage to an ideal integrator is

$$e_1(t) = 12 \cos 6t \qquad \text{for } t > 0$$

Determine the output voltage $e_2(t)$ and sketch the two functions.

2-29. Using phasor diagrams express the following functions in terms of the positive cosine function.
(a) $6 \sin (\omega t + 50°)$ (b) $80 \sin (\omega t - 35°)$
(c) $2 \sin (\omega t - 145°)$ (d) $9 \sin (\omega t + 160°)$

(e) $-8 \sin(\omega t - 25°)$ **(f)** $-4 \sin(\omega t + 140°)$
(g) $-2 \cos (\omega t + 30°)$ **(h)** $-7 \cos (\omega t - 100°)$

2-30. Referring to Problem 2-29, express (a) through (f) in terms of the negative cosine function.

2-31. Using phasor diagrams, express the following functions in terms of the positive sine function.

(a) $6 \cos (\omega t + 50°)$ **(b)** $80 \cos (\omega t - 35°)$
(c) $2 \cos (\omega t - 145°)$ **(d)** $9 \cos (\omega t + 160°)$
(e) $-8 \cos (\omega t - 25°)$ **(f)** $-4 \cos (\omega t + 140°)$
(g) $-2 \sin (\omega t + 30°)$ **(h)** $-7 \sin (\omega t - 100°)$

2-32. Referring to Problem 2-31, express (a) through (f) in terms of the negative sine function.

2-33. Determine the phase angles between the following sets of functions and indicate which one is leading in each case.

(a) $e_1 = 80 \sin \omega t, e_2 = 20 \cos \omega t$
(b) $e_1 = 18 \sin \omega t, e_2 = 9 \sin (\omega t + 120°)$
(c) $i_1 = 2 \cos (\omega t - 10°), i_2 = 4 \cos (\omega t - 60°)$
(d) $i_1 = 5 \sin (\omega t + 30°), i_2 = 3 \sin (\omega t - 80°)$

2-34. Repeat the procedure of Problem 2-33 for the following functions.

(a) $i_1 = 8 \cos (\omega t - 125°), i_2 = 5 \sin (\omega t + 20°)$
(b) $e_1 = 60 \sin (\omega t + 40°), e_2 = 40 \cos (\omega t + 80°)$
(c) $e_1 = 40 \sin (\omega t + 15°), e_2 = -25 \cos (\omega t - 70°)$
(d) $i_1 = -2 \sin (\omega t + 30°), e_2 = -3 \cos (\omega t - 60°)$

2-35. The output voltage v_0 of a certain amplifier is given by the sum of two voltages v_1 and v_2: that is,

$$v_0 = v_1 + v_2$$

Express v_0 as a single sinusoid if

$$v_1 = 3 \sin 100t$$

and

$$v_2 = 4 \cos 100t$$

2-36. Repeat the analysis of Problem 2-35 if

$$v_1 = 8 \sin (200t + 30°)$$

$$v_2 = 5 \cos (200t + 60°)$$

2-37. Express the voltage

$$e(t) = 10 \cos (\omega t - 30°) - 20 \sin (\omega t + 45°) + 12 \cos (\omega t - 10°)$$

as a single sinusoid.

2-38. Express the voltage

$$e(t) = 50 \cos (\omega t + 10°) - 50 \cos (\omega t - 50°) + 50 \cos (\omega t - 20°)$$

as a single sinusoid.

2-39. Describe the following functions for $t > 0$.

(a) $i(t) = 10\epsilon^{-2t} \sin 2\pi t$ **(b)** $v(t) = 50\epsilon^{-t} \sin 200\pi t$

2-40. Describe the following functions for $t > 0$.

(a) $f(t) = 20\epsilon^{-10t} \sin (2t + 45°)$ **(b)** $e(t) = 100\epsilon^{-100t} \cos 2000t$

2-41. A ramp voltage $e(t)$ with a slope of 5 V/s is turned on at $t = 4$ s. Write an equation for the voltage and sketch it.

2-42. A ramp current $i(t)$ with a slope of -2 A/s is turned on at $t = 3$ s. Write an equation for the current and sketch it.

2-43. Sketch the voltage given by the equation

$$v(t) = 8(t - 2)u(t - 2)$$

2-44. Sketch the current given by the equation

$$i(t) = -3(t - 4)u(t - 4)$$

2-45. Write an equation for the voltage shown in Fig. P2-45.

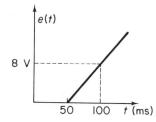

Figure P2-45   **Figure P2-46**

2-46. Write an equation for the current shown in Fig. P2-46.

2-47. A sine-wave current $i(t)$ of amplitude 2 A with a repetition frequency of 50 Hz is turned on at $t = 20$ ms. Write an equation for the current and sketch it.

2-48. A cosine-wave voltage $v(t)$ of amplitude 50 V with a period of 5 ms is turned on at $t = 10$ ms. Write an equation for the voltage and sketch it.

2-49. A certain ideal delay line has a delay of 10 ms. The input voltage $e_1(t)$ is a ramp with a slope of 200 V/s. Write an expression for the output voltage $e_2(t)$ and sketch the two functions.

2-50. For the delay line of Problem 2-49, assume that the input $e_1(t)$ is a sine wave with amplitude 5 V and a period of 20 ms. Write an expression for the output voltage $e_2(t)$ and sketch the two functions.

2-51. Sketch the function given by the equation

$$f(t) = 2u(t - 3) - 2u(t - 7)$$

2-52. Sketch the voltage given by the equation

$$e(t) = 5tu(t) - 10(t - 1)u(t - 1) + 5(t - 2)u(t - 2)$$

2-53. Sketch the current given by the equation

$$i(t) = 5tu(t) - 5(t - 1)u(t - 1) + 5(t - 2)u(t - 2) - 5(t - 3)u(t - 3)$$

2-54. Sketch the function given by the infinite series

$$f(t) = \sum_{n=0}^{\infty} \frac{1}{(2)^n} u(t - nT)$$

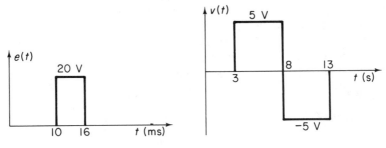

Figure P2-55   **Figure P2-56**

2-55. Write an equation for the pulse waveform of Fig. P2-55.

2-56. Write an equation for the waveform of Fig. P2-56.

2-57. Write an equation for the waveform of Fig. P2-57.

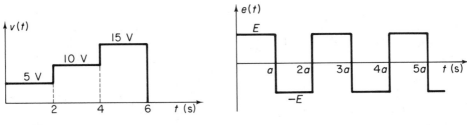

Figure P2-57   **Figure P2-58**

2-58. Write an equation for the square wave of Fig. P2-58. The result will be an infinite series.

2-59. Write an equation for the sawtooth waveform of Fig. P2-59. The result will be an infinite series.

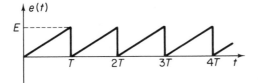

Figure P2-59

2-60. Write an equation for the triangular waveform of Fig. P2-60. The result will be an infinite series.

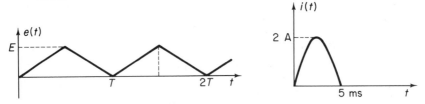

Figure P2-60   **Figure P2-61**

2-61. Write an equation for the single sinusoidal pulse of Fig. P2-61.

2-62. Write an equation for the full-wave rectified voltage of Fig. P2-62. The result will be an infinite series.

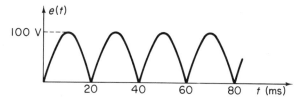

Figure P2-62

2-63. A voltage pulse of amplitude 200 V begins at $t = 0$ and has a width of 50 μs. This time is short compared with the time constants of the system. Write the equation of an impulse voltage $e(t)$ that will approximate the pulse.

2-64. A current pulse of amplitude 80 mA begins at $t = 2$ s and has a width of 50 ms. This width is short compared with the time constants of the system. Write the equation of an impulse current $i(t)$ that will approximate the pulse.

2-65. The input $e_1(t)$ to an ideal infinite range differentiator consists of an approximate step voltage whose final amplitude is 2 V, but whose rise time can be varied. Assume that the input voltage changes linearly until it reaches its final value. Sketch the output voltage $e_2(t)$ for the rise times given below. In addition, determine analytical expressions for input and output in (c).
(a) Rise time $= 1$ s   **(b)** Rise time $= 0.1$ s   **(c)** Rise time $= 0$

2-66. The input $e_1(t)$ to an ideal integrator is a pulse whose width t_1 can be adjusted but whose amplitude changes inversely with the width so as to maintain constant area. The area is set to a value of 12 V $\cdot$ s. Sketch the output voltage $e_2(t)$ for each of the values of t_1 given below. In addition, determine analytical expressions for $e_1(t)$ and $e_2(t)$ in (c).
(a) $t_1 = 1$ s   **(b)** $t_1 = 0.1$ s   **(c)** $t_1 = 0$

2-67. A certain voltage is given by

$$e(t) = \begin{cases} 0 & \text{for } t < 0 \\ 50 & \text{for } t > 0 \end{cases}$$

Write an expression for its derivative $e'(t)$ and sketch the two functions.

2-68. A certain current is given by

$$i(t) = \begin{cases} 0 & \text{for } t < 4 \text{ s} \\ 6 & \text{for } t > 4 \text{ s} \end{cases}$$

Write an expression for its derivative $i'(t)$ and sketch the two functions.

2-69. Write an equation that approximates the definite integral of the voltage of Problem 2-63.

2-70. Write an equation that approximates the definite integral of the current of Problem 2-64.

2-71. Differentiate the pulse of Problem 2-55 both graphically and analytically.

2-72. Differentiate the waveform of Problem 2-56 both graphically and analytically.

2-73. Differentiate the pulse of Example 2-16 (Fig. 2-43) both graphically and analytically.

2-74. Differentiate the triangular waveform of Problem 2-60 both graphically and analytically.

2-75. Integrate the waveform of Problem 2-55 both graphically and analytically.

2-76. Integrate the waveform of Problem 2-56 both graphically and analytically.

2-77. Integrate the waveform of Problem 2-57 both graphically and analytically.

2-78. Integrate the waveform of Problem 2-58 both graphically and analytically.

2-79. One cycle of a certain periodic voltage is shown in Fig. P2-79.
   (a) Determine the average and rms values.
   (b) Determine the average power dissipated by the voltage in a 5-Ω resistor.

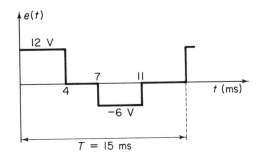

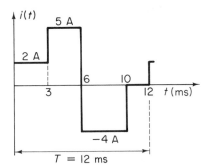

Figure P2-79   **Figure P2-80**

2-80. One cycle of a certain periodic current is shown in Fig. P2-80.
   (a) Determine the average and rms values.
   (b) Determine the average power dissipated by the current in a 2-Ω resistor.

DERIVATION PROBLEMS

2-81. A sinusoidal voltage $e_1(t) = E \sin \omega t$ is applied as the input to an ideal differentiator circuit. The input amplitude is held constant but the frequency is varied. Show that the amplitude of the output voltage $e_2(t)$ is directly proportional to the input frequency.

2-82. A sinusoidal voltage $e_2(t) = E \cos \omega t$ is applied as the input to an ideal integrator circuit. The input amplitude is held constant but the frequency is varied. Show that the amplitude of the output voltage $e_2(t)$ is inversely proportional to the input frequency.

2-83. From the results of Problem 2-81, show that the sinusoidal output of an ideal differentiator leads the input by 90°. Sketch the two functions on the same time scale.

2-84. From the results of Problem 2-82, show that the sinusoidal output of an ideal integrator lags the input by 90°. Sketch the two functions on the same time scale.

2-85. A general combination of a dc component and a single exponential is given by

$$f(t) = A + B\epsilon^{-t/\tau}$$

Show that this equation may be expressed in the form

$$f(t) = f(\infty) + [f(0) - f(\infty)]\epsilon^{-t/\tau}$$

where $f(0)$ represents the initial value of $f(t)$ at $t = 0$ and $f(\infty)$ represents the final value.

2-86. Consider the function switched *off* at $t = 0$ as described by

$$f(t) = \begin{cases} 1 & \text{for } t < 0 \\ 0 & \text{for } t > 0 \end{cases}$$

(a) Show that the derivative is

$$f'(t) = -\delta(t)$$

(b) Sketch the two functions.

2-87. Derive expressions for the average and rms values of the pulse train of Fig. P2-87.

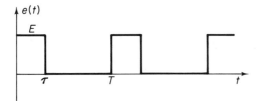

Figure P2-87

2-88. Derive expressions for the average and rms values of the square wave of Fig. P2-88.

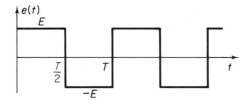

Figure P2-88

2-89. Derive expressions for the average and rms values of the half-wave rectified voltage of Fig. P2-89.

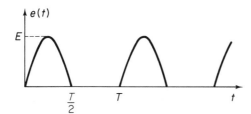

Figure P2-89

2-90. Derive expressions for the average and rms values of the full-wave rectified voltage of Fig. P2-90.

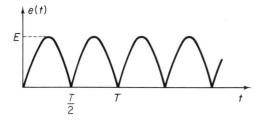

Figure P2-90

COMPUTER PROBLEMS

2-91. Consider a function of the form

$$y = A + B^{-t/\tau}$$

Write a program to compute y over the range from t_1 to t_2 in steps of Δt.

   Input data: $A, B, \tau, t_1, t_2, \Delta t$
   Output data: y

2-92. Consider the sinusoidal function

$$y = A \sin (2\pi ft + \theta)$$

Write a program to compute y over the range from t_1 to t_2 in steps of Δt.

   Input data: $A, f, \theta, t_1, t_2, \Delta t$
   Output data: y

Note: The form of the sinusoidal function accepting radians as the angle will normally be used, and the input angle θ should be expressed in radians.

3

CIRCUIT PARAMETERS

OVERVIEW

The voltage–current relationships for the three basic circuit parameters are studied in detail in this chapter. The three basic circuit parameters are (a) *resistance*, (b) *capacitance*, and (c) *inductance*. The voltage–current relationship for resistance is a simple algebraic equation (Ohm's law), while the relationships for capacitance and inductance involve differential and integral calculus.

The analysis in this chapter utilizes both graphical and analytical techniques. Emphasis is directed toward interpreting the instantaneous relationships between the current into the component and the voltage across it.

OBJECTIVES

After completing this chapter, the reader should be able to

1. Analyze resistive circuits containing one or more time-varying sources.
2. State the voltage–current relationship for a *capacitance*.
3. Apply the relationship of (2), either analytically or graphically, to determine either variable if the other is known.
4. Determine the Thévenin or Norton equivalent circuit of a charged capacitor.
5. State the voltage–current relationship for an *inductance*.
6. Apply the relationship of (5), either analytically or graphically, to determine either variable if the other is known.
7. Determine the Norton or Thévenin equivalent circuit of a fluxed inductor.
8. State the voltage–current relationships for two mutually coupled inductances.
9. Construct a dynamic equivalent circuit to represent the effects of mutual inductance.

63

10. **Determine the equivalent inductance of simple (e.g., series) connections of mutually coupled coils.**
11. **Define the *coefficient of coupling*.**
12. **State the characteristics of an *ideal transformer*.**
13. **State the voltage–current relationships for an ideal transformer.**
14. **Reflect the secondary circuit to the primary, or vice versa, for a circuit containing an ideal transformer.**
15. **Analyze special *RLC* circuit combinations in which each individual element may be isolated for voltage–current analysis.**

3-1 THE RESISTIVE CIRCUIT

The reader is assumed to be familiar with the analysis of linear resistive circuits with dc excitations. To generalize the linear resistive circuit, we need only extend the excitations to include time-varying sources. Although the responses due to such excitations are time varying, we may still use modified dc circuit analysis techniques.

The voltage–current relationship for a single resistance with time-varying voltage and current is

$$v(t) = Ri(t) \tag{3-1}$$

or

$$i(t) = \frac{v(t)}{R} = Gv(t) \tag{3-2}$$

where R is the resistance in *ohms* (Ω) and $G = 1/R$ is the conductance in *siemens* (S).

This idea is illustrated in Fig. 3-1 for some arbitrary waveform. An important point is that *the shapes of the voltage and current waveforms for a resistor are identical.* Of course, the units and scales are different, but the shapes are the same.

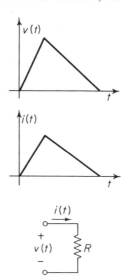

Figure 3-1 Illustration of the fact that the voltage and current waveforms associated with a resistor are identical.

Assume now that a resistive circuit is excited by a single time-varying source. As a result of the linear relationship for a resistor, *all voltages and currents in a purely resistive network will have the same waveform as the source for a single source of excitation.* Clearly, then, for a single excitation, dc circuit analysis can be used to determine the responses in terms of the source, which is perhaps assigned a symbol during the calculations. After the desired response is determined in terms of the source, the value of the symbol is substituted, and the actual time response is obtained. Thus the resistive network simply acts as a scale factor for a single time-varying source.

If there are two or more sources of excitation that have different waveforms, the principle of superposition tells us that any desired response can be considered as the sum of two or more separate responses, each having the waveform of the source that produces it. The result is, then, a linear combination of the sources. Thus *if there are two or more time-varying sources with different waveforms, the responses, in general, will have different waveforms than either individual source.*

In solving such a circuit, we simply use two or more symbols to reperesent the different sources, and solve the problem by dc methods. Any response will then be expressed as a linear combination of the sources, and by substituting the equations for the sources, we may obtain a final expression. In some cases, it is desirable to perform this latter step graphically.

The instantaneous power $p(t)$ in a resistor with time-varying voltage and current is

$$p(t) = v(t)i(t) = Ri^2(t) = \frac{v^2(t)}{R} \tag{3-3}$$

If the voltage and current are *periodic*, the average power over a cycle is

$$P_{av} = \frac{1}{T} \int_0^T p(t)\, dt$$

$$= V_{rms}I_{rms} = RI_{rms}^2 = \frac{V_{rms}^2}{R} \tag{3-4}$$

according to the definitions of Chapter 2.

Example 3-1

The resistive circuit of Fig. 3-2 is excited by the single sinusoidal voltage given. Determine the mesh currents i_1 and i_2. Determine also the average power dissipated in the 2-kΩ resistor.

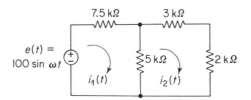

Figure 3-2   Circuit of Example 3-1.

Solution   We shall designate the source as e during the computations. For convenience we will scale the element values. (A complete treatment of scaling is given in Appendix B.) If the resistances are scaled by a factor 1/1000, the currents will be multiplied by 1000 and can thus be considered to be expressed in mA. The mesh equations are

$$12.5i_1 - 5i_2 = e \tag{3-5}$$

$$-5i_1 + 10i_2 = 0 \tag{3-6}$$

resulting in

$$i_1 = 0.1e \text{ mA} \tag{3-7}$$

$$i_2 = 0.05e \text{ mA} \tag{3-8}$$

So far the problem has essentially been identical to a dc resistive problem. However, at this point we must substitute $e = 100 \sin \omega t$ volts into the expressions for i_1 and i_2. We thus obtain

$$i_1 = 10 \sin \omega t \text{ mA} \tag{3-9}$$

$$i_2 = 5 \sin \omega t \text{ mA} \tag{3-10}$$

For this resistive circuit containing only a single excitation, all responses have the same sinusoidal waveform.

The rms value of i_2 is

$$I_{2\text{rms}} = \frac{5}{\sqrt{2}} \text{ mA} \tag{3-11}$$

The average power dissipated in the 2-kΩ resistor is

$$P_{\text{av}} = \left(\frac{5}{\sqrt{2}} \times 10^{-3}\right)^2 \times 2 \times 10^3 = 25 \text{ mW} \tag{3-12}$$

Example 3-2

The resistive circuit of Fig. 3-3a is excited by the sources $e_1(t)$ and $e_2(t)$ whose waveforms are shown in (b) and (c), respectively. Solve for $v_0(t)$ and sketch its waveform.

Solution   We will first solve for v_0 in terms of e_1 and e_2. The mesh current equations will be employed again. They are

$$2i_1 - i_2 = e_1 \tag{3-13}$$

$$-i_1 + 3i_2 = e_2 \tag{3-14}$$

Solution for i_2 yields

$$i_2 = \frac{1}{5}e_1 + \frac{2}{5}e_2 \tag{3-15}$$

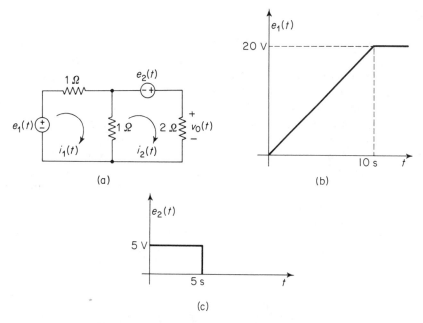

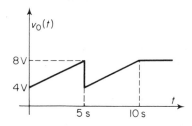

(a)   (b)

(c)

Figure 3-3   Circuit and excitations for Example 3-2.

Figure 3-4   Output voltage of Example 3-2.

and since $v_0 = 2i_2$,

$$v_0 = \frac{2}{5}e_1 + \frac{4}{5}e_2 \qquad (3\text{-}16)$$

We now can substitute expressions for e_1 and e_2 into the expression for v_0. The reader is invited to show that e_1 and e_2 can be expressed as

$$e_1(t) = 2tu(t) - 2(t - 10)u(t - 10) \qquad (3\text{-}17)$$

$$e_2(t) = 5u(t) - 5u(t - 5) \qquad (3\text{-}18)$$

Substitution of e_1 and e_2 into v_0 yields

$$v_0(t) = 4u(t) + \frac{4}{5}tu(t) - 4u(t - 5) - \frac{4}{5}(t - 10)u(t - 10) \qquad (3\text{-}19)$$

This function is shown in Fig. 3-4.

3-2 CAPACITANCE

We now wish to turn our attention to the second of the three basic types of circuit parameters, namely *capacitance*. It will be assumed that the reader has been exposed to the physical phenomena involved, as this topic is adequately covered in many beginning circuit books and most electric and magnetic fields books. Instead, we will concentrate only on the circuit aspects of capacitance. A lumped "package" of capacitance is called a *capacitor*, and the schematic representation is shown in Fig. 3-5. The current flow associated with a capacitor results from a transfer of charge from one plate to the other through the circuit external to the capacitor. However, since the effect is the same as if the current were actually flowing through the capacitor, we shall use the common terminology of current flow *through* the capacitor to represent the actual effect.

It can be determined experimentally that capacitor current is proportional to the rate of change of voltage across it. Thus

$$i(t) \propto \frac{dv(t)}{dt} \tag{3-20}$$

The constant of proportionality is called the *capacitance* of the capacitor and is denoted by the symbol C. The basic capacitance is the *farad* (F), which is a rather enormous unit. More practical units are microfarads (μF) and picofarads (pF), where $1\ \mu\text{F} = 10^{-6}$ farad and $1\ \text{pF} = 10^{-12}$ farad. In order to concentrate on basic principles and eliminate tedious computation, we will frequently assume unrealistic, simple values of capacitance in solving problems. This approach is further justified by the normalization process discussed in Appendix B.

Using the capacitance parameter, the voltage–current relationship is

$$i(t) = C \frac{dv(t)}{dt} \tag{3-21}$$

This relationship is illustrated in Fig. 3-6 for some arbitrary $v(t)$. As can be seen, during intervals when $v(t)$ is changing rapidly (or equivalently, the slope is large), $i(t)$ is larger.

Figure 3-5 Circuit diagram for capacitor.

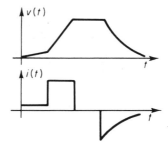

Figure 3-6 Illustration of how capacitor current depends on slope of voltage curve.

On the other hand, when $v(t)$ is constant, $i(t)$ is zero. *Thus, in the dc steady-state, the current through a capacitor is zero, implying that the capacitor acts as an open circuit under this condition.*

Occasionally, we want to calculate the average current through a capacitor, over a short period of time, as a result of a change in voltage whose exact behavior is not readily known or easily calculated. In this case we can approximate Eq. (3-21) as a ratio of differentials and write

$$I_{av} = C \frac{\Delta v}{\Delta t} \tag{3-22}$$

where Δv represents the *change* in voltage, Δt represents the *change* in time, and I_{av} represents the *average value* of the current.

Equation (3-21) provides us with a knowledge of the current if the voltage is known. The inverse relationship is equally important. A basic property of a capacitor is that it holds charge; hence we must consider a possible "history" of the device before the initial point of observation if there is any charge present. The voltage at time t from Eq. (3-21) is

$$v(t) = \frac{1}{C} \int_{-\infty}^{t} i(t) \, dt \tag{3-23}$$

If we are interested only in the behavior of the circuit for $t > 0$, the integral from $-\infty$ to 0 of Eq. (3-23) is actually the initial voltage on the capacitor. Letting V_0 represent this initial voltage, we have

$$V_0 = \frac{1}{C} \int_{-\infty}^{0} i(t) \, dt \tag{3-24}$$

and hence Eq. (3-23) becomes

$$v(t) = \frac{1}{C} \int_{0}^{t} i(t) \, dt + V_0 \tag{3-25}$$

Thus the voltage at any time t $(t > 0)$ is the initial voltage plus the area under the current versus time curve multiplied by $1/C$. This idea is illustrated in Fig. 3-7 for some particular $i(t)$ and an initial voltage V_0.

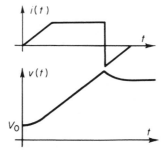

Figure 3-7 Illustration of how capacitor voltage depends on area under current curve.

A resistor *dissipates* the energy that it receives, while a capacitor *stores* energy. It is shown in most basic circuit books that the energy (W) in joules stored in a capacitor charged to V volts is

$$W = \frac{1}{2} CV^2 \qquad (3\text{-}26)$$

This energy is a form of *potential* energy since it is a result of charges at rest.

Example 3-3

The uncharged capacitor shown in Fig. 3-8a is excited at $t = 0$ by the ideal dc voltage shown in (b). Calculate the current.

      Solution   Since the voltage is switched across the capacitor terminals at $t = 0$, we may write

$$v(t) = 10u(t) \qquad (3\text{-}27)$$

The current is

$$i(t) = C \frac{dv(i)}{dt} = 20 \frac{d}{dt} u(t) = 20\delta(t) \qquad (3\text{-}28)$$

Thus, for this ideal, unrealizable case, the current would be an impulse as shown in Fig. 3-8c. Practically, there will always be a short rise time and thus the current pulse, while large, will be finite. Notice that once the pulse of current charges the capacitor, no further current flows, thus indicating a steady-state condition.

Example 3-4

The capacitor shown in Fig. 3-9a is excited at $t = 0$ by the voltage pulse shown in (b). Solve for the current and sketch it.

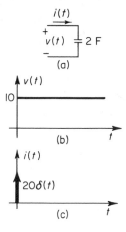

Figure 3-8 Waveforms and capacitor for Example 3-3.

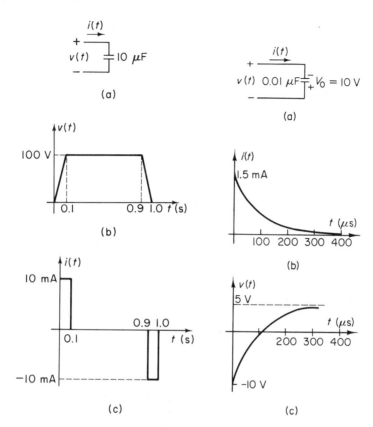

Figure 3-9   Waveforms and capacitor for Example 3-4.

Figure 3-10   Waveforms and capacitor for Example 3-5.

Solution   If desired, we could write the voltage waveform in terms of ramp functions and then perform an analytical differentiation. However, a simpler approach to this problem is a graphical differentiation by inspection. The slope of the voltage curve during the rising portion is $100 \text{ V}/0.1 \text{ s} = 1000 \text{ V/s}$. From Eq. (3-21) the current in this interval is $i = 10 \times 10^{-6} \times 10^{3} = 10 \text{ mA}$. In the interval from $0.1 < t < 0.9$, the slope is zero, and thus no current flows. Finally, during the last interval the slope is -1000 V/s, and thus the current is -10 mA. The resulting waveform of current is shown in Fig. 3-9c.

Example 3-5

The capacitor shown in Fig. 3-10a is initially charged to 10 V in the direction shown. At $t = 0$ the exponential current shown in (b) begins to flow through the circuit. Calculate and sketch the voltage across the capacitor. The time constant of the current is 100 μs.

Solution   This problem is best approached by an analytical scheme. The equation of the current is

$$i(t) = (1.5 \times 10^{-3})\epsilon^{-t/(1 \times 10^{-4})} = (1.5 \times 10^{-3})\epsilon^{-10^4 t} \qquad (3\text{-}29)$$

From Eq. (3-25), the voltage is

$$v(t) = \frac{1.5 \times 10^{-3}}{10^{-8}} \int_0^t \epsilon^{-10^4 t} \, dt - 10$$

$$= -15\epsilon^{-10^4 t} \Big]_0^t - 10 \qquad (3\text{-}30)$$

$$= 5 - 15\epsilon^{-10^4 t}$$

A sketch of the voltage is shown in Fig. 3-10c.

3-3 EQUIVALENT CIRCUITS OF CHARGED CAPACITOR

Let us inspect again the equation in which the voltage across a capacitor is expressed in terms of the current and the initial voltage. We have

$$v(t) = \frac{1}{C} \int_0^t i(t) \, dt + V_0 \qquad (3\text{-}31)$$

Equation (3-31) expresses the total voltage as the sum of two separate voltages. This is precisely the relationship expected for a *series* combination of two separate voltages. The first voltage is simply the voltage that would accumulate on the capacitor as time passed *if the capacitor were uncharged* and *if the current i were the same in the uncharged case*. The second term is simply a dc term of value V_0.

Thus *as far as the external terminals are concerned, a charged capacitor is equivalent, for t > 0, to an uncharged capacitor in series with a dc or step voltage source.* This equivalence is illustrated in Fig. 3-11. The polarity of the voltage source must be observed. Due to the configuration, we may say that this is the *Thévenin equivalent circuit of a charged capacitor.*

We wish to emphasize the importance of the equivalent circuit being true as far as the *external terminals* are concerned. In this case, the external terminals of the capacitor must be interpreted as being terminals a and a' in Fig. 3-11. In other

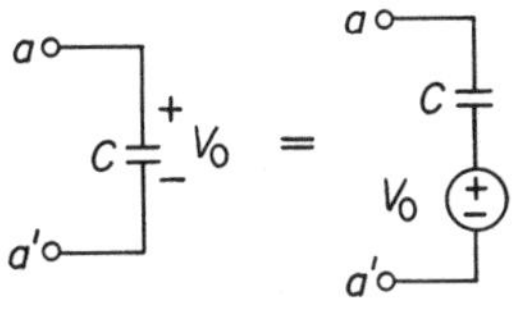

Figure 3-11   Thévenin equivalent circuit of a charged capacitor.

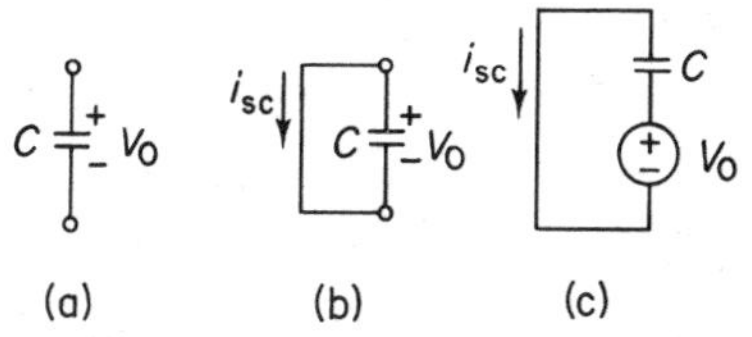

Figure 3-12   Development of Norton equivalent circuit of a charged capacitor.

words, *the dc voltage must be added to the accumulated voltage across the uncharged capacitor to obtain the true voltage across the actual capacitor at any time.*

The beauty of this transformation is that it will allow us, in solving complicated problems later in the text, to work only with uncharged capacitors. It will simply add more sources in a circuit, but this usually adds no unusual complexity to a problem.

The question now arises as to whether or not a Norton equivalent circuit representation can be achieved for a charged capacitor. The answer is yes, if we allow the impulse function to enter into the picture.

Let us see how the Norton equivalent circuit is obtained. Remembering the procedure for obtaining a Norton equivalent circuit, we short-circuit the terminals of the charged capacitor in Fig. 3-12a, as shown in (b). An equivalent circuit is shown in (c) using the Thévenin circuit that we have already derived. In this equivalent representation, the effect of the hypothetical source V_0 is to cause a current to flow through C, since V_0 is connected across C. In any case, the current must be given by

$$i = C \frac{dv}{dt} \tag{3-32}$$

If the short is suddenly placed across the capacitor at $t = 0$, the voltage can be expressed as

$$v(t) = V_0 u(t) \tag{3-33}$$

In the limit, as pointed out in Chapter 2, the derivative of an ideal step function is an impulse function. Thus the short-circuit current, i_{ss}, is

$$i_{ss} = C V_0 \delta(t) \tag{3-34}$$

Hence, the Norton equivalent circuit of a charged capacitor is an impulse current source in parallel with an uncharged capacitor. The impulse is considered to occur at the instant that we wish to begin "recording" a solution, namely $t = 0$. The whole purpose of this impulse current is to establish the initial voltage on the capacitor as required by the initial condition. The Thévenin and Norton equivalent circuits are compared in Fig. 3-13.

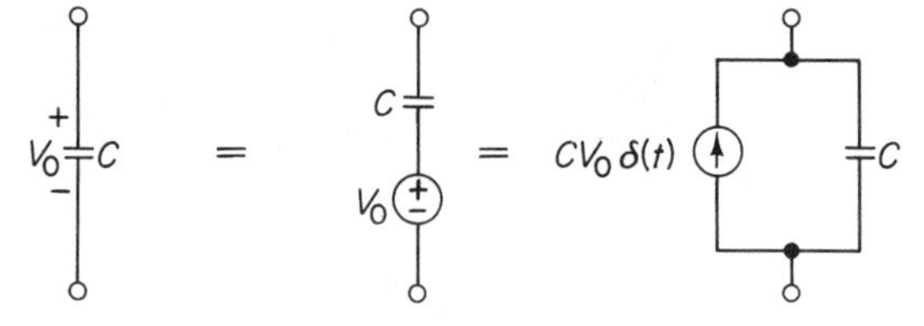

Figure 3-13 Comparison of Thévenin and Norton equivalent circuits of charged capacitor.

When the impulse function was introduced in Chapter 2, it was pointed out that although the ideal impulse was artificial, it frequently arose in conjunction with circuit transformations. Such is the case at hand. Whenever it will be desirable to use current sources in a network containing a charged capacitor (such as in node voltage analysis), we may obtain such a representation if we allow the use of the impulse function. Again, the point of external terminals must be observed.

3-4 SELF-INDUCTANCE

The third and final basic circuit parameter is *inductance*. Inductance is further divided into two types: (a) *self-inductance* and (b) *mutual inductance*. Self-inductance is a property of a single coil, whereas mutual inductance requires that mutual flux link between two or more coils. In this section we will consider only self-inductance. *Whenever no adjective appears in front of the term inductance, it is normally interpreted to mean self-inductance.*

As in the case of capacitance, it will be assumed that the reader has been exposed to the physical phenomena of inductance. We will concentrate only on the circuit aspects of inductance. A lumped "package" of inductance is formally called an *inductor*. Other terms that are often used in certain cases are *coil* and *choke*. The schematic representation of an inductor is shown in Fig. 3-14.

It can be verified experimentally that the voltage induced across an inductor is proportional to the rate of change of current through it. Thus

$$v(t) \propto \frac{di(t)}{dt} \tag{3-35}$$

The constant of proportionality is called the *inductance* and is assigned the symbol L. The basic unit of *inductance* is the *henry* (H). Unlike the farad for capacitors, inductors with inductances in the range of henries will be frequently encountered. In addition, smaller inductances of the order of millihenries (mH) and microhenries (μH) are quite common.

Using the inductance parameter, we have the voltage–current relationship:

$$v(t) = L\frac{di(t)}{dt} \tag{3-36}$$

This relationship is illustrated in Fig. 3-15 for some arbitrary $i(t)$. As can be seen, during periods when $i(t)$ is changing rapidly, $v(t)$ is large. On the other hand, when $i(t)$ is constant, $v(t)$ is zero. Thus, *in the dc steady state, the voltage across a pure*

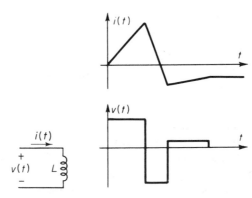

Figure 3-14 Circuit diagram for inductor.

Figure 3-15 Illustration of how inductor voltage is related to slope of current curve.

inductor is zero, implying that the inductor acts as a short circuit under this condition. Since all inductors contain some resistance in the winding, a small resistive voltage drop is present in a practical inductance under operating conditions.

Occasionally, we want to calculate the average voltage across an inductance, over a short period of time, as a result of a change in current whose exact behavior is not readily known or easily calculated. In this case, we can approximate Eq. (3-36) as a ratio of differentials and write

$$V_{av} = L \frac{\Delta i}{\Delta t} \tag{3-37}$$

where Δi represents the *change* in current, Δt represents the *change* in time, and V_{av} represents the *average value* of the voltage.

Equation (3-36) provides us with a knowledge of the voltage across an inductance, if the current through it is known. The inverse relationship is equally important. Since the inductor "stores" current, we must consider a possible "history" of the device prior to the time of observation, if there has been a prior excitation. The current from Eq. (3-36) is

$$i(t) = \frac{1}{L} \int_{-\infty}^{t} v(t) \, dt \tag{3-38}$$

If we are interested only in the behavior of the current for $t > 0$, the integral from $-\infty$ to 0 of Eq. (3-38) is actually the initial current through the inductor. Letting I_0 represent this initial current, we have

$$I_0 = \frac{1}{L} \int_{-\infty}^{0} v(t) \, dt \tag{3-39}$$

Hence Eq. (3-38) becomes

$$i(t) = \frac{1}{L} \int_{0}^{t} v(t) \, dt + I_0 \tag{3-40}$$

Thus the current at any time t $(t > 0)$ is the initial current plus the area under the voltage-vs.-time curve multiplied by $1/L$. This idea is illustrated in Fig. 3-16 for some particular $v(t)$ and an initial current $I = 0$.

A pure inductor, in the same sense as a capacitor, *stores* energy rather than dissipates it. However, while the energy stored in the capacitor is a form of *potential*

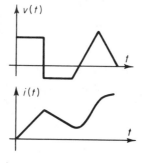

Figure 3-16 Illustration of how inductor current is related to area under voltage curve.

energy and is dependent on the voltage across the capacitor, the energy stored in an inductor is *kinetic* energy and is dependent on the current through the inductor. The stored energy (W) in joules can be shown to be

$$W = \frac{1}{2} LI^2 \qquad (3\text{-}41)$$

where I is the current through the inductor. This energy is kinetic since the current depends on charge in motion. Both capacitors and inductors are referred to as *energy-storage* devices.

As we have already seen, a capacitor containing an initial voltage is said to be *charged* since the voltage is related to electric charge. We shall refer to an inductor carrying an initial current as a *fluxed* inductor since the current is related to magnetic flux.

Example 3-6

The unfluxed inductor shown in Fig. 3-17a is excited at $t = 0$ by the current shown in (b). Solve for the voltage across this inductance and sketch it.

     Solution   This problem is probably best handled by a graphical procedure. The basic relationship is

$$v = L \frac{di}{dt} \qquad (3\text{-}42)$$

At $t = 0$, the current jumps instantaneously from 0 to 2 A. In the ideal case, the slope (or derivative) is an impulse function of area 2 A · s; hence $v(0)$ is an impulse given by

$$v(0) = L \frac{di(0)}{dt} = 5 \times 2\delta(t) = 10\delta(t) \qquad (3\text{-}43)$$

During the interval $0 < t < 1$, the current is constant and $v = 0$. During the interval from $1 < t < 2$, the slope of the current is 2 A/s, and hence

$$v(t) = 10 \text{ V} \qquad \text{for } 1 < t < 2 \qquad (3\text{-}44)$$

The reader should be able to complete the response, which is shown in Fig. 3-17c.

Example 3-7

The 2-H inductance shown in Fig. 3-18a is fluxed to 5 A in the direction shown. (The current is actually flowing into some external circuit not shown.) For $t > 0$, the voltage is given by

$$v(t) = 10\epsilon^{-t/4} \qquad (3\text{-}45)$$

as shown in (b). Calculate the current through the inductor for $t > 0$.

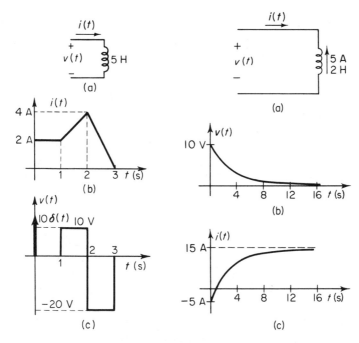

Figure 3-17 Inductor and waveforms of Example 3-6.

Figure 3-18 Inductor and waveforms of Example 3-7.

Solution The current in amperes is given by

$$i(t) = \frac{1}{L} \int_0^t v(t) \, dt + I_0 = \frac{1}{2} \int_0^t 10\epsilon^{-t/4} \, dt - 5$$

$$= -\frac{4}{2} \times 10\epsilon^{-t/4} \Big]_0^t - 5 = 15 - 20\epsilon^{-t/4} \qquad (3\text{-}46)$$

The current is shown in Fig. 3-18c.

3-5 EQUIVALENT CIRCUITS OF FLUXED INDUCTOR

We have seen how a capacitor with an initial electric charge could be represented by either a Thévenin or a Norton equivalent circuit, each containing a source and an equivalent uncharged capacitor. We now wish to show that an inductor containing an initial magnetic flux can also be represented by similar forms.

Consider again the expression for current through an inductor. The expression is

$$i(t) = \frac{1}{L} \int_0^t v(t) \, dt + I_0 \qquad (3\text{-}47)$$

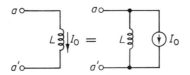

Figure 3-19 Norton equivalent circuit of a fluxed inductor.

Equation (3-47) expresses the total current as the sum of two separate currents. This is exactly the relationship that exists for a *parallel* combination of two separate currents. The first current is simply the current that would flow through the inductor, as time passed, *if the inductor were unfluxed* and *if the voltage v were the same in the unfluxed case*. The second term assures that this condition is met. It is simply a dc current of value I_0.

Thus, *as far as the external terminals are concerned, a fluxed inductor is equivalent, for t > 0, to an unfluxed inductor in parallel with a dc or step-current source*. This equivalence is illustrated in Fig. 3-19. The direction of the current source must be observed. Due to the configuration, this circuit may be called the *Norton equivalent circuit of a fluxed inductor*. In all cases, the external terminals of the inductor are interpreted to mean a-a' in Fig. 3-19. In other words, the dc current must be added to the current through the unfluxed inductor to obtain the true current through the actual inductor at any time.

The Thévenin equivalent circuit of the fluxed inductor may also be obtained. Consider the fluxed inductor shown in Fig. 3-20a whose Norton equivalent circuit is shown in (b). To determine the Thévenin equivalent, we must open the loop at $t = 0$ and measure the open-circuit voltage as shown in (c). From the equivalent circuit, the current must establish itself instantaneously through the equivalent inductance. The voltage is, of course, given by

$$v = L\frac{di}{dt} \tag{3-48}$$

As a result of the switching action, the current through L can be expressed as

$$i(t) = I_0 u(t) \tag{3-49}$$

Substituting Eq. (3-49) into Eq. (3-48) and considering the ideal limit, we obtain the open-circuit voltage:

$$v_{oc} = LI_0\delta(t) \tag{3-50}$$

Hence the Thévenin equivalent circuit of a fluxed inductor is an unfluxed inductor in series with an impulse voltage source that occurs at $t = 0$. The impulse voltage source serves to establish an instantaneous current through the inductor at $t = 0$ to satisfy the initial condition. The Norton and Thévenin equivalent circuits are compared in Fig. 3-21.

(a)

(b)

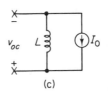
(c)

Figure 3-20 Development of the Thévenin equivalent circuit of a fluxed inductor.

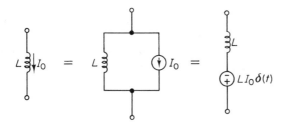

Figure 3-21 Comparison of Norton and Thévenin equivalent circuits of fluxed inductor.

3-6 MUTUAL INDUCTANCE

We now wish to consider the phenomena of *mutual inductance*. Mutual inductance exists whenever the magnetic flux of one coil links with the flux of another coil. Such coupling may exist deliberately as in the case of a transformer, or it may be accidental as a result of improper circuit placement. In either case, the effect of mutual coupling must be considered if it is present in a given circuit. As in the case of the previous parameters, we concentrate only on the circuit aspects of mutual inductance.

To consider mutual inductance, we must have at least two coils. As an example, consider the two coils wound on a common core as shown in Fig. 3-22. The two coils each possess self-inductance which we shall designate as L_1 and L_2, respectively. If a current flowing in either winding is changing with time, a voltage is induced in that respective winding as a result of the self-inductance. In addition, the changing flux linking with the other coil also induces a voltage in that particular coil.

First assume that coil 2 is open and a current i_1 is flowing in coil 1. It can be determined experimentally that the open-circuit voltage induced in L_2 is proportional to the time rate of change of i_1. Thus

$$v_{2oc} \propto \frac{di_1}{dt} \tag{3-51}$$

If we now open-circuit coil 1 and feed a current i_2 into L_2, an open-circuit voltage across coil 1 can be measured. Thus

$$v_{1oc} \propto \frac{di_2}{dt} \tag{3-52}$$

The constants of proportionality for both cases can be shown to be equal by the principle of reciprocity. This constant is called the *mutual inductance* and is designated by the symbol M. The basic unit is the same as for self-inductance, namely

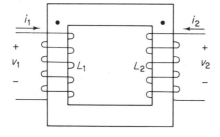

Figure 3-22 Two coils with mutual coupling.

Figure 3-23 Static equivalent circuit of mutually coupled coils.

the *henry*. Hence

$$v_{2oc} = M \frac{di_1}{dt} \tag{3-53}$$

$$v_{1oc} = M \frac{di_2}{dt} \tag{3-54}$$

The polarity of the mutual voltage is determined by the relative directions of the windings. Since it is not always possible to "look inside" a pair of coils and determine the manner in which the coils are wound, a dot convention is employed to specify the relative directions from a circuit point of view. Schematic representations of two pairs of mutual coils with two possible relative dot locations are shown in Fig. 3-23. The dot convention is defined as follows: *If the positive reference direction of either current is assumed into the dot, the voltage induced in the other winding is positive at the dot end of that coil.* For the coils considered in Fig. 3-22, if the dot of coil 2 is also at the top. If either winding were wound in the opposite direction, the dot of one coil would be reversed. However, if both coils were wound in opposite directions, the dots would retain the first-mentioned directions.

Let us designate either schematic drawing of Fig. 3-23 as a *static* equivalent circuit for a pair of mutually coupled coils. To solve a portion of a circuit containing a mutual inductance, we must define a representation that provides us with a quantitative relationship for the coils. We shall designate such a representation as a *dynamic* equivalent circuit.

Consider, for example, the static circuit of the pair of coils shown in Fig. 3-24a connected to some external circuit. To obtain the dynamic equivalent circuit shown in (b), we treat the circuit as follows:

1. Currents are assumed to be flowing in the meshes containing L_1 and L_2. The choice of directions is arbitrary.

2. As a result of i_1, a generator of value $M(di_1/dt)$ is placed in series with L_2. If i_1 is *entering* the dot of L_1, the generator polarity is such that it tends to force current *out* of the dot of L_2.

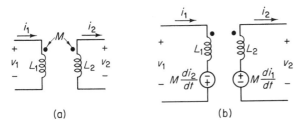

(a)   (b)

Figure 3-24 Development of dynamic equivalent circuit of mutually coupled coils.

3. As a result of i_2, a generator of value $M(di_2/dt)$ is placed in series with L_1. The same convention applies as in step 2.

4. The circuit may now be solved by any method in which the currents i_1 and i_2 appear as variables (e.g., mesh current analysis). Of course, we have not considered solutions involving more than one kind of element yet, so the reader may not be able to perform step 4 at this time. Nevertheless, the first three steps may be done for a given pair of coils in anticipation of step 4.

Example 3-8

Write mesh equations for the two mutually coupled meshes considered in Fig. 3-24.

   Solution   If we sum the voltage drops as positive, the equation for mesh 1 is

$$-v_1 + L_1 \frac{di_1}{dt} - M \frac{di_2}{dt} = 0 \tag{3-55}$$

Around mesh 2 we have

$$-M \frac{di_1}{dt} + L_2 \frac{di_2}{dt} + v_2 = 0 \tag{3-56}$$

Example 3-9

Determine the equivalent inductance of the two coils connected in series as shown in Fig. 3-25a.

   Solution   As a result of the mutual coupling, we cannot simply add the two inductances as can be done for noncoupled coils. Instead, we must write a voltage–circuit equation for the circuit. First we draw a dynamic equivalent circuit as shown in Fig. 3-25b. Note that since the same current flows through both coils, two generators are produced in series. Assuming a current i, the

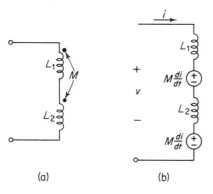

(a)   (b)   **Figure 3-25**   Coils of Example 3-9.

voltage across the combination is

$$v = L_1 \frac{di}{dt} + M \frac{di}{dt} + L_2 \frac{di}{dt} + M \frac{di}{dt}$$

$$= (L_1 + L_2 + 2M) \frac{di}{dt} \qquad (3\text{-}57)$$

The equivalent inductance must satisfy the equation

$$v = L_{eq} \frac{di}{dt} \qquad (3\text{-}58)$$

Comparing the last two equations, we have

$$L_{eq} = L_1 + L_2 + 2M \qquad (3\text{-}59)$$

Thus, for this *aiding* connection, the equivalent inductance is greater than the sum of the two inductances. In one of the exercises to follow, the reader can show that for the *opposing* connection, the inductance is smaller than the sum of the two inductances.

3-7 IDEAL TRANSFORMER

We now wish to consider an ideal case of two mutually coupled coils arranged in an optimum fashion. The device that we will describe is called a *transformer*. An *ideal transformer* possesses the following characteristics:

1. All the flux created by the first winding links with the second winding.
2. The self-inductances of the two windings are very large and approach infinity in the ideal limit.

Although the ideal transformer is a hypothetical model, practical transformers with ferromagnetic cores may be made to approach the ideal model very closely. Thus, for many practical problems, it is not unrealistic to assume an ideal transformer.

The relationships stated in this section are derived in most basic circuits texts, and we will not repeat the proofs. Instead, we will simply state the important results and show their applications to more general circuits problems.

A property characterizing a pair of mutually coupled coils is the inequality

$$M \leq \sqrt{L_1 L_2} \qquad (3\text{-}60)$$

As the coupling increases, this inequality approaches an equality. A measure of the degree of coupling can be achieved by defining a constant K called the *coefficient of coupling*. This quantity is defined as

$$K = \frac{M}{\sqrt{L_1 L_2}} \qquad 0 \leq K \leq 1 \qquad (3\text{-}61)$$

Thus for $K = 0$ there is no coupling at all between the two coils, and at the other extreme, when $K = 1$, the coils constitute one of the requirements of an ideal transformer. While it is essentially correct to speak of any two coupled coils as constituting a transformer, a popular convention has been to use the term transformer in reference to devices with values of K near unity (the nearly perfect category).

To consider any transformer as a nearly perfect device, we must assume that the voltages and currents are varying reasonably rapidly with time. In the limit of dc excitation, the transformer acts essentially as a short circuit, and no transformer action occurs. Furthermore, for very slow variations of voltage or current, the transformer may not function properly. Transformers may be designed primarily for use with a single-frequency sinusoidal excitation such as the common 60-Hz power transformers, or they may be designed for use with pulse-type waveforms such as are encountered in electronic circuits.

Under conditions approaching the ideal transformer, the values of the self- and mutual inductances become less significant in describing the circuit phenomena. Instead, the *turns ratio* serves to describe the transformer from an external point of view. Transformer windings are often designated as *primary* and *secondary* windings, respectively. With a single excitation, the primary winding is that winding nearest to the excitation. However, the convention depends on the type of application. The schematic representation of an ideal transformer with voltages and currents present is shown in Fig. 3-26. The turns ratio is specified in the form $1:n$, where n is the ratio of turns on the right-hand side to turns on the left-hand side. (Some authors prefer a notation of the form $n_1:n_2$.) Note that we have also specified a dot convention indicating the relative polarity of the voltages.

The voltage–current equations for the ideal transformer are

$$\frac{v_2}{v_1} = n \tag{3-62}$$

$$\frac{i_2}{i_1} = \frac{1}{n} \tag{3-63}$$

Thus *for the ideal transformer the voltage–current relationships are merely scale factors.*

A circuit involving a close approximation to an ideal transformer may be solved by the usual methods of analysis, with special relationships around the transformer expressed by Eqs. (3-62) and (3-63). On the other hand, it is often convenient to "reflect" the network across a transformer so that an equivalent network can be solved without considering the transformer. The reflected network has the same topology as the original network, but element values are scaled. If the desired response has to be shifted back across the transformer again, it is multiplied by the appropriate turns ratio.

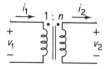

Figure 3-26 Schematic representation of ideal transformer.

Assume that the given transformer has turns ratio $1:n$ from primary to secondary. To reflect any circuit from primary to secondary, we first redraw the part of the circuit to be shifted on the secondary side. Unshifted parts retain their original values. Shifted values are modified as follows:

1. Voltages are multiplied by n.
2. Currents are multiplied by $1/n$.
3. Resistances are multiplied by n^2.
4. Inductances are multiplied by n^2.
5. Capacitances are multiplied by $1/n^2$.

To reflect a circuit from secondary to primary, we first redraw the part of the circuit to be shifted on the primary side. Unshifted parts retain their original values. Shifted values are modified as follows:

1. Voltages are multiplied by $1/n$.
2. Currents are multiplied by n.
3. Resistances are multiplied by $1/n^2$.
4. Inductances are multiplied by $1/n^2$.
5. Capacitances are multiplied by n^2.

An example will best clarify the procedures.

Example 3-10

The circuit shown in Fig. 3-27a contains an ideal transformer. Determine i_1, i_2, and i_3 by each of the following methods.
    (a) Direct solution of the circuit equations.
    (b) Reflecting the circuit to the secondary.
    (c) Reflecting the circuit to the primary.
      Solution   (a) The circuit equations for the three meshes are

$$20i_1 + v_1 = e \tag{3-64}$$

$$-v_2 + 250i_2 + 1000(i_2 - i_3) = 0 \tag{3-65}$$

$$1000(i_3 - i_2) + 1000i_3 = 0 \tag{3-66}$$

In addition, we have the relationships

$$v_2 = 5v_1 \tag{3-67}$$

and

$$i_1 = 5i_2 \tag{3-68}$$

Simultaneous solution of the five preceding equations yields

$$i_1 = \frac{1}{50}e = 2\sin\omega t \tag{3-69}$$

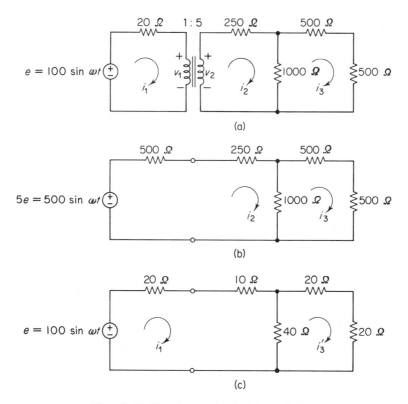

Figure 3-27   Transformer circuit of Example 3-10.

$$i_2 = \frac{1}{250} e = 0.4 \sin \omega t \tag{3-70}$$

$$i_3 = \frac{1}{500} e = 0.2 \sin \omega t \tag{3-71}$$

(b) We will now reflect the primary circuit to the secondary. Following the rules outlined in the text, we obtain the circuit shown in Fig. 3-27b. Note that we "lose" one mesh in this particular problem as a result of the configuration. The circuit from the 250-Ω resistor to the right is unmodified; thus, all responses in this region are correct as predicted from the reflected circuit. Therefore, we need only to use currents i_2 and i_3 in writing our equations. The circuit may be solved by any standard dc method. The reader can verify that the results for i_2 and i_3 are the same as before. The current i_1 is then found by reflecting back across the transformer.

(c) The circuit reflected to the primary side is shown in Fig. 3-27c. Note that we have now used the symbol i_3' to describe the primary reflected value of i_3. The reader can solve this circuit and verify that the reflection of i_1 back to the secondary yields i_2 and that the reflection of i_1 back to the secondary yields i_2 and that the reflection of i_3' back to the secondary yields i_3.

3-8 SPECIAL RLC COMBINATIONS

Thus far in this chapter we have considered the complete solutions of circuits for only those circuits containing one type of element, with the exception of resistive circuits containing an ideal transformer. We shall see in the next chapter that general *RLC* circuits present new concepts which must be investigated before solutions are possible. However, there is a special class of *RLC* problems that can be solved by a simple extension of the methods of this chapter.

If the voltage across a parallel combination of single elements is given, the individual currents may be calculated by the methods of this chapter, and the total current may be determined by application of Kirchhoff's current law, regardless of the type of components. Similarly, if the current through a series combination of single elements is known, the individual voltages can be calculated, and the total voltage may be determined by application of Kirchhoff's voltage law. The key to applying such an approach is that we must be able to work with each individual element at a time.

Often, this approach is useful in cases where the response is known and the excitation must be determined. In such cases, we work "backwards" from the desired response to obtain a source that will produce it. Let us illustrate with an example.

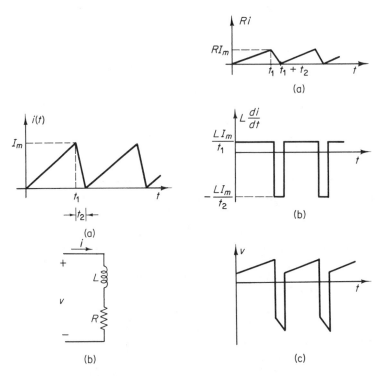

Figure 3-28 Desired current and circuit of Example 3-11.

Figure 3-29 Waveforms of Example 3-11.

Example 3-11

The current waveform shown in Fig. 3-28a is of the form desired in a television deflection coil. The equivalent circuit of the coil is shown in (b). Determine the waveform of voltage required to produce the given current.

    Solution   This problem is best handled by a graphical approach. Let us look at the first cycle. There are two time intervals to be considered: $0 < t < t_1$ and $t_1 < t < t_1 + t_2$. The equation that must govern the circuit at all times is

$$v = iR + L\frac{di}{dt} \tag{3-72}$$

    We first sketch iR as shown in Fig. 3-29a. We next calculate di/dt for each of the two intervals, multiply by L, and sketch as shown in (b). The sum of these two curves at all points is the required voltage as shown in (c).

GENERAL PROBLEMS

3-1. Solve for the currents $i_1(t)$ and $i_2(t)$ in the circuit of Fig. P3-1 using mesh current analysis.

3-2. Repeat the analysis of Problem 3-1 if the two sources are

$$e_1(t) = 10 \sin \omega t$$

$$e_2(t) = 12 \cos \omega t$$

Express each current as a single sinusoid.

3-3. Solve for the voltages $v_1(t)$ and $v_2(t)$ in the circuit of Fig. P3-3 using mesh current analysis.

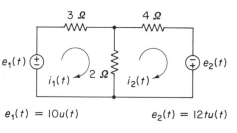

$e_1(t) = 10u(t)$ $e_2(t) = 12tu(t)$ $i_1(t) = 4u(t)$ $i_2(t) = 5tu(t)$

Figure P3-1   **Figure P3-3**

3-4. Repeat the analysis of Problem 3-3 if the two sources are

$$i_1(t) = 4 \sin \omega t$$
$$i_2(t) = 5 \cos \omega t$$

Express each voltage as a single sinusoid.

3-5. Using the voltage-divider rule, determine the voltage $v_o(t)$ in the circuit of Fig. P3-5.

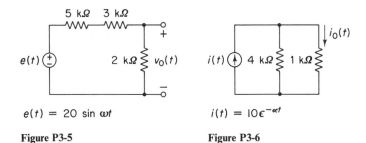

$$e(t) = 20 \sin \omega t \qquad\qquad i(t) = 10\epsilon^{-\alpha t}$$

Figure P3-5   **Figure P3-6**

3-6. Using the current-divider rule, determine the current $i_o(t)$ in the circuit of Fig. P3-6.

3-7. Determine the Thévenin equivalent circuit for the circuit of Fig. P3-7.

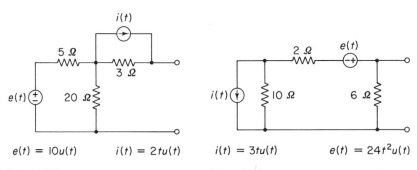

$$e(t) = 10u(t) \qquad i(t) = 2tu(t) \qquad\quad i(t) = 3tu(t) \qquad\quad e(t) = 24t^2u(t)$$

Figure P3-7   **Figure P3-8**

3-8. Determine the Thévenin equivalent circuit for the circuit of Fig. P3-8.

3-9. During an interval of 50 μs, the voltage across a 0.2-μF capacitor changes by 6 V. Determine the average current during this interval.

3-10. During an interval of 20 ms, the voltage across a 5-μF capacitor changes by 200 mV. Determine the average current during this interval.

3-11. During an interval of 40 ms, the average current in a 20-μF capacitor is 8 mA. Determine the change in voltage across the capacitor.

3-12. During an interval of 120 μs, the average current in a 0.02-μF capacitor is 3 mA. Determine the change in voltage across the capacitor.

3-13. The voltage across a 0.2-F capacitor is given by the equation

$$v(t) = 4(1 - \epsilon^{-10t}) \qquad \text{for } t > 0$$

Determine an expression for the current $i(t)$.

3-14. The voltage across a 50-μF capacitor is given by the equation

$$v(t) = 40 \sin 2000t \qquad \text{for } t > 0$$

Determine an expression for the current $i(t)$.

3-15. The current in an initially uncharged 0.5-F capacitor is given by the equation

$$i(t) = 10 + 12\epsilon^{-4t} \qquad \text{for } t > 0$$

Determine an expression for the voltage $v(t)$.

3-16. The current in an initially uncharged 0.2-μF capacitor is given by the equation

$$i(t) = 0.06 \cos 10^4 t \qquad \text{for } t > 0$$

Determine an expression for the voltage $v(t)$.

3-17. Repeat the analysis of Problem 3-15 if the capacitor is initially charged to 25 V.

3-18. Repeat the analysis of Problem 3-16 if the capacitor is initially charged to 40 V.

3-19. The voltage across a 0.5-F capacitor is shown in Fig. P3-19. Determine the current $i(t)$ *graphically*, and label all significant points.

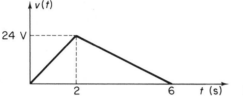

Figure P3-19

3-20. The voltage across a 3-μF capacitor is shown in Fig. P3-20. Determine the current $i(t)$ *graphically*, and label all significant points.

Figure P3-20

3-21. Repeat the analysis of Problem 3-19 by first expressing the voltage according to the methods of Section 2-12 and performing an *analytical* analysis.

3-22. Repeat the analysis of Problem 3-20 by first expressing the voltage according to the methods of Section 2-12 and performing an *analytical* analysis.

3-23. The current in an initially uncharged 0.2-F capacitor is shown in Fig. P3-23. Determine the voltage $v(t)$ *graphically*, and label all significant points.

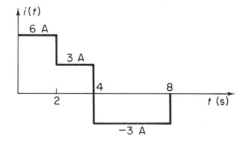

Figure P3-23

3-24. The current in an initially uncharged 0.5-μF capacitor is shown in Fig. P3-24. Determine the voltage $v(t)$ *graphically*, and label all significant points.

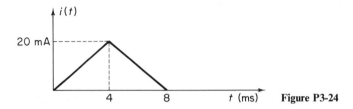

Figure P3-24

3-25. Repeat the analysis of Problem 3-23 by first expressing the current according to the methods of Section 2-12 and performing an *analytical* analysis.

3-26. Repeat the analysis of Problem 3-24 by first expressing the current according to the methods of Section 2-12 and performing an *analytical* analysis.

3-27. Determine the Thévenin equivalent circuit of the charged capacitor shown in Fig. P3-27.

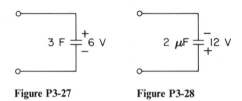

Figure P3-27   Figure P3-28

3-28. Determine the Thévenin equivalent circuit of the charged capacitor shown in Fig. P3-28.

3-29. Determine the Norton equivalent circuit of the charged capacitor of Problem 3-27 (Fig. P3-27).

3-30. Determine the Norton equivalent circuit of the charged capacitor of Problem 3-28 (Fig. P3-28).

3-31. During an interval of 50 ms, the current through a 200-mH inductor changes by 3 A. Determine the average voltage during this interval.

3-32. During an interval of 400 μs, the current through an 80-μH inductor changes by 120 mA. Determine the average voltage during this interval.

3-33. During an interval of 20 μs, the average voltage across an 80-μH inductor is 12 V. Determine the change in current during this interval.

3-34. During an interval of 5 s, the average voltage across a 4-H inductor is -8 V. Determine the change in current during this interval.

3-35. The current through a 2-H inductor is given by the equation

$$i(t) = 4 \sin 3t \qquad \text{for } t > 0$$

Determine an expression for the voltage $v(t)$.

3-36. The current through a 50-mH inductor is given by the equation

$$i(t) = 3(1 - \epsilon^{-40t}) \qquad \text{for } t > 0$$

Determine an expression for the voltage $v(t)$.

3-37. The voltage across an initially unfluxed 0.5-H inductor is given by the equation

$$v(t) = 20 \cos 4000t \qquad \text{for } t > 0$$

Determine an expression for the current $i(t)$.

3-38. The voltage across an initially unfluxed 40-mH inductor is given by the equation

$$v(t) = 12\epsilon^{-2000t} \qquad \text{for } t > 0$$

Determine an expression for the current $i(t)$.

3-39. The current through a 0.1-H inductor is shown in Fig. P3-39. Determine the voltage $v(t)$ *graphically*, and label all significant points.

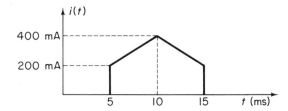

Figure P3-39

3-40. The voltage across an initially unfluxed 4-H inductor is shown in Fig. P3-40. Determine the current $i(t)$ *graphically*, and label all significant points.

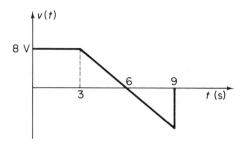

Figure P3-40

3-41. Repeat the analysis of Problem 3-39 by first expressing the current according to the methods of Section 2-12 and performing an *analytical* analysis.

3-42. Repeat the analysis of Problem 3-40 by first expressing the voltage according to the methods of Section 2-12 and performing an *analytical* analysis.

3-43. Determine the Norton equivalent circuit of the fluxed inductor shown in Fig. P3-43.

3-44. Determine the Norton equivalent circuit of the fluxed inductor shown in Fig. P3-44.

3-45. Determine the Thévenin equivalent circuit of the fluxed inductor of Problem 3-43 (Fig. P3-43).

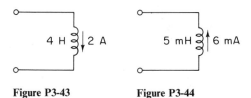

Figure P3-43   Figure P3-44

3-46. Determine the Thévenin equivalent circuit of the fluxed inductor of Problem 3-44 (Fig. P3-44).

3-47. A certain pair of mutually coupled coils has the static equivalent circuit shown in Fig. P3-47. With coil 2 open ($i_2 = 0$), determine expressions for v_1 and v_2 if the input current is

$$i_1 = 4 \sin 20t \qquad \text{for } t > 0$$

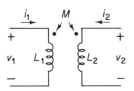

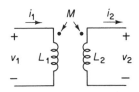

$L_1 = 0.2$ H, $L_2 = 0.8$ H, $M = 0.3$ H $L_1 = 40$ mH, $L_2 = 10$ mH, $k = 0.2$

Figure P3-47   **Figure P3-48**

3-48. A certain pair of mutually coupled coils has the static equivalent circuit shown in Fig. P3-48. With coil 2 open ($i_2 = 0$), determine expressions for v_1 and v_2 if the input current is

$$i_1 = 40t \qquad \text{for } t > 0$$

3-49. For the circuit of Problem 3-47, assume that i_1 is the same as before but that coil 2 is connected to a circuit that establishes the current

$$i_2 = 2 \sin 20t \qquad \text{for } t > 0$$

Determine expressions for v_1 and v_2. Note that direction of i_2 in Fig. P3-47.

3-50. For the circuit of Problem 3-48, assume that i_1 is the same as before but that coil 2 is connected to a circuit that establishes the current

$$i_2 = 20t \qquad \text{for } t > 0$$

Determine expressions for v_1 and v_2. Note the direction of i_2 in Fig. P3-48.

3-51. For the mutually coupled coils of Problem 3-47 (Fig. P3-47), determine the coefficient of coupling K.

3-52. For the mutually coupled coils of Problem 3-48 (Fig. P3-48), determine the mutual inductance M.

3-53. For the circuit of Fig. P3-53, solve for the currents i_1 and i_2 and the voltages v_1 and v_2 by writing equations for the circuit as given and solving them simultaneously.

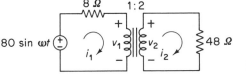

Figure P3-53

3-54. For the circuit of Fig. P3-54, solve for the currents i_1 and i_2 and the voltages v_1 and v_2 by writing equations for the circuit as given and solving them simultaneously.

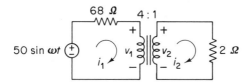

Figure P3-54

3-55. Analyze the circuit of Problem 3-53 by reflecting the circuit to the *primary*.

3-56. Analyze the circuit of Problem 3-54 by reflecting the circuit to the *primary*.

3-57. Analyze the circuit of Problem 3-53 by reflecting the circuit to the *secondary*.

3-58. Analyze the circuit of Problem 3-54 by reflecting the circuit to the *secondary*.

3-59. Reflect the circuit of Fig. P3-59 to the *primary*.

3-60. Reflect the circuit of Problem 3-59 to the *secondary*.

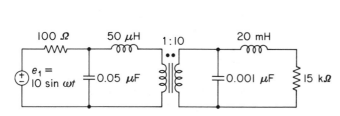

Figure P3-59

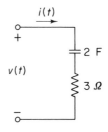

Figure P3-61

3-61. The current in the series circuit of Fig. P3-61 is

$$i(t) = 4u(t)$$

Determine an expression for the voltage $v(t)$.

3-62. The current in the series circuit of Fig. P3-62 is

$$i(t) = 2 \sin 100t$$

Determine an expression for the voltage $v(t)$.

3-63. The voltage across the parallel circuit of Fig. P3-63 is

$$v(t) = 12 \sin 2t$$

Determine an expression for the current $i(t)$.

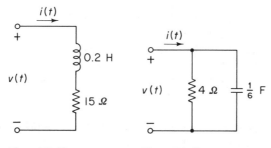

Figure P3-62   **Figure P3-63**

3-64. The voltage across the parallel circuit of Fig. P3-64 is

$$v(t) = 12tu(t)$$

Determine an expression for the current $i(t)$.

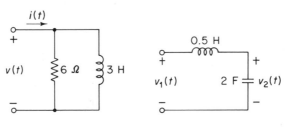

Figure P3-64   **Figure P3-65**

3-65. The circuit shown in Fig. P3-65 is required to yield the output voltage

$$v_2(t) = 2t^2u(t)$$

Determine the required input voltage $v_1(t)$.

3-66. The circuit shown in Fig. P3-66a is required to yield the output voltage $v_2(t)$ shown in Fig. P3-66b. Determine the required input voltage $v_1(t)$.

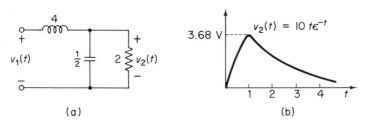

(a)   (b)

Figure P3-66

DERIVATION PROBLEMS

3-67. Prove the validity of the following statements with respect to a capacitor.
  (a) A ramp voltage produces a step current.
  (b) A step voltage produces an impulse current.

3-68. Prove the validity of the following statements with respect to a capacitor.
  (a) An impulse current produces a step voltage.
  (b) A step current produces a ramp voltage.

3-69. Prove the validity of the following statements with respect to an inductor.
  (a) A ramp current produces a step voltage.
  (b) A step current produces an impulse voltage.

3-70. Prove the validity of the following statements with respect to an inductor.
  (a) An impulse voltage produces a step current.
  (b) A step voltage produces a ramp current.

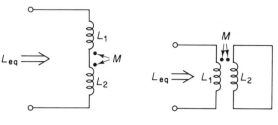

Figure P3-71   Figure P3-72

3-71. Show that the equivalent inductance of the series-opposing connection of Fig. P3-71 is

$$L_{eq} = L_1 + L_2 - 2M$$

3-72. Show that the equivalent inductance of the connection of Fig. P3-72 is

$$L_{eq} = L_1(1 - K^2)$$

3-73. Consider that the current in a capacitor is sinusoidal and expressed by

$$i(t) = I \cos(\omega t + \theta)$$

By means of Euler's formula, $\epsilon^{j\phi} = \cos\phi + j\sin\phi$, we may write

$$i(t) = \text{real part} \left[I\epsilon^{j(\omega t + \theta)}\right] \stackrel{\text{def}}{=} I\epsilon^{j(\omega t + \theta)}$$
$$= I\epsilon^{j\theta + j\omega t} = \bar{I}\epsilon^{j\omega t}$$

where the real part designation is understood and $\bar{I}$ is a complex phasor

$$\bar{I} = I\epsilon^{j\theta}$$

Show that if the voltage is assumed to be sinusoidal of the form

$$v(t) = \bar{V}\epsilon^{j\omega t}$$

then

$$\bar{V} = \frac{-j}{\omega C}\bar{I}$$

Sketch a typical phasor diagram showing $\bar{V}$ and $\bar{I}$, and interpret its meaning.

3-74. Consider that the current through an inductor is sinusoidal and expressed by

$$i(t) = I \cos(\omega t + \theta)$$

Using the ideas and notation expressed in Problem 3-73 show that

$$\bar{V} = j\omega L\bar{I}$$

Sketch a typical phasor diagram showing $\bar{V}$ and $\bar{I}$, and interpret its meaning.

COMPUTER PROBLEMS

3-75. The process of integration can be approximated with a digital computer or calculator by breaking up the function into narrow segments of area and summing the individual areas. Under favorable conditions, the approximation may be very close to the exact area.

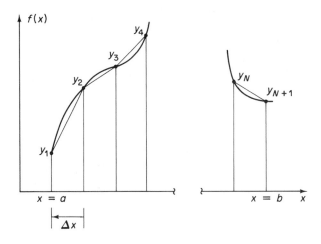

There are a number of complex algorithms for integration available in the software libraries of computer systems, but the purpose here is to investigate the trapezoidal integration algorithm, which can be readily programmed on a personal computer or calculator. While limited in scope, it is capable of reasonable accuracy for "well-behaved" functions if the sampling interval is sufficiently small.

Assume that the area under a curve $y = f(x)$ from a to b is approximated by a series of N trapezoids, each of width Δx, as shown in Fig. P3-75. The left-hand height of the first trapezoid is y_1, and the right-hand height is y_2. The left-hand height of the second trapezoid is y_2, and the right-hand height is y_3, etc.

(a) Show that the total area A_{ab} from $x = a$ to $x = b$ of the N trapezoids is

$$A_{ab} = \left(\frac{1}{2} y_1 + \frac{1}{2} y_{N+1} + \sum_{n=2}^{N} y_n \right) \Delta x$$

(b) Assume that the current (in A) flowing into an initially uncharged $\frac{1}{3}$-F capacitor is given by

$$i(t) = 2t \qquad \text{for } t > 0$$

By first calculating the value of the current at 0.5-s intervals, use the trapezoidal algorithm to compute the voltage at $t = 4$ [i.e., $v(4)$].

(c) Now calculate $v(4)$ using conventional analytical calculus, and compare the result with (b). What can you conclude and why?

(d) Next assume that

$$i(t) = 2\epsilon^{-t} \qquad \text{for } t > 0$$

Again, calculate the current at 0.5-s intervals, and use the trapezoidal rule to compute $v(4)$.

(e) Repeat (d) but with the current calculated at 0.05-s intervals.

(f) Now, calculate $v(4)$ using conventional analytical calculus, and compare the result with (d) and (e). What can you conclude?

3-76. The process of approximating differentiation on a digital computer or calculator is often more difficult than integration because the differentiation operation tends to accentuate noise. Consequently, accurate differentiation algorithms are quite involved and require a background in numerical analysis for full understanding.

One simplified differentiation scheme, which is useful for certain noncritical purposes, will be investigated here. The function $y = f(x)$ is represented in the same manner as was done in Problem 3-75 (Fig. P3-75). Let y'_n represent the slope of the straight line between points on the curve at $x = n\,\Delta x$ and $(n + 1)\,\Delta x$. This quantity may, under certain conditions, be a reasonable estimate of the derivative at $x = n\,\Delta x$.

(a) Show that

$$y'_n = \frac{y_{n+1} - y_n}{\Delta x}$$

(b) Assume that the current (in amperes) flowing in a 3-H inductor is given by

$$i(t) = 2t \qquad \text{for } t > 0$$

By first calculating the value of the current at a few 0.05-s intervals, use the result of (a) to compute the voltage at $t = 1$ [i.e., $v(1)$].

(c) Now calculate $v(1)$ using conventional analytical calculus, and compare the result with (b). What can you conclude, and why?

(d) Next, assume that

$$i(t) = 2t^2 \qquad \text{for } t > 0$$

Again, calculate the current at a few 0.5-s intervals, and use the result of (a) to compute $v(1)$.

(e) Repeat (d) but with the current calculated at a few 0.05-s intervals.

(f) Repeat (d) but with the current calculated at a few 0.01-s intervals.

(g) Calculate $v(1)$ using conventional analytical calculus, and compare the results with (d), (e), and (f). What can you conclude?

4

THE BASIC
TIME-DOMAIN CIRCUIT

OVERVIEW

The voltage–current relationship and general behavior of the basic circuit parameters were considered in Chapter 3. For the most part, emphasis was on the response of a single element when analysis was performed.

The emphasis in this chapter is toward a broader viewpoint in which complete circuits as combinations of different types of elements are considered. The final steady-state behavior of complete circuits with dc excitations is considered, and the initial behavior under transient switching conditions is studied. Techniques for determining complete solutions of single time constant (first-order) circuits are developed.

OBJECTIVES

After completing this chapter, the reader should be able to

1. Discuss and compare the concepts of the *time domain* and the *transform domain*.
2. State the form of the differential equation of a lumped, linear network with constant parameters.
3. Write the integrodifferential node or mesh equations for a circuit containing energy storage elements.
4. Define the concepts of *transient response* and *steady-state response*, and determine these functions from a given total response.
5. Construct the dc steady-state equivalent circuit for a given general circuit and analyze it.
6. Construct the initial equivalent circuit for a given general circuit and analyze it.

7. **Write the form of the general response of a single time constant circuit with dc excitations.**

8. **Determine the equation of any complete voltage or current response in a single time constant circuit by determining the boundary conditions and matching them to the general response form.**

9. **Discuss the different approaches in which initially charged capacitors and initially fluxed inductors are dealt with in formulating network equations.**

4-1 GENERAL DISCUSSION

The mesh or node equations of a circuit containing resistance, capacitance, and inductance are called *integrodifferential* equations due to the presence of both integrals and derivatives. Consider, for example, the mesh shown in Fig. 4-1. The mesh current equation expressing Kirchhoff's voltage law for this mesh is

$$L \frac{d}{dt} (i_2 - i_1) + Ri_2 + \frac{1}{C} \int_{-\infty}^{t} (i_2 - i_3)\, dt = 0 \tag{4-1}$$

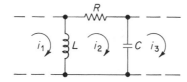   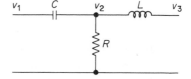

Figure 4-1 Typical *RLC* mesh.   **Figure 4-2** Typical *RLC* node.

As another example, consider the middle node shown in Fig. 4-2. The node voltage equation expression Kirchhoff's current law for this node is

$$C \frac{d}{dt} (v_2 - v_1) + \frac{v_2}{R} + \frac{1}{L} \int_{-\infty}^{t} (v_2 - v_3)\, dt = 0 \tag{4-2}$$

Thus, in the general circuit, the equation for each mesh or node is an integrodifferential equation. In general, we may eliminate the integral terms by differentiating all terms of the resulting equation with respect to time. In this case, the resulting equation is called a *differential equation* since it contains no integral terms.

To solve for some desired response, a series of simultaneous integrodifferential equations that completely characterize the circuit can first be written. Next, all the variables but the desired one can be eliminated by some successive elimination scheme. In general, this is more complicated than in the case of algebraic simultaneous equations. After proper elimination, a single equation expressing the desired variable in the form of a differential equation is obtained. For a lumped, linear network with constant parameters, the differential equation can always be expressed in the form

$$b_n \frac{d^n y}{dt^n} + b_{n-1} \frac{d^{n-1} y}{dt^{n-1}} + \cdots + b_0 y = f(t) \tag{4-3}$$

where $y(t)$ represents the desired response (either voltage or current) and $f(t)$ represents some combination of excitations.

A differential equation of the form of Eq. (4-3) is called an ordinary, linear, constant-coefficient differential equation of order n. More will be said about the properties of such equations in later chapters, but at this point we will merely point out a few basic properties. The order (n) of the differential equation is the order of the highest derivative when all integral terms have been removed. The linearity property of such an equation permits the use of superposition in obtaining the response due to several excitations.

After the differential equation is obtained, it still must be solved for $y(t)$. There are a number of different methods for solving differential equations. Two important approaches used in electrical engineering are (a) the so-called "classical" differential equation methods and (b) the transform methods.

Transform methods have become quite popular in electrical engineering in recent years. Engineers and technologists have learned to think in terms of such techniques when viewing circuits and systems. However, it is necessary to obtain a certain familiarity with basic classical methods before introducing transform methods. Our purpose in discussing classical methods is to provide the reader with some knowledge as to "why," rather than "how." The transform methods are more concerned with "how" rather than "why."

Actually, the single-element cases of Chapter 3 are good, relatively simple examples of the classical methods. In such cases, we usually needed only to differentiate or integrate an individual excitation to obtain the subsequent response.

In general, the classical methods are characterized by the fact that the solution is obtained directly from the differential equation by appropriate mathematical methods. Since the variables involved (currents and voltages) are functions of time, we say that the differential equations of a circuit are the representation of the circuit in the *time domain*.

Transform methods are characterized by the fact that a mathematical transformation is made on the original circuit and its excitations in such a manner that a new system is obtained that is no longer a function of time. Instead, the quantities in the new system are functions of some new variable, with the resulting property that the responses can be determined by algebraic manipulations. We say that the transformed equations of the circuit are the representations of the circuit in the *transform domain*.

For single-element circuits, it is usually much easier to work in the time domain as was demonstrated in Chapter 3. In addition, there are also a few combinations of two-element circuits in which the solution can be obtained much easier in the time domain, as will be demonstrated later in this chapter.

Regardless of whether time-domain or transform techniques are to be used, it is necessary to make certain inspections of the network in the time domain. The purpose of such inspections is to determine certain initial and final conditions that exist within a network. To illustrate this idea, the reader should recall the process of integration. An arbitrary constant must be added to the indefinite integral, and this constant is usually evaluated by specifying an initial condition. In general, it can be shown that for a differential equation of order n, it is necessary to determine n initial conditions.

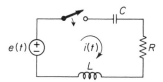

Figure 4-3   Circuit of Example 4-1.

In time-domain solutions, it is usually necessary to specify the desired function and its first $(n - 1)$ derivatives at $t = 0$ (or whenever the response begins). In circuit transform methods, the determination of arbitrary constants, in a manner of speaking, takes care of itself if certain initial inspections and procedures are followed. Furthermore, the process of eliminating all but the desired variable from the circuit equations can be done by algebra if transform methods are employed.

Let us illustrate with some examples how the differential equations of circuits are obtained. All the examples considered in this section assume initially *relaxed* energy-storage components, as some further discussion is necessary in relation to initial energy storage. A relaxed energy-storage device is one containing no initial energy.

Example 4-1

The single-mesh relaxed *RLC* circuit of Fig. 4-3 is excited at $t = 0$ by the voltage source shown. Obtain a differential equation describing the circuit from a mesh current analysis.

Solution   A single response $i(t)$ is sufficient to characterize the solution since, if $i(t)$ can be determined, the voltages across the components could be determined by the methods of the preceding chapter. The mesh current equation for $t > 0$ is

$$L \frac{di}{dt} + Ri + \frac{1}{C} \int_0^t i\, dt = v(t) \qquad (4\text{-}4)$$

To remove the integral term, we differentiate all terms with respect to t. This results in

$$L \frac{d^2 i}{dt^2} + R \frac{di}{dt} + \frac{i}{C} = v_1(t) \qquad (4\text{-}5)$$

where $v_1(t)$ is a new function of time given by

$$v_1(t) = \frac{dv(t)}{dt} \qquad (4\text{-}6)$$

Equation (4-5) can be seen to be in the form of Eq. (4-3). Thus the differential equation for this circuit is a *second-order* differential equation. The general solution of this equation would contain two arbitrary constants, which could be evaluated from a knowledge of the initial value of the current and its first derivative.

4-2 STEADY-STATE AND TRANSIENT SOLUTIONS

When a linear network is first excited by an arbitrary number of time-varying excitations, an initial transfer of energy begins to take place around the network. Eventually, the balance of energy in the circuit will usually adjust to an equilibrium or semiequilibrium position in which a particular response may be either constant or more uniform in its behavior. The initial "adjustment" response is called the *transient response*, and the final equilibrium or semiequilibrium response is called the *steady-state response*.

The reader who has studied dc and ac circuits has already been exposed to two important cases of steady-state responses. As we shall see in the next section, the steady-state response of an *RLC* network excited by dc sources can be obtained from dc resistive circuit analysis. In such a case, all responses become constant in the steady state as the effects of inductors and capacitors "disappear" from the scene.

In general, the transient response is created by the adjustment process within the network and will usually approach zero after a sufficiently long time. The steady-state response is the response that remains after the transient disappears.

In some cases, the steady-state response is identically zero. Consider, for example, the uncharged capacitor shown in Fig. 4-4a excited at $t = 0$ by the step voltage source shown in (b). In the ideal limit of zero rise time and perfect capacitance, the current would approach an impulse at the point of switching, as shown in (c). (The reader can verify this result.) In this case, the transient response is an impulse, and the steady-state response is zero.

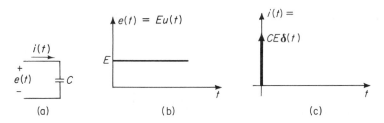

Figure 4-4   Response of an ideal capacitor excited by an ideal step voltage.

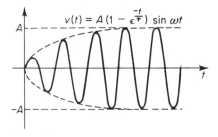

Figure 4-5   Waveform of Example 4-2.

In some cases, the transient response is identically zero. Such is the case of a purely resistive circuit regardless of the type of excitation. This is due to the fact that there are no energy-storage elements within a resistive circuit, and hence the final forms of the responses are established immediately.

In a few special cases, there is some ambiguity as to whether a portion of a response should be termed a transient response or a steady-state response. We shall say more about this subject whenever it occurs in later problems.

In general, the form and duration of the transient response are determined by the elements within the circuit and not by the excitation. On the other hand, the form of the steady-state solution is determined by the excitation, but its magnitude is determined by the circuit elements.

Electrical engineers and technologists in earlier years were trained primarily only in steady-state concepts such as dc and sinusoidal ac. However, modern technology requires that an engineer or engineering technologist be able to think in terms of both steady-state and transient conditions. In many applications, the transient phenomena are more important than the steady-state situation.

Example 4-2

A certain voltage response is given by $v(t) = A(1 - \epsilon^{-t/\tau}) \sin \omega t$ for $t > 0$. Sketch the function and identify the transient and steady-state portions of the response. Write an equation for the steady-state response.

Solution   The simplest way to sketch the function is to first draw the envelope $A(1 - \epsilon^{-t/\tau})$ as shown by the dashed lines in Fig. 4-5. This function simply acts as a multiplier for the sinusoidal function $\sin \omega t$. The transient portion of the response is the initial part of the response as the amplitude changes on each cycle. In about five time constants ($t = 5\tau$), the function reaches a semi-equilibrium state of constant-amplitude oscillations. Thus the portion of the response for $t > 5\tau$ is the steady-state response.

We obtain the equation for the steady-state response by assuming that $\epsilon^{-t/\tau}$ approaches zero. Letting $v_{ss}(t)$ represent the steady-state response, we have

$$v_{ss}(t) = A \sin \omega t \qquad (4\text{-}7)$$

4-3 THE STEADY-STATE DC CIRCUIT

A special case of considerable importance is the steady-state response of the general linear network with one or more dc excitations. It is important not only for predicting the final values of the responses for such a network, but also the methods of analysis are useful in predicting initial values for many transient problems.

Whenever an *RLC* circuit is excited by one or more dc sources, an initial transient occurs, as we have already predicted. However, since the sources are all constants, eventually, the currents through inductors and the voltages across capacitors will approach equilibrium constant values. Since no voltage appears across an inductance whenever the current through it is constant, it acts as a short circuit in the dc steady state. Since no current flows in a capacitor whenever the voltage across it is constant, it acts as an open circuit in the dc steady state. *Thus, in the dc steady state, a pure inductance appears as a short circuit, and a pure capacitance appears as an open circuit.* This idea is illustrated in Fig. 4-6. On the other hand, there can indeed be currents through inductors and voltages across capacitors in the dc steady state, representing energy stored in these components.

From a circuit analysis point of view, to solve for the various steady-state voltages and currents in a circuit excited by dc sources, we simply replace inductors by short circuits, replace capacitors by open circuits, and solve for the various responses that are obtained. It is often desirable to determine currents through inductors and voltages across capacitances, so it is usually necessary to keep in mind the locations of capacitors and inductors in the equivalent circuit.

We wish to emphasize again that the responses calculated from the steady-state portions of the total responses tell us nothing about the initial transient. The steady-state dc response is often referred to as the dc circuit at infinity (∞). In this context, infinity is interpreted to mean any time sufficiently large such that the transient response has disappeared. The use of the symbol (∞) as an argument for a dc response is convenient as a means of identifying the steady-state dc value of the function. Thus

$$y(\infty) = y_{ss}(t) = \text{constant} \tag{4-8}$$

for the dc steady-state case, where y is any voltage or current response.

In summary, the steady-state responses of a circuit excited by dc sources are all constant regardless of the number of energy-storage elements present. The various responses are determined by replacing inductors by short circuits, capacitors by open circuits, and solving for the desired response by dc circuit analysis.

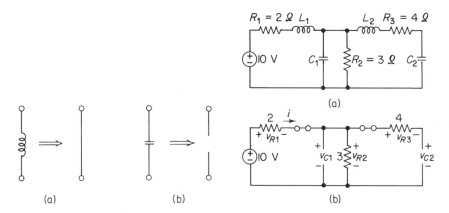

Figure 4-6 Dc steady-state behavior of inductance and capacitance.

Figure 4-7 Circuit of Example 4-3.

Example 4-3

Solve for the steady-state dc voltages and currents in the circuit shown in Fig. 4-7a.

  Solution   We first replace the inductors by short circuits and capacitors as open circuits as shown in Fig. 4-7b. At the risk of being redundant, we will tabulate the voltage and current associated with each component in the circuit. Since we are interested in the response of this circuit only in the steady-state region, we will omit any subscripts or arguments identifying the responses as steady-state since this fact is understood. The solution of the circuit can be obtained from the single current i, since the current in the right-hand mesh is zero as a result of the open circuit. The current i is

$$i = \frac{10 \text{ V}}{2 \Omega + 3 \Omega} = 2 \text{ A} \tag{4-9}$$

All other quantities are tabulated below.

$$i_{R1} = i_{L1} = i_{R2} = i = 2 \text{ A} \tag{4-10}$$

$$i_{C1} = i_{C2} = i_{L2} = i_{R3} = 0 \tag{4-11}$$

$$v_{R1} = 2 \text{ A} \times 2 \Omega = 4 \text{ V} \tag{4-12}$$

$$v_{L1} = v_{L2} = v_{R3} = 0 \tag{4-13}$$

$$v_{R2} = v_{C1} = v_{C2} = 2 \text{ A} \times 3 \Omega = 6 \text{ V} \tag{4-14}$$

4-4 THE INITIAL CIRCUIT

We saw in Section 4-3 that the final or steady-state values of the responses of a circuit with dc excitations can be obtained by considering the limiting behavior of inductors and capacitors. An analogous situation occurs when a circuit is first excited, but the behavior of inductors and capacitors will be shown to be opposite to their behavior in the dc steady state.

  Consider, first, the case of an uncharged capacitor excited at $t = 0$ by an arbitrary current $i(t)$. Let us look at the voltage an infinitesimally small time later at $t = 0^+$. The voltage $v_C(0^+)$ is

$$v_C(0^+) = \frac{1}{C} \int_0^{0^+} i \, dt \tag{4-15}$$

If we exclude the possibility of an impulse current, the net area under the current curve gained during this infinitesimally short period is essentially zero. Hence

$$v_C(0^+) = 0 \tag{4-16}$$

for an uncharged capacitor.

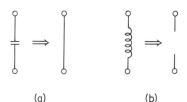

Figure 4-8 Initial behavior of uncharged capacitor and unfluxed inductor.

(a)   (b)

Thus *an uncharged capacitor initially appears as a short circuit, irrespective of the type of excitation, as long as the current is not impulsive in nature.* This idea is illustrated in Fig. 4-8a.

Now consider the case of an unfluxed inductor excited at $t = 0$ by an arbitrary voltage $v(t)$. Let us look at the current at $t = 0^+$. The current $i_L(0^+)$ is

$$i_L(0^+) = \frac{1}{L} \int_0^{0^+} v \, dt \tag{4-17}$$

Excluding the possibility of an impulse voltage, the current is

$$i_L(0^+) = 0 \tag{4-18}$$

Thus *an unfluxed inductor initially appears as an open circuit, irrespective of the type of excitation, as long as the voltage is not impulsive in nature.* This idea is illustrated in Fig. 4-8b.

The next question concerns the initial response of energy-storage devices whenever the components contain initial energy. Rather than treating these cases as separate concepts, let us employ the equivalent circuits of Chapter 3.

Consider, first, the charged capacitor shown in Fig. 4-9a. This is equivalent, for $t > 0$, to a voltage source in series with an uncharged capacitor. At $t = 0^+$, the equivalent uncharged capacitor appears as a short circuit, leaving only the voltage source as shown. Thus a *charged capacitor initially acts as a voltage source equal to the initial value of the voltage.*

Consider next the inductor with initial current as shown in Fig. 4-9b. This is equivalent, for $t > 0$, to a current source in parallel with an unfluxed inductor. At $t = 0^+$, the equivalent unfluxed inductor appears as an open circuit, leaving only the current source as shown. Thus a *fluxed inductor initially acts as a current source equal to the initial value of current.*

The reader may note that we used the Thévenin equivalent circuit of the charged capacitor and the Norton equivalent circuit of the fluxed inductor as these are the representations involving step functions. The Norton equivalent circuit of the charged capacitor and the Thévenin equivalent of the fluxed inductor both involve impulse

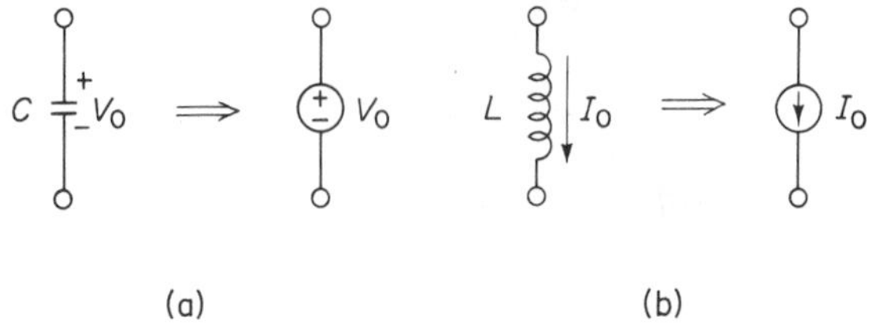

Figure 4-9 Initial behavior of charged capacitor and inductor.

(a)   (b)

functions, and it seemed advisable to avoid these forms for the purpose at hand. In general, one should be careful when using the impulse function concept since there are certain mathematical problems which often arise. As we proceed, we will point out the cases where the impulse function can be used without introducing difficulties.

As an extension of the properties we have discussed in this section, it can be implied that *a capacitor tends to oppose a change in voltage across it, and an inductor tends to oppose a change in current through it*. This concept is associated with the fact that energy stored in a capacitor is dependent on the voltage across the capacitor, and the energy stored in an inductor is dependent on the current through it. Since work must be done both to establish or to change this energy, the natural tendency is to oppose change.

A mathematical statement concerning the initial circuit can be made. If a circuit is excited or subjected to a new change in conditions at a general time $t = t_0$, and *if there are no impulsive conditions present*,

$$v_C(t_0^-) = v_C(t_0^+) \tag{4-19}$$

$$i_L(t_0^-) = i_L(t_0^+) \tag{4-20}$$

where v_C is the voltage across a capacitor, i_L is the current through an inductor, t_0^- is the instant just before the change, and t_0^+ is the instant just after the change. These equations are called *boundary conditions* for capacitance and inductance.

The impulsive conditions arise whenever ideal energy-storage components and ideal sources are considered. For example, consider the pure inductor of Fig. 4-10a excited at $t = 0$ by the step current shown in (b). In the ideal limit, an impulse of voltage would appear across the inductor, as shown in (c), which would establish an instantaneous current through the inductor as required by the assumed ideal current source. Clearly, Eq. (4-20) is violated in this case, a direct consequence of the impulse voltage. Impulse sources may also arise as a result of suddenly changing a circuit configuration.

In summary of the past few paragraphs, we may state:

1. With nonimpulsive conditions, the voltage across a capacitor and the current through an inductor cannot be changed instantaneously.

2. An impulse current can establish an instantaneous change in the voltage across a capacitor, and an impulse voltage can establish an instantaneous change in the current through an inductor.

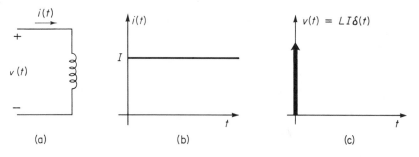

Figure 4-10   Response of an ideal inductor excited by an ideal step current.

The reader is not expected at this point to be able to identify all those situations that constitute impulsive action. These situations will be considered from time to time within the book, but the reader should remember the basic ideas involved.

Example 4-4

The circuit of Fig. 4-11a is excited at $t = 0$ by the dc source shown. C_1 is uncharged, L contains an initial current of 5 A, and C_2 is charged to 4 V as shown.

   (a) Determine all initial voltages and currents in the circuit ($t = 0^+$).

   (b) Since the excitation is dc, determine the dc steady-state voltages and currents ($t = \infty$).

   Solution   (a) The initial circuit at $t = 0^+$ is shown in Fig. 4-11b with C_1 replaced by a short, L replaced by a 5-A current source, and C_2 replaced by a 4-V voltage source. The solution of this circuit can now proceed by mesh analysis techniques as shown in (c). The equations are

$$4i_1(0^+) - 2i_2(0^+) = 10 \tag{4-21}$$

$$i_2(0^+) = 5 \text{ A} \tag{4-22}$$

$$-4i_2(0^+) + 8i_3(0^+) = -4 \tag{4-23}$$

yielding

$$i_1(0^+) = 5 \text{ A} \tag{4-24}$$

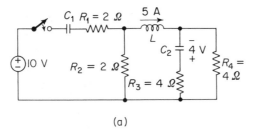

(a)

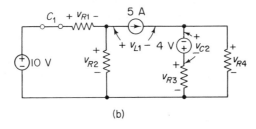

(b)

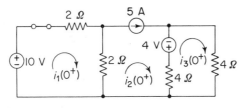

(c)

Figure 4-11   Circuit of Example 4-4.

and

$$i_3(0^+) = 2 \text{ A} \qquad (4\text{-}25)$$

The various branch currents at $t = 0^+$ are

$$i_{R1}(0^+) = i_{C1}(0^+) = i_1(0^+) = 5 \text{ A} \qquad (4\text{-}26)$$

$$i_{R2}(0^+) = i_1(0^+) - i_2(0^+) = 0 \qquad (4\text{-}27)$$

$$i_L(0^+) = 5 \text{ A} \qquad (4\text{-}28)$$

$$i_{C2}(0^+) = i_{R3}(0^+) = i_2(0^+) - i_3(0^+) = 3 \text{ A} \qquad (4\text{-}29)$$

$$i_{R4}(0^+) = i_3(0^+) = 2 \text{ A} \qquad (4\text{-}30)$$

The various branch voltages at $t = 0^+$ are

$$v_{C1}(0^+) = 0 \qquad (4\text{-}31)$$

$$v_{R1}(0^+) = 2 \times 5 = 10 \text{ V} \qquad (4\text{-}32)$$

$$v_{R2}(0^+) = 0 \qquad (4\text{-}33)$$

$$v_{C2}(0^+) = -4 \text{ V} \qquad (4\text{-}34)$$

$$v_{R3}(0^+) = 3 \times 4 = 12 \text{ V} \qquad (4\text{-}35)$$

$$v_L(0^+) = v_{R2}(0^+) - v_{C2}(0^+) - v_{R3}(0^+) = -8 \text{ V} \qquad (4\text{-}36)$$

$$v_{R4}(0^+) = 2 \times 4 = 8 \text{ V} \qquad (4\text{-}37)$$

(b) The dc steady-state circuit is shown in Fig. 4-12. Note that we have retained the initial condition generators. In many cases the initial condition generators have no effect at $t = \infty$, but when in doubt, it is best to retain them. In a few cases the final values reached in a circuit will depend in part on initial conditions.

Solution of the steady-state dc circuit results in zero voltage and current for every variable except $v_{C1}(\infty)$. We have

$$v_{C1}(\infty) = 10 \text{ V} \qquad (4\text{-}38)$$

The reader is invited to verify that the omission of the intial condition generators results in the same values for this example.

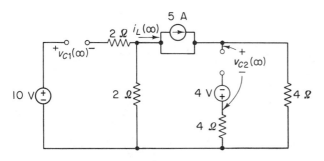

Figure 4-12 Steady-state equivalent circuit of Example 4-4.

Example 4-5

The circuit of Fig. 4-13a is given. The switch has been in position 1 for a long time. At $t = 0$, the switch is thrown to position 2. Determine all voltages and currents in the circuit for $t = 0^+$.

   Solution   We must first determine the initial voltage across C and the initial currents through L_1 and L_2. Since the switch has been in position 1 for a long time and since the previous excitation has been dc, we can use the dc steady-state concept. Thus the circuit for $t = 0^-$ is shown in Fig. 4-13b. The solution follows immediately.

$$i(0^-) = 2 \text{ A} = i_{L1}(0^-) = i_{L2}(0^-) \tag{4-39}$$

$$v_C(0^-) = 6 \text{ V} \tag{4-40}$$

Thus, the inductors each possess initial currents of 2 A, and the capacitor is initially charged to 6 V. At $t = 0^+$, the equivalent circuit is now constructed according to the principles of the initial circuit. Although the excitation in this circuit is time varying, at $t = 0^+$, it acts as a 20-V source. The circuit for $t = 0^+$ is shown in Fig. 4-13c. Some of the quantities desired are labeled in (d). The branch currents are

$$i_{L1}(0^+) = i_{R1}(0^+) = 2 \text{ A} \tag{4-41}$$

$$i_{R2}(0^+) = i_{L2}(0^+) = 2 \text{ A} \tag{4-42}$$

$$i_C(0^+) = 2 - 2 = 0 \tag{4-43}$$

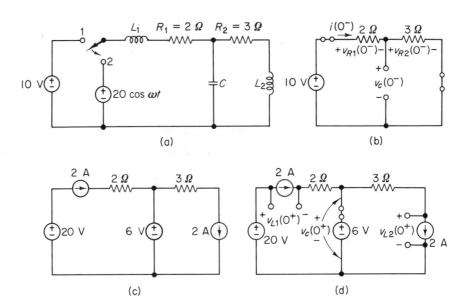

Figure 4-13   Circuit of Example 4-5.

The branch voltages are

$$v_{R1}(0^+) = 4 \text{ V} \tag{4-44}$$

$$v_C(0^+) = 6 \text{ V} \tag{4-45}$$

$$v_{L1}(0^+) = 20 - 4 - 6 = 10 \text{ V} \tag{4-46}$$

$$v_{R2}(0^+) = 6 \text{ V} \tag{4-47}$$

$$v_{L2}(0^+) = 6 - 6 = 0 \tag{4-48}$$

Since the excitation in this circuit is periodic, the responses will never reach a constant value but will always vary with time. However, since the excitation is sinusoidal, a steady-state sinusoidal condition will be reached in which all responses are sinusoidal. This state could be determined from steady-state ac analysis if ω were specified.

4-5 SINGLE TIME CONSTANT CIRCUITS

In the past few sections, we have considered various limiting properties of the circuit elements. In a sense, we have considered the limiting behavior of the differential equations at the times when they reduce to ordinary algebraic equations. But not once have we actually attempted the complete solution of a general circuit problem. We have seen in Chapter 3 that there are a number of fairly simple cases that can be readily solved by a direct application of the voltage-current characteristics in the time domain. We wish to consider in this section an important type of circuit that is more easily solved in the time domain than by transform methods. The circuits that we will consider are called *single time constant circuits. A single time constant circuit is one whose response (either voltage or current) can be described by a first-order linear differential equation.*

As we shall see shortly, to satisfy the single time constant condition, the circuit must be capable of being reduced to one containing a single resistor and a single capacitor, or a single resistor and a single inductor. Thus, we will not allow both an inductor and a capacitor to be present. As a further restriction to enhance the development of a general solution, *we will restrict all external excitations to be dc.* Initial energy storage will be permitted.

Let us look at a specific example. Consider the *RL* circuit shown in Fig. 4-14a. The inductor is assumed to be initially unfluxed, and the switch is closed at $t = 0$. The

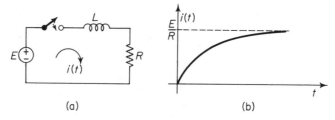

(a)   (b)

Figure 4-14   *RL* circuit and its response.

equation governing the response is

$$L\frac{di}{dt} + iR = E \qquad \text{for } t > 0 \tag{4-49}$$

which is recognized to be a first-order differential equation as discussed in Section 4-1. The general solution of this equation can be shown to be

$$i(t) = \frac{E}{R} + K\epsilon^{-Rt/L} \tag{4-50}$$

where K is an arbitrary constant.

As we might have suspected, since this is a first-order differential equation, there is one arbitrary constant. Thus a knowledge of $i(0^+)$ is sufficient to determine the constant, and since our previous studies tell us that an unfluxed inductor appears as an open circuit at $t = 0^+$, we know that $i(0^+) = 0$. Substituting $t = 0$ into Eq. (4-50) results in

$$0 = \frac{E}{R} + K \tag{4-51}$$

or

$$K = -\frac{E}{R} \tag{4-52}$$

Thus

$$i(t) = \frac{E}{R}(1 - \epsilon^{-Rt/L}) \qquad \text{for } t > 0 \tag{4-53}$$

Applying the definition of time constant (τ) from Chapter 2, we have

$$\tau = \frac{L}{R} \tag{4-54}$$

and hence

$$i(t) = \frac{E}{R}(1 - \epsilon^{-t/\tau}) \tag{4-55}$$

A sketch of $i(t)$ is shown in Fig. 4-14b. We note from the curve that $i(0^+) = 0$ as required. We further note that $i(\infty) = E/R$. This latter fact corresponds to the dc steady state in which the inductor acts as a short circuit, and the current is opposed only by the resistor. We could, of course, have obtained both these limiting values from the material of the preceding few sections without solving a differential equation. The solution of the differential equation has "filled in the middle" and has shown us the exact manner in which the response changes. We can also note the relative "inertia" of the inductor in allowing current to change rather slowly. Only impulse-type excitations can change the current through an inductor instantaneously.

Let us now look at the RC circuit of Fig. 4-15a. In contrast to the previous example, we will assume that the capacitor is initially charged to E_0 volts with polarity as shown. The excitation is E_1 volts applied at $t = 0$, and we will assume that $E_1 > E_0$. Replacing the charged capacitor by its equivalent circuit, we obtain the

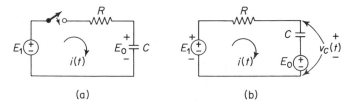

Figure 4-15 *RC circuit with initially charged capacitor and its equivalent circuit.*

circuit shown in Fig. 4-15b. The mesh equation for this loop is

$$Ri + \frac{1}{C}\int_0^t i\,dt = E_1 - E_0 \qquad \text{for } t > 0 \tag{4-56}$$

The integral sign may be removed if we differentiate once with respect to time. We obtain

$$R\frac{di}{dt} + \frac{i}{C} = 0 \qquad \text{for } t > 0 \tag{4-57}$$

for which the general solution is known to be

$$i(t) = K\epsilon^{-t/RC} \tag{4-58}$$

A few readers may pose a question at this point. Since the dc voltage was applied at $t = 0$, suppose that we had written the equation as

$$Ri + \frac{1}{C}\int_0^t i\,dt = (E_1 - E_0)u(t) \tag{4-59}$$

in terms of the unit step function. Then, in differentiating the equation, would we have "differentiated" the step function and obtained the impulse function on the right-hand side? The reader may recall that in Chapter 2 we pointed out that the step function notation can usually be dropped when differentiating, if we ignore the time $t = 0$. On the other hand, we will later show that the step function and the resulting impulse function obtained have certain advantages in some cases.

We will postpone detailed consideration of this point until the next section. Meanwhile, we will again point out that the impulse function must be treated cautiously. For this case, we will avoid it altogether at this time with the restriction, $t > 0$, as noted beside Eqs. (4-56) and (4-57). Thus, as long as $t > 0$, the derivative of the constant is clearly zero, and we will not need to concern ourselves with the impulse function.

Returning to Eq. (4-58), the reader may verify that since the charged capacitor acts as a voltage source at $t = 0^+$,

$$i(0^+) = \frac{E_1 - E_0}{R} \tag{4-60}$$

Solution for the constant K yields

$$K = \frac{E_1 - E_0}{R} \tag{4-61}$$

and hence

$$i(t) = \frac{E_1 - E_0}{R} \epsilon^{-t/\tau} \qquad (4\text{-}62)$$

In this case, the time constant (τ) is recognized to be

$$\tau = RC \qquad (4\text{-}63)$$

A sketch of the current is shown in Fig. 4-16a. Note that $i(\infty) = 0$, implying the fact that the capacitor approaches an open circuit in the steady state.

As another point of interest, let us calculate the voltage $v_C(t)$ across the capacitor. Since the equivalent voltage source E_0 must be considered as a part of the capacitor terminals, we have

$$\begin{aligned} v_C(t) &= \frac{1}{C} \int_0^t i(t)\, dt + E_0 \\ &= \frac{1}{C} \int_0^t \frac{(E_1 - E_0)}{R} \epsilon^{-t/RC}\, dt + E_0 \qquad (4\text{-}64) \\ &= (E_0 - E_1)\epsilon^{-t/RC}\Big]_0^t + E_0 \\ &= E_1 + (E_0 - E_1)\epsilon^{-t/RC} \end{aligned}$$

This response is shown in Fig. 4-16b.

Let us now write a general expression for the response of a single time constant circuit with dc excitations. The general solution of the differential equation for such a circuit always yields an equation of the form

$$y(t) = y(\infty) + [y(0^+) - y(\infty)]\epsilon^{-t/\tau} \qquad \text{for } t > 0 \qquad (4\text{-}65)$$

where $y(t)$ may represent either a voltage or a current, and

$$y(\infty) = \text{final steady-state value of } y$$

$$y(0^+) = \text{initial value of } y$$

$$\tau = RC \text{ for } RC \text{ circuits}$$

$$\tau = \frac{L}{R} \text{ for } RL \text{ circuits}$$

The quantities R, L, and C represent the equivalent values when the network is reduced to the required form.

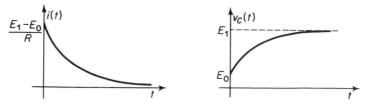

Figure 4-16   Response of the RC circuit of Fig. 4-15.

Since $y(0^+)$, $y(\infty)$, and τ may be determined by inspection, the entire solution reduces to a simple inspection procedure. In the event that $t = a$ is the starting point, a shift in scale of Eq. (4-65) results in

$$y(t) = y(\infty) + [y(a^+) - y(\infty)]\epsilon^{-(t-a)/\tau} \qquad \text{for } t > a \qquad (4\text{-}66)$$

The reader should go back and verify that the two examples considered can be solved very quickly by the use of the general equation.

Example 4-6

The switch in the circuit shown in Fig. 4-17a is instantaneously switched from position 1 to position 2 at $t = 0$. Steady-state conditions existed at $t = 0^-$. Determine $i_L(t)$ and $v_L(t)$ for $t > 0$.

Solution   To determine the initial values of the desired quantities, we first construct the steady-state dc circuit at $t = 0^-$ as shown in Fig. 4-17b. (The right-hand side of the circuit is not connected, so we have omitted it at this time.) The entire available current of 15 A is flowing through the short circuit represented by the inductor. Thus

$$i_L(0^-) = 15 \text{ A} \qquad (4\text{-}67)$$

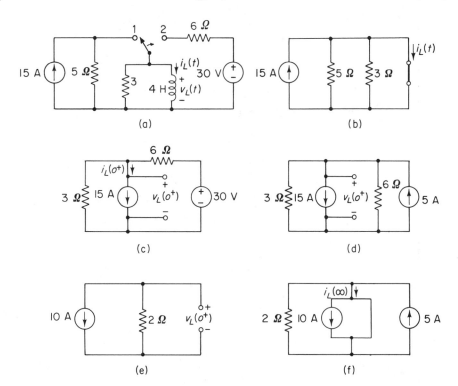

Figure 4-17   Circuit of Example 4-6.

Next, we construct the equivalent circuit at $t = 0^+$ as shown in Fig. 4-17c. The inductor now acts as a current source of 15 A. Thus

$$i_L(0^+) = 15 \text{ A} \tag{4-68}$$

which, of course, could have been predicted by the boundary condition requirements since

$$i_L(0^-) = i_L(0^+) \tag{4-69}$$

We can use at least two approaches to obtain $v_L(t)$. We could continue our present calculation to obtain $i_L(t)$, and then use the voltage–current relationship for an inductor, or we can solve for $v_L(t)$ from initial and final conditions in the same way as for $i_L(t)$. Although the former approach is probably simpler, we will choose the latter approach due to its instructive nature. Thus we must calculate $v_L(0^+)$ next. For the circuit at hand, this is probably more easily done by converting the voltage source in series with the 6-Ω resistor to a current source in parallel with the resistor, as shown in Fig. 4-17d. This reduces the entire network to a simple parallel arrangement as shown in (e). Thus

$$v_L(0^+) = -20 \text{ V} \tag{4-70}$$

Next, we must construct the equivalent circuit at $t = \infty$ (steady-state circuit). As in the preceding step, the current source model of the external generator is more convenient. Thus the circuit at $t = \infty$ is shown in Fig. 4-17e. The final values of i_L and v_L are

$$i_L(\infty) = 5 \text{ A} \tag{4-71}$$

$$v_L(\infty) = 0 \tag{4-72}$$

Finally, the time constant is

$$\tau = \frac{L}{R} = \frac{4}{2} = 2 \text{ s} \tag{4-73}$$

Thus

$$\begin{aligned} i_L(t) &= 5 + (15 - 5)\epsilon^{-t/2} \\ &= 5 + 10\epsilon^{-t/2} \end{aligned} \tag{4-74}$$

and

$$\begin{aligned} v_L(t) &= 0 + (-20 - 0)\epsilon^{-t/2} \\ &= -20\epsilon^{-t/2} \end{aligned} \tag{4-75}$$

The reader can verify that

$$v_L(t) = L \frac{di_L(t)}{dt} \tag{4-76}$$

The functions i_L and v_L are sketched in Fig. 4-18.

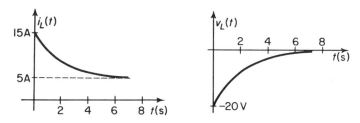

Figure 4-18   Responses of circuit of Example 4-6.

Example 4-7

The capacitor in the circuit of Fig. 4-19a is initially charged to 2 V as shown. At $t = 0$, the switch is closed. Solve for the voltage across the capacitor, $v_0(t)$, for $t > 0$.

   Solution   As far as the desired response is concerned, everything to the left of the switch may be replaced by its Thévenin equivalent circuit as shown in Fig. 4-19b. The equivalent circuit at $t = 0^+$ is shown in (c). Thus

$$v_0(0^+) = v(0^-) = 2 \text{ V} \qquad (4\text{-}77)$$

The equivalent circuit at $t = \infty$ is shown in (d). Thus

$$v_0(\infty) = 10 \text{ V} \qquad (4\text{-}78)$$

implying that the capacitor will eventually charge to 10 V. The time constant is

$$\tau = R_{eq}C = 15 \times 5 = 75 \text{ s} \qquad (4\text{-}79)$$

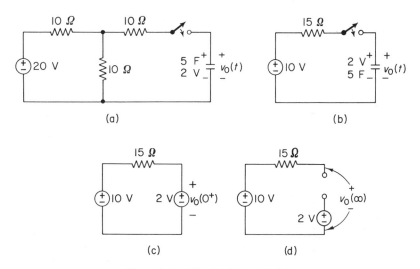

Figure 4-19   Circuit of Example 4-7.

Substitution into the general form yields

$$v_0(t) = 10 + [2 - 10]\epsilon^{-t/75}$$
$$= 10 - 8\epsilon^{-t/75}$$

(4-80)

A sketch of this function is shown in Fig. 4-20.

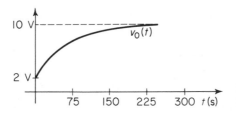

Figure 4-20   Response of circuit of Example 4-7.

Example 4-8

The *RC* circuit shown in Fig. 4-21a is excited by the voltage pulse shown in (b). Solve for the voltage $v_0(t)$ across the capacitor for the following ratios of pulse width to time constant.

(a) $T/\tau = 1000$
(b) $T/\tau = 1$
(c) $T/\tau = 0.001$

The capacitor is initially uncharged.

  Solution   Although the excitation is not quite a simple dc excitation, we may still treat it as such if we look at the solution in two intervals. First, during the interval $0 < t < T$, the excitation *e* is nothing more than a simple dc voltage. (The circuit does not "know" that the voltage will become zero at $t = T$.) Next, for $t > T$, *e* becomes zero and thus the voltage that has accumulated on the capacitor will discharge through the circuit.

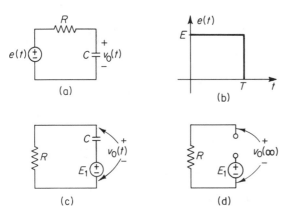

Figure 4-21   Circuit and excitation of Example 4-8.

First, let us look at the interval, $0 < t < T$. In determining $v_0(\infty)$, we look at the response *as if* e would remain at E volts forever. Thus

$$v_0(\infty) = E \qquad (4\text{-}81)$$

$$v_0(0^+) = 0 \qquad (4\text{-}82)$$

The voltage is

$$v_0(t) = E(1 - \epsilon^{-t/\tau}) \qquad \text{for } 0 < t < T \qquad (4\text{-}83)$$

Notice that for this particular response, we did not need to actually know R and C, but merely their product. At time $t = T$, the voltage accumulated on the capacitor is

$$v_0(T) = E(1 - \epsilon^{-T/\tau}) \qquad (4\text{-}84)$$

which is some constant value. For convenience, let

$$E_1 = v_0(T) = E(1 - \epsilon^{-T/\tau}) \qquad (4\text{-}85)$$

Thus, for $t > T$, the problem reduces to the circuit shown in Fig. 4-21c. At $t = \infty$, we obtain the equivalent circuit shown in (d). Thus

$$v_0(\infty) = 0 \qquad (4\text{-}86)$$

We now write the response in this interval, noting that the time scale is now shifted to the right by T units. Thus

$$v_0(t) = E_1\epsilon^{-(t-T)/\tau} \qquad \text{for } t > T \qquad (4\text{-}87)$$

where E_1 is defined by Eq. (4-85). The complete solution is

$$v_0(t) = \begin{cases} E(1 - \epsilon^{-t/\tau}) & \text{for } 0 < t < T \\ E_1\epsilon^{-(t-T)/\tau} & \text{for } t > T \end{cases} \qquad (4\text{-}88)$$

Before looking at the results, let us look at an alternative method for solving this problem. First, we note that $e(t)$ may be considered to be the algebraic sum of two dc voltages (step functions) in the following manner.

$$e(t) = Eu(t) - Eu(t - T) \qquad (4\text{-}89)$$

By the principle of superposition, the total response may be considered to be due to the sum of the responses obtained from the separate step functions. We shall designate the response due to the first step function as $v_0'(t)$. This response is identical with Eq. (4-83), but with the step function attached to clarify the beginning time of the response. Thus

$$v_0'(t) = E(1 - \epsilon^{-t/\tau})u(t) \qquad (4\text{-}90)$$

We shall designate the response due to the second step function as $v_0''(t)$. This response is identical in form with the first response, with the exception of a negative sign, a shift in scale of T units, and the delayed step-function factor. Thus

$$v_0''(t) = -E[1 - \epsilon^{-(t-T)/\tau}]u(t - T) \qquad (4\text{-}91)$$

The reader may wonder why we did not consider any initial voltage in determining $v_0''(t)$. The fact is that, when we use the principle of superposition, we determine each individual response *as if* the excitation under consideration were the only excitation. The voltage on the capacitor for $t > \tau$ is taken care of by the fact that both responses, (4-90) and (4-91), are considered to extend to $t = \infty$ even though the actual pulse is only of finite width.

The total response is the sum of the two responses $v_0'(t)$ and $v_0''(t)$.

$$v_0(t) = v_0'(t) + v_0''(t)$$
$$= E(1 - \epsilon^{-t/\tau})u(t) - E[1 - \epsilon^{-(t-T)/\tau}]u(t - T) \qquad (4\text{-}92)$$

Let us demonstrate that Eq. (4-92) is equivalent to Eq. (4-88). First, for $t < T$, the second response is zero, and hence

$$v_0(t) = \begin{cases} E(1 - \epsilon^{-t/\tau})u(t) & \text{for } t < T \\ E(1 - \epsilon^{-t/\tau}) & \text{for } 0 < t < T \end{cases} \qquad (4\text{-}93)$$

as expected. For $t > T$, both step functions are unity, and hence

$$\begin{aligned} v_0(t) &= E - E\epsilon^{-t/\tau} - E + E\epsilon^{-(t-T)/t} \\ &= E[\epsilon^{-(t-T)/\tau} - \epsilon^{-t/\tau}] \\ &= E[1 - \epsilon^{-T/\tau}]\epsilon^{-(t-T)/\tau} \\ &= E_1\epsilon^{-(t-T)/\tau} \qquad \text{for } t > T \end{aligned} \qquad (4\text{-}94)$$

according to the definition of E_1 from Eq. (4-85). The result is in agreement with Eq. (4-88). For this particular example, the two approaches were comparable in complexity. In a problem involving many sudden steps, the second approach is easier. Now let us look at the responses for different ratios of pulse width to time constant.

(a) $T/\tau = 1000$. In this case, the time constant is so short in comparison to the pulse width that the capacitor essentially charges to E volts very early in the pulse width. This response is shown in Fig. 4-22a.

(b) $T/\tau = 1$. In this case, the time constant and pulse width are equal, so the output voltage reaches about 63% of the input voltage during the pulse width. This response is shown in Fig. 4-22b.

(c) $T/\tau = 0.001$. In this case, the time constant is so long in comparison to the pulse width that the output charges up to only a small fraction of the input voltage as shown in Fig. 4-22c. (Note that the scale has been magnified.)

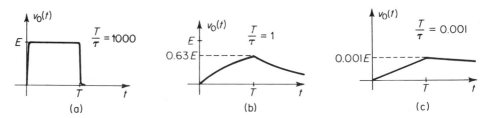

Figure 4-22   Responses of circuit of Example 4-8.

The response is almost a straight line in this interval. This fact may be proved from the power series expansion of the exponential function. Expansion of ϵ^{-x} reads

$$\epsilon^{-x} = 1 - x + \frac{x^2}{2!} - \frac{x^3}{3!} + \cdots \tag{4-95}$$

Thus,

$$1 - \epsilon^{-t/\tau} = \frac{t}{\tau} - \frac{1}{2!}\left(\frac{t}{\tau}\right)^2 + \frac{1}{3!}\left(\frac{t}{\tau}\right)^3 + \cdots \tag{4-96}$$

If $t/\tau \ll 1$, we may neglect all terms after the first term in Eq. (4-96). Thus

$$E(1 - \epsilon^{-t/\tau}) \approx \frac{Et}{\tau} \qquad \text{for } \frac{t}{\tau} \ll 1 \tag{4-97}$$

The approximate voltage at $t = T$ is

$$v_0(T) \approx \frac{ET}{\tau} = 0.001E \tag{4-98}$$

Since the integral of a step function is a ramp function, this circuit is producing an output voltage proportional to the integral of the input voltage. For that reason, this circuit is often referred to as an *integrator circuit*. To qualify for this description, the time constant must be very large in comparison to the duration of the waveform, as was true in part (c) of this example (see Problem 4-45 for more details).

4-6 INITIAL CONDITIONS AND DIFFERENTIAL EQUATIONS

Let us investigate the manner in which initial voltages on capacitors or initial currents through inductors are treated in expressing the differential equations of electric circuits. First, consider an initially charged capacitor as shown in Fig. 4-23a. The Thévenin equivalent circuit for the charged capacitor is shown in (b) as it was developed in Chapter 3. As far as a mesh current analysis is concerned, this voltage source acts along with any other sources in the loop to determine the net current and, hence, is treated as a regular voltage source in writing the mesh equations. Thus for the

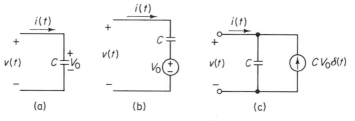

Figure 4-23 Thévenin and Norton equivalent circuits of charged capacitor.

portion of a circuit containing the charged capacitor, the voltage in Fig. 4-23 is

$$v = \frac{1}{C} \int_0^t i \, dt + V_0 \qquad (4\text{-}99)$$

as previously established, with the proper polarity of V_0 maintained.

A somewhat different problem occurs, however, in writing node voltage equations in the presence of a charged capacitor. Since it is desirable to express all sources as equivalent current sources when writing node voltage equations, we obtain the Norton equivalent circuit of Fig. 4-23c as was developed in Chapter 3. The question now arises as to what to do with the impulse current source. One possibility is to write the equation at the node as

$$i = C \frac{dv}{dt} - CV_0 \, \delta(t) \qquad (4\text{-}100)$$

However, this question is not particularly desirable due to certain difficulties that arise in dealing with the impulse function in a differential equation. Perhaps we can resolve this difficulty by asking ourselves what the purpose of this impulse function really is anyway. The whole purpose in developing an equivalent circuit for a charged capacitor is so that we may work with uncharged capacitors. In other words, by considering the capacitor as if it were initially uncharged, we put the requirement on Eq. (4-100) that $v_C(0) = 0$. On the other hand, we know that for $t = 0^+$ the capacitor must be charged to V_0 volts. We have also noted before that an impulse current could actually produce an instantaneous charge in the voltage across a capacitor. Thus *the purpose of the impulse current source is to bring the uncharged capacitor instantaneously to its required value of V_0 volts at $t = 0^+$*.

On the other hand, since the solution of a differential equation always involves arbitrary constants, we can achieve the same results by placing an appropriate initial condition on $v_C(0^+)$ and avoiding the impulse function. Thus, if we ignore $t = 0$ and consider only $t = 0^+$, we may write instead of Eq. (4-100)

$$i = C \frac{dv_C}{dt} \qquad (4\text{-}101)$$

with the additional specification that

$$v_C(0^+) = V_0 \qquad (4\text{-}102)$$

Thus, theoretically, there are two approaches to writing a node voltage equation at a node involving a charged capacitor. The first approach involves writing a node equation similar to Eq. (4-100) containing the impulse-current source and with the specification $v_C(0) = 0$. This form does not lend itself easily to "classical" methods and is usually avoided when one is working strictly in the time domain. However, as we shall see in our later work, this form is quite adaptable to transform methods.

The second approach involves writing a node equation similar to Eq. (4-101) in which the specification $v_C(0^+) = V_0$ is made. The two methods yield identical results, and the latter method is more suitable for time-domain considerations. The quantity $t = 0^+$ is usually treated like $t = 0$ when substituting into an equation, except in the case of functions with a "jump" at $t = 0$. We use the 0^+ concept to emphasize more rigorously the ideas involved.

Recall from Section 4-5 that a question arose regarding the differentiation of an apparent step function [refer back to Eq. (4-59)]. By ignoring $t = 0$ and considering only the region for $t > 0$, we considered the derivative of the step as being zero, with the initial requirement that $i(0^+) = (E_1 - E_0)/R$ for that particular example. If we had chosen to include $t = 0$ in the range, and thus had considered the derivative of the equivalent step function as being an impulse function, we would have required that $i(0) = 0$. The impulse function would have created the instantaneous voltages required to produce the necessary initial current if we had known how to solve the equation containing the impulse.

Let us now consider the manner in which initial conditions are handled with inductors. First consider the initially fluxed inductor shown in Fig. 4-24a. The Norton equivalent circuit for the fluxed inductor is shown in (b) as it was developed in Chapter 3. As far as a node voltage analysis is concerned, this current source acts along with any other currents feeding the node to determine the net voltage; hence it is treated as a regular current source. Thus for a node involving a fluxed inductor, the current for Fig. 4-24 is

$$i = \frac{1}{L} \int_0^t v \, dt + I_0 \tag{4-103}$$

as previously established, with the proper direction of I_0 maintained.

A somewhat different problem occurs, however, when we write mesh current equations around a fluxed inductor. We shall not elaborate through the details here, as this situation is analogous to the node voltage equations of a charged capacitor.

Consider the Thévenin equivalent circuit of the fluxed inductor as shown in Fig. 4-24c. If the impulse source is considered, the equation for this branch is

$$v = L \frac{di}{dt} - LI_0 \, \delta(t) \tag{4-104}$$

with the requirement that

$$i(0) = 0 \tag{4-105}$$

The effect of the impulse voltage source is to establish the current I_0 through the inductor at $t = 0^+$. On the other hand, the form more appropriate for time-domain analysis is to ignore the point $t = 0$ and to consider only the region $t > 0$. In this region, the equation is

$$v = L \frac{di}{dt} \tag{4-106}$$

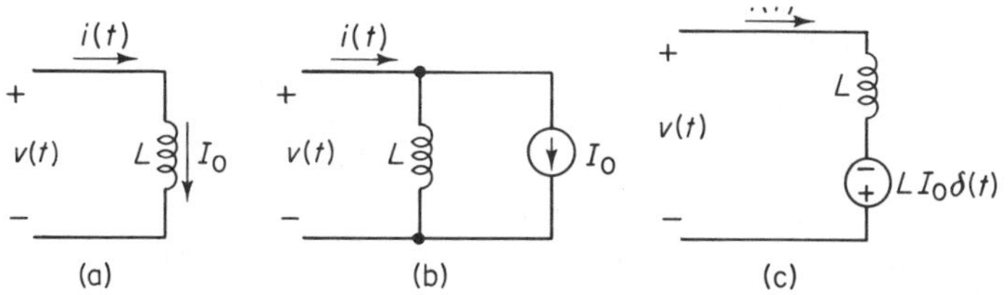

Figure 4-24   Norton and Thévenin equivalent circuits of fluxed inductor.

with the requirement that

$$i(0^+) = I_0 \tag{4-107}$$

Both forms yield identical results. However, as previously stated, the impulse form is more suitable for the transform methods of later work, whereas the method avoiding impulses is more suitable for time-domain classical methods.

To summarize this section, we may state:

1. In expressing Kirchhoff's voltage law around a closed loop containing a charged capacitor (e.g., mesh current analysis), we use the Thévenin equivalent circuit of the charged capacitor, and we treat the voltage source as any other voltage source. (Of course, the source must be considered to be part of the capacitor terminals.)

2. In expressing Kirchhoff's current law at a node involving a fluxed inductor (e.g., node voltage analysis), we use the Norton equivalent circuit of the fluxed inductor and we treat the current source as any other current source. (Again, the source must be considered to be part of the inductor terminals.)

3. In expressing Kirchhoff's current law at a node involving a charged capacitor, one of two approaches is possible:
   (a) The Norton equivalent circuit involving an impulse function can be used, in which case $v_C(0) = 0$ is specified.
   (b) The impulse source may be ignored, in which case $v_C(0^+) = V_0$ is specified.

4. In expressing Kirchhoff's voltage law around a closed loop containing a fluxed inductor, one of two approaches is possible:
   (a) The Thévenin equivalent circuit involving an impulse function can be used, in which case $i_L(0) = 0$ is specified.
   (b) The impulse source may be ignored, in which case $i_L(0^+) = I_0$ is specified.

For 3 and 4, approach (a) will be seen to be the most convenient approach in transform analysis, whereas approach (b) is more suitable in time-domain analysis.

Example 4-9

The circuit shown in Fig. 4-25a is excited at $t = 0$ by the sinusoidal source shown. All energy-storage components are initially charged or fluxed with values and directions as shown.

(a) Draw an equivalent circuit suitable for mesh current analysis, with initial conditions represented by sources. Write the mesh equations ignoring impulse functions, and specify the appropriate initial conditions on the three mesh currents.

(b) Repeat the procedure of (a) using node voltage analysis.

Solution (a) For writing mesh current equations, we wish to use the Thévenin equivalent circuits for the initial conditions. The resulting equivalent circuit is shown in Fig. 4-25b, with the three mesh currents i_a, i_b, and i_c assumed. However, since we will be following the approach more suitable for time-domain

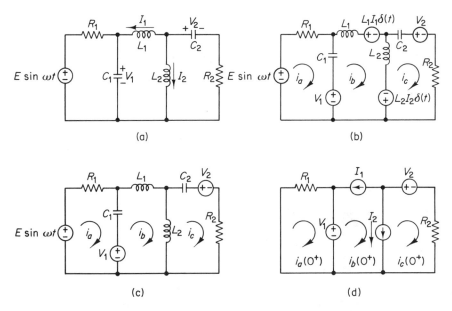

Figure 4-25   Circuit of Example 4-9 (mesh current analysis).

analysis, namely the approach of neglecting impulse voltage sources, we redraw
the circuit in (c) with the impulse generators removed. (The reader is probably
wondering why we even bothered drawing the circuit first with the impulse
generators. As we have said before, with transform methods, we will find the
impulse approach to be the most useful. Therefore, it is a good habit to at
least think about the impulse idea at this point, even though we will not actually
use it for a while.) The equations for $t > 0$ are

$$R_1 i_a + \frac{1}{C_1} \int_0^t (i_a - i_b) \, dt = E \sin \omega t - V_1 \tag{4-108}$$

$$\frac{1}{C_1} \int_0^t (i_b - i_a) \, dt + L_1 \frac{di_b}{dt} + L_2 \frac{d}{dt} (i_b - i_c) = V_1 \tag{4-109}$$

$$L_2 \frac{d}{dt} (i_c - i_b) + \frac{1}{C_1} \int_0^t i_c \, dt + R_2 i_c = -V_2 \tag{4-110}$$

To obtain the initial values of the three mesh currents, we draw an equivalent circuit at $t = 0^+$ according to the methods of this chapter. The circuit
is shown in Fig. 4-25d. Note that the input voltage is zero at $t = 0$. From the
equivalent circuit, we note that

$$i_a(0^+) = -\frac{V_1}{R_1} \tag{4-111}$$

$$i_b(0^+) = -I_1 \tag{4-112}$$

$$i_c(0^+) = -I_1 - I_2 \tag{4-113}$$

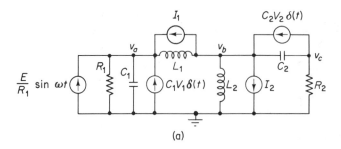

(a)

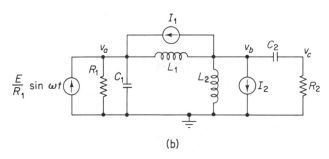

(b)

Figure 4-26   Circuit of Example 4-9 (node voltage analysis).

(b) Now let us characterize this network from the node voltage equations. We first draw an equivalent circuit in which all initial condition generators are expressed by their Norton equivalent circuits as shown in Fig. 4-26a. Note that the external generator has also been converted to an equivalent current source. (We could treat it as a node whose voltage is known if desired.) Next, we redraw the circuit with the impulse current sources omitted as shown in (b). The node voltage equations for $t > 0$ are

$$\frac{v_a}{R_1} + C_1 \frac{dv_a}{dt} + \frac{1}{L_1} \int_0^t (v_a - v_b)\, dt = \frac{E}{R_1} \sin \omega t + I_1 \qquad (4\text{-}114)$$

$$\frac{1}{L_1} \int_0^t (v_b - v_a)\, dt + \frac{1}{L_2} \int_0^t v_b\, dt + C_2 \frac{d}{dt}(v_b - v_c) = -I_1 - I_2 \quad (4\text{-}115)$$

$$C_2 \frac{d}{dt}(v_c - v_b) + \frac{v_c}{R_2} = 0 \qquad (4\text{-}116)$$

To obtain the initial values of the three voltages, we refer back to the equivalent circuit at $t = 0^+$, shown in Fig. 4-25d. From the circuit, we note that

$$v_a(0^+) = V_1 \qquad (4\text{-}117)$$

$$v_c(0^+) = -R_2(I_1 + I_2) \qquad (4\text{-}118)$$

$$v_b(0^+) = v_c(0^+) + V_2 = V_2 - R_2(I_1 + I_2) \qquad (4\text{-}119)$$

GENERAL PROBLEMS

The energy-storage elements in Problems 4-1 through 4-4 are initially relaxed. In each case the excitations are assumed to be applied at $t = 0$.

4-1. Write a mesh current equation for the circuit of Fig. P4-1. Remove the integral term and classify the resulting differential equation.

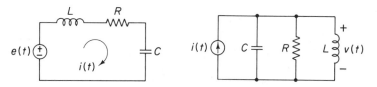

Figure P4-1   **Figure P4-2**

4-2. Write a node voltage equation for the circuit of Fig. P4-2. Remove the integral term and classify the resulting differential equation.

4-3. Write the simultaneous mesh curent equations for the circuit of Fig. P4-3. Eliminate i_1 and obtain a differential equation involving only i_2. What is its order?

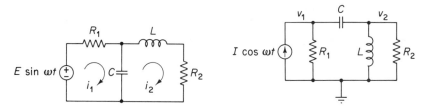

Figure P4-3   **Figure P4-4**

4-4. Write the simultaneous node voltage equations for the circuit of Fig. P4-4. Eliminate v_1 and obtain a differential equation involving only v_2. What is its order?

In Problems 4-5 through 4-10, sketch the response given for $t > 0$. Write an equation for the steady-state response in each case.

4-5. $v(t) = A(1 + \epsilon^{-t/\tau}) \sin \omega t$ **4-6.** $i(t) = A(1 - 0.5\epsilon^{-t/\tau}) \sin \omega t$

4-7. $v(t) = A(1 - \epsilon^{-t/\tau})$ **4-8.** $i(t) = A(1 + \epsilon^{-t/\tau})$

4-9. $v(t) = A \cos \omega t$ **4-10.** $i(t) = A\epsilon^{-t/\tau}$

In Problems 4-11 through 4-16, draw the steady-state dc equivalent circuit, and determine the steady-state dc voltage and current associated with each component.

4-11. See Fig. P4-11.   **4-12.** See Fig. P4-12.

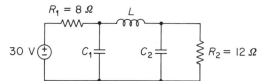

Figure P4-11

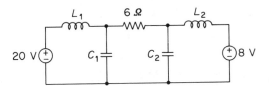

Figure P4-12

4-13. See Fig. P4-13.   **4-14.** See Fig. P4-14.

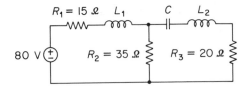

Figure P4-13

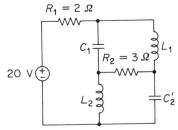

Figure P4-14

4-15. See Fig. P4-15.   **4-16.** See Fig. P4-16.

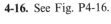

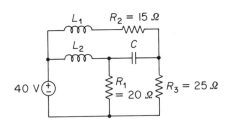

Figure P4-15

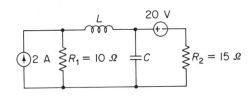

Figure P4-16

In Problems 4-17 through 4-22, the switches are either closed (Problems 4-17 through 4-20), opened (Problem 4-22), or changed from position 1 to 2 (Problem 4-21) at $t = 0$. Any initial capacitor voltages or inductor currents are labeled on the respective components. (a) For each case, draw the equivalent circuit at $t = 0^+$, and calculate the initial values of voltage and current associated with each component. (b) For each case, draw the steady-state dc equivalent circuit and calculate the steady-state voltages and currents.

4-17. See Fig. P4-17.   **4-18.** See Fig. P4-18.   **4-19.** See Fig. P4-19.
4-20. See Fig. P4-20.   **4-21.** See Fig. P4-21.   **4-22.** See Fig. P4-22.

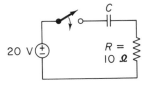

Figure P4-17

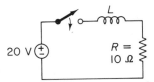

Figure P4-18

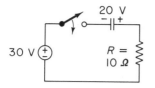

Figure P4-19

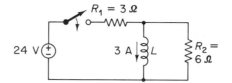

Figure P4-20

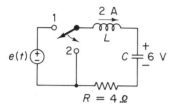

Figure P4-21

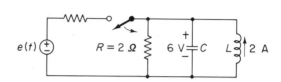

Figure P4-22

In Problems 4-23 through 4-28, the circuits have reached steady-state conditions prior to $t = 0$. At $t = 0$, the switch is instantaneously changed from position 1 to position 2. (a) For each case, draw an equivalent circuit at $t = 0^-$, and calculate the currents through inductors and the voltages across capacitors. (b) For each case, draw an equivalent circuit at $t = 0^+$, and calculate the initial voltage and current associated with each component. (c) For each case, draw the steady-state dc equivalent circuit, and calculate the steady-state voltages and currents.

4-23. See Fig. P4-23.   **4-24.** See Fig. P4-24.   **4-25.** See Fig. P4-25.
4-26. See Fig. P4-26.   **4-27.** See Fig. P4-27.   **4-28.** See Fig. P4-28.

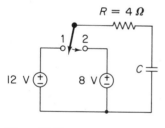

Figure P4-23

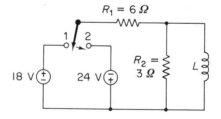

Figure P4-24

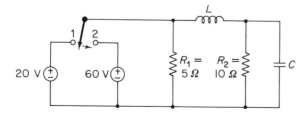

Figure P4-25

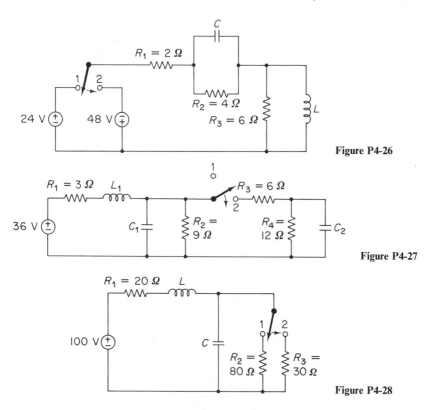

Figure P4-26

Figure P4-27

Figure P4-28

In Problems 4-29 through 4-34, the switches are closed at $t = 0$. Any initial capacitor voltages or inductor currents are labeled on the respective components. For each circuit, determine equations for the voltages and currents indicated for $t > 0$ and sketch them.

4-29. See Fig. P4-29.   **4-30.** See Fig. P4-30.   **4-31.** See Fig. P4-31.
4-32. See Fig. P4-32.   **4-33.** See Fig. P4-33.   **4-34.** See Fig. P4-34.

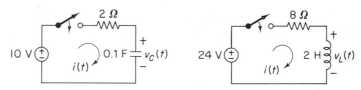

Figure P4-29   Figure P4-30

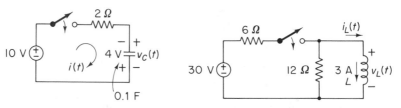

Figure P4-31   Figure P4-32

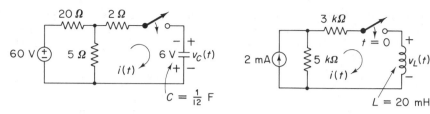

Figure P4-33   Figure P4-34

4-35. The circuit of Fig. P4-35 has reached a steady-state condition at $t = 0^-$. At $t = 0$, the switch is closed. Determine an equation for $v_C(t)$ for $t > 0$ and sketch it.

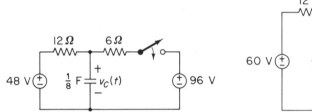

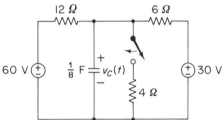

Figure P4-35   Figure P4-36

4-36. The circuit of Fig. P4-36 has reached a steady-state condition at $t = 0^-$. At $t = 0$, the switch is closed. Determine an equation for $v_C(t)$ for $t > 0$ and sketch it.

In Problems 4-37 through 4-39, (a) draw an equivalent circuit for $t > 0$ appropriate for mesh current analysis; that is, retain initial condition step voltage sources, but not impulse voltage sources. (b) Write the mesh current equation(s). (c) Determine the initial value of each mesh current.

4-37. See Fig. P4-37.   **4-38.** See Fig. P4-38.   **4-39.** See Fig. P4-39.

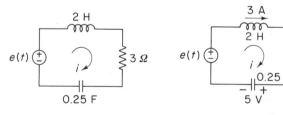

Figure P4-37   Figure P4-38

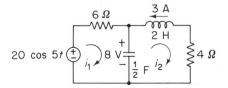

Figure P4-39

In Problems 4-40 through 4-42, (a) draw an equivalent circuit for $t > 0$ appropriate for node voltage analysis; that is, retain initial condition step current sources, but not impulse current sources. (b) Write the node voltage equation(s). (c) Determine the initial value of each node voltage.

4-40. See Fig. P4-40.   **4-41.** See Fig. P4-41.   **4-42.** See Fig. P4-42.

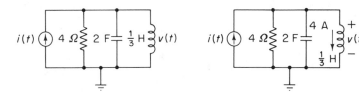

Figure P4-40   **Figure P4-41**

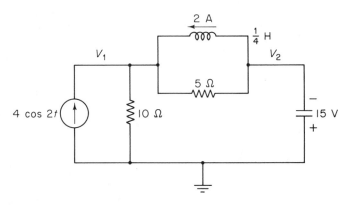

Figure P4-42

DERIVATION PROBLEMS

4-43. In working with the response of a single time constant circuit as given by Eq. (4-65), it is often necessary to "invert" the equation, that is, to determine the value of time t_1 at which the response y reaches a level y_1. Show that this time can be expressed as

$$t_1 = \tau \ln \frac{y(\infty) - y(0^+)}{y(\infty) - y_1}$$

4-44. The effect of shunt capacitance or other frequency-limiting effects is to introduce a *rise time* for any dc or pulse-type input v_1 to a transmission system. To minimize errors and uncertainties in initial and final levels, the IEEE definition of rise time T_R is

$$T_R = t_{90} - t_{10}$$

where t_{90} represents the time required for the output v_2 to reach 90% of the final level and t_{10} is the time required to reach 10% of the final level. For the basic *RC* circuit of Fig. P4-44, show that the rise time is approximately

$$T_R \simeq 2.2RC$$

4-45. The simple *RC* circuit of Fig. P4-45 will be investigated as it relates to a possible *integrator circuit* as discussed in Example 4-8.

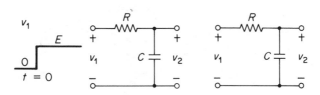

Figure P4-44   **Figure P4-45**

(a) Write a node voltage equation at the output node, and show that the following differential equation is obtained:

$$\frac{dv_2}{dt} + \frac{1}{RC} v_2 = \frac{1}{RC} v_1$$

(b) Under certain conditions in which RC is large, the second term on the left in the equation above is small compared to the first term. Under these conditions, show that the voltage $v_2(t)$ is approximately

$$v_2(t) \simeq \frac{1}{RC} \int_0^t v_1(t)\, dt$$

4-46. The simple RC circuit of Fig. P4-46 will be investigated as it relates to a possible *differentiator circuit*.

(a) Write a node voltage equation at the output node, and show that the following differential equation is obtained:

$$RC \frac{dv_2}{dt} + v_2 = RC \frac{dv_1}{dt}$$

(b) Under certain conditions in which RC is small, the first term on the left in the equation above is small compared to the second term. Under these conditions, show that the voltage $v_2(t)$ is approximately

$$v_2(t) \simeq RC \frac{dv_1(t)}{dt}$$

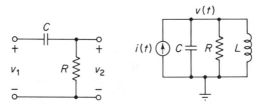

Figure P4-46   **Figure P4-47**

4-47. For the circuit of Fig. P4-47, show that the following differential equation can be obtained:

$$\frac{d^2v}{dt^2} + \frac{1}{RC} \frac{dv}{dt} + \frac{v}{LC} = \frac{1}{C} \frac{di}{dt}$$

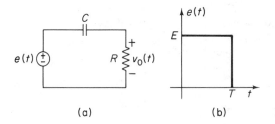

(a)   (b)   **Figure P4-48**

4-48. The RC circuit shown in Fig. P4-48a is excited by the voltage pulse shown in (b). Solve for the voltage across the resistor, $v_0(t)$, for each of the following ratios of pulse width to time constant.

(a) $T/\tau = 1000$ **(b)** $T/\tau = 1$ **(c)** $T/\tau = 0.001$

Illustrate that condition (a) results in an output that is, in some sense, the *derivative* of the input voltage (see Problem 4-46 for further discussion of the differentiator effect).

COMPUTER PROBLEMS

4-49. Write a computer program to determine $y(t)$ of Eq. (4-65) for any arbitrary single time constant circuit. Provision should be made for varying the independent variable t from t_1 to t_2 in steps of Δt.

> Input data: $y(0^+)$, $y(\infty)$, τ, t_1, t_2, Δt
>
> Output data: t, $y(t)$

4-50. Write a computer program to determine the time t_1 for a first-order response to reach the level y_1 (see Problem 4-43).

> Input data: $y(0^+)$, $y(\infty)$, τ, y_1
>
> Output data: t_1

5

LAPLACE TRANSFORM

OVERVIEW

In this chapter we pause momentarily from our consideration of circuit analysis in order to develop some important mathematical techniques. The Laplace transform and its basic mathematical properties will be studied extensively. The Laplace transform permits the solution of complex integrodifferential equations using algebraic methods. In particular, actual circuit models can be represented directly using Laplace techniques.

Both the transforms of time functions and the inverse transforms of Laplace functions are considered. The waveforms of Chapter 2 appear extensively in these developments since they are widely employed in circuit applications.

OBJECTIVES

After completing this chapter, the reader should be able to

1. Discuss the advantages of the Laplace transform approach in solving differential equations.
2. State the definition of the Laplace transform.
3. Derive the Laplace transforms of some of the common functions of Table 5-1.
4. Use Table 5-1 to determine the Laplace transforms of common waveforms encountered in circuit analysis.
5. Apply the operation pairs of Table 5-2 for appropriate waveforms.
6. Write the form of a polynomial, identify its *degree* (or *order*), and indicate the number of roots.
7. Express a polynomial in factored form.
8. State the various classifications of roots.

135

9. **Apply Table 5-1 to determine inverse transforms of easily identifiable transform functions.**

10. **Define the terms *zero* and *pole* as they relate to Laplace functions.**

11. **State the four ways of classifying poles for the purposes of inverse transformation.**

12. **Perform a partial fraction expansion for first-order real poles.**

13. **Determine inverse transforms corresponding to first-order complex poles.**

14. **Determine inverse transforms corresponding to multiple-order poles.**

5-1 GENERAL DISCUSSION

The French mathematician Laplace (1749–1827) developed a mathematical method for solving differential equations by the use of algebraic techniques. The method of Laplace utilizes a so-called *transform* or *operational* procedure. The British physicist Oliver Heaviside (1850–1925) is credited with early usage of operational methods in circuit analysis.

As an aid in understanding transform analysis, let us consider the process of multiplication by logarithms as illustrated in Fig. 5-1. The first step in performing the multiplication is to replace the original numbers by their logarithms. Next, the logarithms are added, and the result, of course, is the logarithm of the quantity desired. As a final step, the antilogarithm (or inverse logarithm) is found, and the desired result is obtained. Thus, the process of multiplication has been replaced by transformation, addition, and inverse transformation.

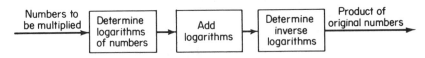

Figure 5-1   Steps involved in multiplication by logarithms.

An analogous situation exists with transform circuit methods. A general description of the process is shown in Fig. 5-2. We first begin with a circuit or system that is described by a set of differential equations. We then replace the system by a so-called *transform system*. We speak of the original system as a representation in the *time domain* and the new system as a representation in the *transform domain*. (The transform domain is often referred to as the *complex-frequency domain* for reasons that will be clearer in later chapters.) The transform system has the property that any solution may be obtained by algebraic procedures. After the desired response is obtained in the transform domain, we invert or calculate the *inverse* transform to obtain the desired response in the time domain.

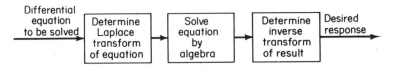

Figure 5-2   Steps involved in Laplace transform analysis.

In performing the transformation to the transform domain, two approaches are possible:

1. The differential equation may be written in the time domain, followed by transformation of the differential equation.
2. The circuit or system may first be transformed directly, followed by direct expression of the equations in the transform domain.

The engineering analyst will have occasion to follow both procedures, but the second method is usually simpler for circuit applications. Since we are primarily interested in circuit applications, we concentrate more heavily on the second approach, but we also discuss the first approach in some detail.

In this chapter we learn to calculate the Laplace transforms of the waveforms of interest and to determine the appropriate inverse transforms. We also investigate the transforms of basic mathematical operations. The actual evaluation of circuit responses is considered in Chapter 6.

5-2 LAPLACE TRANSFORMS

The Laplace transform of a function of time $f(t)$ is a function of a new variable s and is designated as $F(s)$. The process of performing the Laplace transformation is designated as

$$\mathscr{L}[f(t)] = F(s) \tag{5-1}$$

The process of inverse transformation is designated as

$$\mathscr{L}^{-1}[F(s)] = f(t) \tag{5-2}$$

Thus, Eqs. (5-1) and (5-2) describe symbolically the process of transformation and inverse transformation. The reader is probably beginning to wonder exactly what this new variable s is. We discuss this in greater length in a later chapter. Let us say at the moment that it is, in a sense, an artificial variable or *operator* that must be carried in all expressions involving transforms in order to produce the desired results.

Note that we have used a lowercase f for $f(t)$ and a capital F for $F(s)$. It is necessary to make this distinction, as $f(t)$ and $F(s)$ are *not* equal. In this context, we emphatically state that

$$f(t) \neq F(s) \tag{5-3}$$

It is a common occurrence to see beginners tend to write expressions of the form of Eq. (5-3) as an equality. At the risk of redundancy, we again emphasize that the pair of functions $f(t)$ and $F(s)$ correspond to each other in a transform sense, but are not equal. The correct way to express an equality is by Eqs. (5-1) and (5-2).

Now let us see what is actually implied by the Laplace transformation. The definition of the Laplace transform of $f(t)$ is

$$\mathscr{L}[f(t)] = F(s) = \int_0^\infty f(t)\epsilon^{-st}\, dt \tag{5-4}$$

We thus multiply $f(t)$ by ϵ^{-st} and integrate from $t = 0$ to $t = \infty$ to obtain $F(s)$. This means that the given function times ϵ^{-st} is integrated over all positive time. This may require breaking up the function into several different integrals in cases where $f(t)$ is defined in a different sense over different intervals, or it may require integrating only over a short period of time in cases where $f(t)$ is identically zero over most of positive time.

Before calculating any Laplace transforms, we will state two fairly simple but important relationships that will be useful in determining all transforms.

1. The Laplace transform of a constant times a given function of time is equal to the constant times the transform of the original function of time. Thus

$$\mathscr{L}[Kf(t)] = K\mathscr{L}[f(t)] = KF(s) \tag{5-5}$$

2. The Laplace transform of the sum of several functions is equal to the sum of the individual transforms. Thus,

$$\mathscr{L}[f_1(t) + f_2(t) + \cdots + f_n(t)] = \mathscr{L}[f_1(t)] + \mathscr{L}[f_2(t)] + \cdots + \mathscr{L}[f_n(t)]$$
$$= F_1(s) + F_2(s) + \cdots + F_n(s) \tag{5-6}$$

TABLE 5-1   COMMON LAPLACE TRANSFORM PAIRS

$f(t)$	$F(s) = \mathscr{L}[f(t)]$	
$\delta(t)$	1	(T-1)
1 or $u(t)$	$\dfrac{1}{s}$	(T-2)
t	$\dfrac{1}{s^2}$	(T-3)
$\epsilon^{-\alpha t}$	$\dfrac{1}{s + \alpha}$	(T-4)
$\sin \omega t$	$\dfrac{\omega}{s^2 + \omega^2}$	(T-5)
$\cos \omega t$	$\dfrac{s}{s^2 + \omega^2}$	(T-6)
$\epsilon^{-\alpha t} \sin \omega t$	$\dfrac{\omega}{(s + \alpha)^2 + \omega^2}$ [a]	(T-7)
$\epsilon^{-\alpha t} \cos \omega t$	$\dfrac{s + \alpha}{(s + \alpha)^2 + \omega^2}$ [a]	(T-8)
t^n	$\dfrac{n!}{s^{n+1}}$	(T-9)
$\epsilon^{-\alpha t} t^n$	$\dfrac{n!}{(s + \alpha)^{n+1}}$	(T-10)

[a] Complex roots.

On the other hand, the reader should be cautioned about one common mistake made by beginners. The Laplace transform of the product of two functions is *not* the product of the two individual transforms. Thus

$$\mathcal{L}[f_1(t)f_2(t)] \neq F_1(s)F_2(s) \qquad (5\text{-}7)$$

A list of transform pairs for the most common basic circuit functions is given in Table 5-1. The examples that follow will demonstrate the procedure for calculating transforms by deriving a few of the pairs in the table; a few examples will employ the results of the table. Some pairs will be left for the student as exercises. The (T-*n*) designation in the table is merely an arbitrary system for later reference. The table is repeated in Appendix D.

Example 5-1

Derive the Laplace transform of the unit step function.
    Solution By definition, we have

$$\mathcal{L}[u(t)] = \mathcal{L}[1] = \int_0^\infty (1)\epsilon^{-st}\, dt = \left.\frac{\epsilon^{-st}}{-s}\right]_0^\infty \qquad (5\text{-}8)$$

We interpret the *s* to be positive in the foregoing equation so that $\lim\limits_{t \to \infty} \epsilon^{-st} = 0$. Thus

$$F(s) = \frac{1}{s} \qquad (5\text{-}9)$$

Example 5-2

Derive the Laplace transform pair (T-5). We may use "simpler" transform pairs to accomplish the task.
    Solution Although we could integrate $\epsilon^{-st} \sin \omega t$ if we cared to do so, we can accomplish the same task by accepting, for the moment, pair (T-4) and employing the exponential definition of the sine function. This definition reads

$$\sin \omega t = \frac{\epsilon^{j\omega t} - \epsilon^{-j\omega t}}{2j} \qquad (5\text{-}10)$$

Thus, by Eqs. (5-5), (5-6), and (T-4),

$$\mathcal{L}[\sin \omega t] = \mathcal{L}\left[\frac{\epsilon^{j\omega t} - \epsilon^{-j\omega t}}{2j}\right] = \frac{1}{2j}\left\{\mathcal{L}[\epsilon^{j\omega t}] - \mathcal{L}[\epsilon^{-j\omega t}]\right\}$$

$$= \frac{1}{2j}\left[\frac{1}{s - j\omega} - \frac{1}{s + j\omega}\right] = \frac{(s + j\omega) - (s - j\omega)}{2j(s^2 + \omega^2)} \qquad (5\text{-}11)$$

$$= \frac{\omega}{s^2 + \omega^2}$$

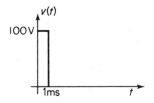

Figure 5-3 Voltage pulse of Example 5-3.

Example 5-3

The narrow voltage pulse shown in Fig. 5-3 is used to excite a network whose time constants are very long compared to the width of the pulse. Define an impulse that approximates the given pulse, and find its Laplace transform.

    Solution The area of the pulse is $100 \text{ V} \times 10^{-3} \text{ s} = 0.1 \text{ V·s}$. Thus, according to the methods of Chapter 2, we may approximate $v(t)$ as

$$v(t) \approx 0.1\delta(t) \tag{5-12}$$

The transform of $v(t)$ is $V(s)$. From (T-1),

$$V(s) = \mathscr{L}[v(t)] = \mathscr{L}[0.1\delta(t)] = 0.1\mathscr{L}[\delta(t)] = 0.1 \tag{5-13}$$

We might note in passing that, while the impulse function may be somewhat strange in the time domain, its transform is the simplest of all functions because it is merely a constant!

Example 5-4

Determine the Laplace transform of the current
$$i(t) = 10 \cos (20t + 30°) \tag{5-14}$$

    Solution In order to use the table, we must decompose $i(t)$ into the sum of a sine function and a cosine function. The reader can verify that

$$i(t) = 5\sqrt{3} \cos 20t - 5 \sin 20t \tag{5-15}$$

Letting $I(s)$ represent the transform of $i(t)$, we have from (T-5) and (T-6),

$$\begin{aligned}
I(s) &= 5\sqrt{3} \, \mathscr{L}[\cos 20t] - 5\mathscr{L}[\sin 20t] \\
&= \frac{5\sqrt{3}\,s}{s^2 + 400} - \frac{100}{s^2 + 400} \\
&= \frac{5\sqrt{3}\,s - 100}{s^2 + 400}
\end{aligned} \tag{5-16}$$

5-3 LAPLACE TRANSFORM OPERATIONS

In Section 5-2 we presented a few basic transform pairs for some common types of functions encountered in circuit theory. We now wish to consider a few operations on time functions and how they affect the transforms of the respective functions. The

TABLE 5-2   A FEW LAPLACE TRANSFORM OPERATIONS

$f(t)$	$F(s)$	
$f'(t)$	$sF(s) - f(0)$	(O-1)
$\int_0^t f(t)\,dt$	$\dfrac{F(s)}{s}$	(O-2)
$f(t - a)u(t - a)$	$\epsilon^{-as}F(s)$	(O-3)
$\epsilon^{-\alpha t}f(t)$	$F(s + \alpha)$	(O-4)

proofs of these operations are presented in Appendix C and not in the text. Here we concentrate only on what they mean and how to use them. The operations are summarized in Table 5-2 and again in Appendix D for later reference. Let us look at each operation individually. In all operations to follow, we assume as a reference a time function $f(t)$ and its transform $F(s)$.

Derivative of a Time Function

$$\mathcal{L}[f'(t)] = sF(s) - f(0) \tag{5-17}$$

According to this theorem, if we start out with a function $f(t)$ whose transform is $F(s)$, and then differentiate $f(t)$, we obtain the transform of the new function by simply multiplying the original transform by s and subtracting the initial value of the time function at $t = 0$, which is, of course, a constant. This theorem does not apply to functions that are indeterminant at $t = 0$ (e.g., impulse function). If we, for the moment, consider time functions that are zero at $t = 0$, we can say that *differentiation in the time domain corresponds to multiplication by* s *in the transform domain.*

Let us illustrate this theorem by using an example from the transform pair table. Accepting for the moment the transform of the ramp function, we have

$$f(t) = t \qquad \text{for } t > 0 \tag{5-18}$$

The derivative of the ramp function is the step function.

$$\begin{aligned} f'(t) &= 1 \qquad \text{for } t > 0 \\ &= u(t) \end{aligned} \tag{5-19}$$

We know that $F(s) = 1/s^2$ for the ramp function.
From Eq. (5-17) we have

$$\begin{aligned} \mathcal{L}[u(t)] &= sF(s) - f(0) \\ &= s\frac{1}{s^2} - 0 \\ &= \frac{1}{s} \end{aligned} \tag{5-20}$$

as expected.

Integral of a Function

$$\mathcal{L}\left[\int_0^t f(t)\, dt\right] = \frac{F(s)}{s} \tag{5-21}$$

This theorem states that if we start out with a function $f(t)$ whose transform is known, and then integrate $f(t)$, we obtain the transform of the new function by simply dividing the original transform by s. Thus we can say that *integration in the time domain corresponds to dividing by* s *in the transform domain.*

Let us again use an example. This time we will operate on pair (T-6) in order to derive pair (T-5). Thus

$$f(t) = \cos \omega t \tag{5-22}$$

Its integral from 0 to t is

$$\int_0^t f(t)\, dt = \frac{1}{\omega} \sin \omega t \tag{5-23}$$

We start with the fact that

$$F(s) = \mathcal{L}[\cos \omega t] = \frac{s}{s^2 + \omega^2} \tag{5-24}$$

According to Eq. (5-21), the transform of $\int_0^t f(t)\, dt$ is

$$\mathcal{L}\left[\int_0^t f(t)\, dt\right] = \mathcal{L}\left[\frac{1}{\omega} \sin \omega t\right] = \frac{F(s)}{s} = \frac{1}{s^2 + \omega^2} \tag{5-25}$$

Hence

$$\mathcal{L}[\sin \omega t] = \frac{\omega}{s^2 + \omega^2} \tag{5-26}$$

Shifting Theorem

$$\mathcal{L}[f(t-a)u(t-a)] = \epsilon^{-as}\,\mathcal{L}[f(t)]$$
$$= \epsilon^{-as} F(s) \tag{5-27}$$

This theorem states that if the original time function is shifted to the right by a, its transform is the original transform multiplied by ϵ^{-as}. In a sense, the ϵ^{-as} factor is an identification factor which "tags" a particular transform to indicate that it is the transform of a shifted function.

For example, let us calculate the transform of the shifted step function $u(t-a)$. Let

$$f(t) = u(t) \tag{5-28}$$

$$g(t) = u(t-a) \tag{5-29}$$

The transform of $g(t)$ is

$$G(s) = \mathcal{L}[u(t-a)] = \epsilon^{-as}\,\mathcal{L}[u(t)]$$
$$= \frac{\epsilon^{-as}}{s} \tag{5-30}$$

We might at this point discuss a quite simple, but nevertheless confusing, point on notation. The tables employ f and F to designate the basic functions. A natural tendency might have been to write $f(t) = u(t - a)$ for the unit step function above. However, this is not the best notation since $f(t)$, in reference to the table, would imply the unshifted step function. Thus, we avoided this difficulty by using the symbol g in Eq. (5-29). In general, to avoid confusion, it is necessary to be somewhat careful about notation in considering these operations.

Multiplication by $\epsilon^{-\alpha t}$

$$\mathscr{L}[\epsilon^{-\alpha t} f(t)] = F(s + \alpha) \tag{5-31}$$

This theorem states that if the original time function is multiplied by $\epsilon^{-\alpha t}$, we obtain the transform of the resulting product by replacing s by $s + \alpha$ in the original transform. For example, let us derive pair (T-8) from (T-6) by this operation. We have

$$f(t) = \cos \omega t \tag{5-32}$$

$$F(s) = \frac{s}{s^2 + \omega^2} \tag{5-33}$$

We wish to determine

$$\mathscr{L}[\epsilon^{-\alpha t} \cos \omega t]$$

According to Eq. (5-31), we simply replace s by $s + \alpha$ in (5-33). Thus,

$$\mathscr{L}[\epsilon^{-\alpha t} \cos \omega t] = \frac{s + \alpha}{(s + \alpha)^2 + \omega^2} \tag{5-34}$$

A few other operations are included in the table of Appendix D, but they will not be discussed at this point. We are now equipped to determine a more general set of transforms by use of the transform pairs in conjunction with the operations.

Example 5-5

(a) Determine the exact Laplace transform of the unit-area square pulse shown in Fig. 5-4. (In other words, initially assume that the width of the pulse is *not* small in comparison to any time constants under consideration, and the impulse approximation cannot be used.)

(b) Show that as a approaches zero, the correct transform of the impulse function is obtained.

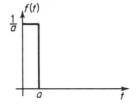

Figure 5-4   Pulse of Example 5-5.

Solution   (a) The pulse may be represented as

$$f(t) = \frac{1}{a}[u(t) - u(t - a)] \qquad (5\text{-}35)$$

The transform according to (T-2) and (O-3) is

$$F(s) = \frac{1}{as}[1 - \epsilon^{-as}] \qquad (5\text{-}36)$$

(b) Let $G(s)$ represent the limiting value of $F(s)$ as a approaches zero. Thus,

$$G(s) = \lim_{a \to 0} \frac{1}{as}[1 - \epsilon^{-as}] \qquad (5\text{-}37)$$

Since this yields an indeterminate form, we employ L'Hôpital's rule and take the ratio of the derivatives with respect to a. Thus

$$G(s) = \lim_{a \to 0} \frac{s\epsilon^{-as}}{s} = 1 \qquad (5\text{-}38)$$

which is recognized as being the transform of the impulse function.

Example 5-6

Determine the Laplace transform of the waveform shown in Fig. 5-5.
   Solution   The function can be expressed by

$$f(t) = 2tu(t) - 2(t - 1)u(t - 1) - 2(t - 3)u(t - 3) + 2(t - 4)u(t - 4) \quad (5\text{-}39)$$

With the aid of (T-3) and (O-3), we have

$$F(s) = \frac{2}{s^2} - \frac{2}{s^2}\epsilon^{-s} - \frac{2}{s^2}\epsilon^{-3s} + \frac{2}{s^2}\epsilon^{-4s}$$

$$= \frac{2}{s^2}[1 - \epsilon^{-s} - \epsilon^{-3s} + \epsilon^{-4s}] \qquad (5\text{-}40)$$

We note that the exponential terms in s serve as identification factors for the times at which the various components of the time function begin.

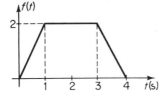

Figure 5-5   Waveform of Example 5-6.

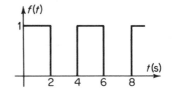

Figure 5-6   Waveform of Example 5-7.

Example 5-7

Determine the Laplace transform of the periodic square wave shown in Fig. 5-6.
   Solution   The time function can be expressed as

$$f(t) = u(t) - u(t - 2) + u(t - 4) - u(t - 6) + \cdots + (-1)^n u(t - 2n)$$

$$= \sum_{n=0}^{\infty} (-1)^n u(t - 2n)$$

(5-41)

From (T-2) and (O-3), the transform is

$$F(s) = \frac{1}{s} \left[1 - \epsilon^{-2s} + \epsilon^{-4s} - \epsilon^{-6s} + \cdots + (-1)^n \epsilon^{-2ns} + \cdots \right]$$

$$= \frac{1}{s} \sum_{n=0}^{\infty} (-1)^n \epsilon^{-2ns}$$

(5-42)

It can be shown by the theory of geometric series that this function may also be expressed in closed form. However, for most purposes, the infinite series form is more suitable; therefore, we shall simply leave $F(s)$ in that form.

5-4 SOME DEFINITIONS AND CONCEPTS

Now that we have seen how to determine the Laplace transforms of time functions, the next logical question concerns how to determine the time function if we are given the transform function. However, before considering this problem, we need to pause briefly and introduce a few definitions and concepts. Some of these ideas will probably be familiar to many readers, while others may seem entirely new. The definitions and concepts will help to provide a deeper insight into the theory of transform methods.

   Polynomial.   A polynomial is the sum of a number of terms of different integer degrees in some variable. (We will be primarily concerned with the variable s.) A polynomial in s is a series of terms of the form

$$P(s) = a_n s^n + a_{n-1} s^{n-1} + a_{n-2} s^{n-2} + \cdots + a_1 s + a_0$$

(5-43)

where the a's are all constants.

   Degree of a polynomial.   The *degree* of a polynomial is the highest degree present. Thus, for Eq. (5-43), the degree is n. The term *order* is frequently used synonymously with the term *degree*.

   Fundamental theorem on the number of roots.   A fundamental theorem of algebra states that the number of *roots* of a polynomial is equal to the degree of the polynomial. Thus a polynomial of order n has n roots. We determine the *roots* by solving for the values of s that satisfy the equation

$$a_n s^n + a_{n-1} s^{n-1} + a_{n-2} s^{n-2} + \cdots + a_1 s + a_0 = 0 \qquad (5\text{-}44)$$

Factored form of a polynomial.   A polynomial may be written in *factored form* as follows: Assume that the n roots of an nth-degree polynomial are known. Let us designate them as $s_1, s_2, \ldots, s_n$. Then the polynomial can be written in the factored form

$$P(s) = a_n(s - s_1)(s - s_2) \cdots (s - s_n) \qquad (5\text{-}45)$$

where a_n is the coefficient of the leading term [see Eq. (5-45)]. Thus, *except for a constant multiplier, a polynomial is completely specified by a knowledge of its roots.*

From either Eq. (5-44) or (5-45) it is seen that a root of a polynomial is a value such that, when substituted into the polynomial, it results in the polynomial being identically zero. For this reason, the roots of a polynomial are often called the *zeros* of the polynomial.

Classification of roots.   The roots of a polynomial may be classified in two ways:

1. A root may be an ordinary real number (either positive or negative), in which case it is said to be a *real root*. A root may be purely imaginary (e.g., $j5$), in which case it is said to be an *imaginary root*. Finally, a root may consist of both a real and an imaginary part, in which case it is said to be a *complex root* (e.g., $4 + j5$). A complex or imaginary root is always accompanied by its *complex conjugate*.

    The *complex conjugate* of a complex or imaginary number is another complex number formed from the original by retaining the same real part (which is zero in the case of an imaginary number) and reversing the sign of the imaginary part. Thus the complex conjugate of $a + jb$ is $a - jb$, and if $a + jb$ is a root of a polynomial, $a - jb$ is a root also. Similarly, $-jc$ is the conjugate of jc, and if one is a root, so is the other. Thus complex and imaginary roots occur in conjugate pairs.

2. Roots may be classified according to their *order*, which is the number of times a root is repeated in a given solution. The most common root is the *first-order* or *simple-order* root. Higher-order roots are referred to as *multiple-order* roots. A multiple-order root may be further designated by its particular order.

    The determination of the roots of a polynomial involves solving Eq. (5-44). For first- and second-degree equations, a solution is always straightforward. For third- and higher-degree equations, numerical methods must often be used. Fortunately, most of the problems in this text will not require the solution of higher-degree equations. Computer programs are readily available for determining the roots of such equations.

Quadratic formula.   In high school, one is usually taught the quadratic formula in the following manner: Using s as the variable, consider a quadratic of the form

$$As^2 + Bs + C = 0 \tag{5-46}$$

The values of the roots are then given by

$$s = \frac{-B \pm \sqrt{B^2 - 4AC}}{2A} \tag{5-47}$$

It has been the experience of the author that this equation is not the most suitable form for working with transform problems. Instead, we present the following form. First, assume that the factor of s^2 is unity. [In many cases this is true anyway; otherwise, we may divide through Eq. (5-47) by the leading term.] We then have an equation of the form

$$s^2 + bs + c = 0 \tag{5-48}$$

The roots are then given by

$$s = -\frac{b}{2} \pm \sqrt{\left(\frac{b}{2}\right)^2 - c} \tag{5-49}$$

As a further step, this can be remembered as a procedure. First, we look at the middle term factor (b) and divide it by 2. The negative of this term is written in front $-b/2$. Next, the number that was written in front is squared and placed as the first term under the radical. Then, the last term is subtracted from the previous quantity, and the square root is taken. From here on, the procedure is that of combining the numbers according to Eq. (5-49). Of course, if the roots are complex, we could express this in the form

$$s = -\frac{b}{2} \pm j \sqrt{c - \left(\frac{b}{2}\right)^2} \tag{5-50}$$

The test as to what type of roots we have is:

1. If $(b/2)^2 > c$, the roots are real and of simple order.
2. If $(b/2)^2 < c$, the roots are complex and of simple order.
3. If $(b/2)^2 = c$, the roots are real and of second order.

Now let us consider an example.

Example 5-8

Classify the following polynomials and their roots. Write each in factored form.
    (a) $P(s) = 2s^2 + 3s$
    (b) $P(s) = 2s^2 + 6s + 4$
    (c) $P(s) = 5s^2 + 10s + 25$
    (d) $P(s) = s^2 + 8s + 16$
    (e) $P(s) = 8s^2 + 128$
    (f) $P(s) = s^3 + 7s^2 + 31s + 25$ ($s = -1$ is known to be a root.)

Solution   Polynomials (a) through (e) are all second degree, while the polynomial of (f) is third degree. For the most part, we will leave the determination of the roots to the reader in the steps to follow.

(a) $P(s) = 2s^2 + 3s$. The roots are $s_1 = 0$ and $s_2 = -\frac{3}{2}$, which are both real and simple. The factored form is

$$P(s) = 2s^2\left[s - \left(-\frac{3}{2}\right)\right] = 2s\left(s + \frac{3}{2}\right) \tag{5-51}$$

(b) $P(s) = 2s^2 + 6s + 4$. The roots are $s_1 = -1$ and $s_2 = -2$ which are both real and simple. The factored form is

$$P(s) = 2(s + 1)(s + 2) \tag{5-52}$$

(c) $P(s) = 5s^2 + 10s + 25$. The roots are $s_1 = -1 + j2$ and $s_2 = -1 - j2$ which are both complex, but of first order. The factored form is

$$\begin{aligned} P(s) &= 5[s - (-1 + j2)][s - (-1 - j2)] \\ &= 5(s + 1 - j2)(s + 1 + j2) \end{aligned} \tag{5-53}$$

(d) $P(s) = s^2 + 8s + 16$. The roots are $s_1 = s_2 = -4$. Thus, we say that $s = -4$ is a real root of *second order*. The factored form is

$$P(s) = (s + 4)^2 \tag{5-54}$$

(e) $P(s) = 8s^2 + 128$. The roots are $s_1 = +j4$ and $s_2 = -j4$, which are both imaginary and simple. The factored form is

$$P(s) = 8(s - j4)(s + j4) \tag{5-55}$$

(f) $P(s) = s^3 + 7s^2 + 31s + 25$. Although this is a third-degree equation, the knowledge that $s = -1$ is a root allows us to factor it since this fact implies that $(s + 1)$ is an exact factor of the equation. The reader can verify that division by $s + 1$ yields

$$P(s) = (s + 1)(s^2 + 6s + 25) \tag{5-56}$$

The roots of the quadratic can then be determined. The three roots of the whole equation are $s_1 = -1$, $s_2 = -3 + j4$, and $s_3 = -3 - j4$. The first root given is real and of simple order, whereas the latter two roots are both complex and of simple order. The factored form is

$$P(s) = (s + 1)(s + 3 - j4)(s + 3 + j4) \tag{5-57}$$

5-5 INVERSE TRANSFORMS BY IDENTIFICATION

In this section we begin considering the problem of determining the time function from a given transform function. This process is called *inverse Laplace transformation*, and $f(t)$ can be considered to be the *inverse transform* of $F(s)$.

We saw that we could always determine $F(s)$ from $f(t)$ by using Eq. (5-4), the basic definition of the transform. The inverse integral that could be used to determine $f(t)$ from $F(s)$ is derived in many advanced mathematical books. The reader not familiar with complex variable theory will not be expected to understand or appreciate this integral; we show it only as a matter of interest. The inverse transform is

$$f(t) = \frac{1}{2\pi j} \int_{\gamma - j\infty}^{\gamma + j\infty} F(s)\epsilon^{st}\, ds \qquad (5\text{-}58)$$

where γ is a real number chosen to satisfy certain restrictions. The integral involves *complex integration* which is not within the scope of this book.

However, all is not lost, for if we recall from the previous two sections, we depended somewhat on the tables once we had discussed the basic transform method. If the tables were suitable for determining $F(s)$ from $f(t)$, they certainly should be adequate in many cases for determining $f(t)$ from $F(s)$ if the appropriate pair can be identified. In this section we consider inverse transformation by use of the tables for functions that are easily identified. We can best illustrate by means of examples.

Example 5-9

Determine the inverse transform of

$$I(s) = \frac{100}{s^2 + 25} \qquad (5\text{-}59)$$

Solution   In checking Table 5-1 we observe that (T-5) appears to be of the appropriate form. However, (T-5) seems to require that we have $\omega/(s^2 + \omega^2)$ which is not quite true in our case. This discrepancy is readily resolved if we write $F(s)$ in the form

$$I(s) = 20\left[\frac{(5)}{s^2 + 25}\right] \qquad (5\text{-}60)$$

The quantity in the brackets is now the exact Laplace transform of $\sin 5t$. Thus

$$i(t) = 20 \sin 5t \qquad (5\text{-}61)$$

With some practice, it will not be necessary to write the intermediate step given by Eq. (5-60). We might compare this procedure with that of determining the differential in integration. For example, consider

$$y = 100 \int \epsilon^{2x}\, dx \qquad (5\text{-}62)$$

We could rewrite this in the form

$$y = 50 \int \epsilon^{2x} 2\, dx \qquad (5\text{-}63)$$

in which the quantity under the integral is now an exact differential of the form $u\, du$.

Example 5-10

Determine the inverse transform of

$$V(s) = \frac{s + 4}{s^2 + 3s + 2} \tag{5-64}$$

Solution   At first glance, this function appears to be in the form of (T-7) or (T-8). However, *a second-degree polynomial in the denominator should first be factored to determine if the roots are real or complex.* As noted beside (T-7) and (T-8), these forms apply for complex roots of the denominator. In this case, the roots of the denominator polynomial are determined to be $s_1 = -1$ and $s_2 = -2$ which are both real. Thus since the roots are real, we should not use (T-7) or (T-8). How, then, can we determine the inverse transform of this function since it is not given in the table? The answer is that in the next section we will show that it may be expanded into the sum of two functions of the form of (T-4). Thus we will abandon this problem for the moment and return later.

Example 5-11

Determine the inverse transform of

$$V(s) = \frac{10s - 2}{s^2 + 6s + 13} \tag{5-65}$$

Solution   As stated previously, we should first determine the roots of the denominator. The roots are $s_1 = -3 + j2$ and $s_2 = -3 - j2$. Since the roots are complex, we will investigate (T-7) and (T-8) as likely candidates. First we must place the denominator in the proper form $(s + \alpha)^2 + \omega^2$. This can be done by completing the square. Recalling this procedure, we take half the middle factor and square it. This quantity is added and subtracted as follows:

$$D(s) = s^2 + 6s + 13 = s^2 + 6s + (3)^2 + 13 - 9 = s^2 + 6s + (3)^2 + 4 \tag{5-66}$$

where $D(s)$ represents the denominator polynomial. The first three terms now form a perfect square. Thus

$$D(s) = (s + 3)^2 + (2)^2 \tag{5-67}$$

We might note that this latter operation could have been done by inspection from a knowledge of the roots. Thus, the quantity α is the negative of the real part of the roots, and ω is the imaginary part of the roots without regard to sign. We now write $V(s)$ as

$$V(s) = \frac{10s - 2}{(s + 3)^2 + (2)^2} \tag{5-68}$$

The s in the numerator implies that we need (T-8). However, we need $s + 3$ in the numerator. We achieve this by adding and subtracting $32/D(s)$ from the function. Thus

$$
\begin{aligned}
V(s) &= \frac{10s - 2}{(s + 3)^2 + (2)^2} + \frac{32}{(s + 3)^2 + (2)^2} - \frac{32}{(s + 3)^2 + (2)^2} \\
&= \frac{10(s + 3)}{(s + 3)^2 + (2)^2} - \frac{32}{(s + 3)^2 + (2)^2}
\end{aligned} \tag{5-69}
$$

The first term is now in the proper form. The second term is of the form of (T-7) if we mentally factor 32 into 16×2. Thus

$$
\begin{aligned}
v(t) &= 10\epsilon^{-3t} \cos 2t - 16\epsilon^{-3t} \sin 2t \\
&= 2\epsilon^{-3t}[5 \cos 2t - 8 \sin 2t] \\
&= 2\sqrt{89}\,\epsilon^{-3t} \cos (2t + \theta)
\end{aligned} \tag{5-70}
$$

where $\theta = \tan^{-1} \frac{8}{5}$.

Example 5-12

Determine the inverse transform of

$$
I(s) = \frac{5}{s^2} [1 - 2\epsilon^{-3s} + 2\epsilon^{-6s} - 2\epsilon^{-9s} + 2\epsilon^{-12s} + \cdots] \tag{5-71}
$$

Solution   The transform given is an infinite series in ϵ^{-s}. Whenever we observe exponential terms in s, we immediately recognize shifted functions according to (O-3). We may multiply through by $5/s^2$ if desired, although it is just as easy to mentally visualize $1/s^2$ as a factor for all internal terms. Since

$$
\mathscr{L}^{-1}\left[\frac{1}{s^2}\right] = t \tag{5-72}
$$

and

$$
\mathscr{L}^{-1}[\epsilon^{-as}F(s)] = f(t - a)u(t - a) \tag{5-73}
$$

we deduce that

$$
\mathscr{L}^{-1}\left[\frac{\epsilon^{-as}}{s^2}\right] = (t - a)u(t - a) \tag{5-74}
$$

Therefore, on a term-by-term basis,

$$
\begin{aligned}
i(t) = 5[tu(t) &- 2(t - 3)u(t - 3) + 2(t - 6)u(t - 6) \\
&- 2(t - 9)u(t - 9) + 2(t - 12)u(t - 12) - \cdots]
\end{aligned} \tag{5-75}
$$

This function is shown in Fig. 5-7.

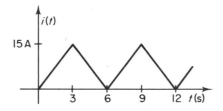

Figure 5-7 Waveform of Example 5-12.

5-6 INVERSE TRANSFORMS BY PARTIAL FRACTION EXPANSION

We saw in Section 5-5 how the inverse transforms of many functions could be immediately recognized in the forms of the tables and thus determined mainly by inspection. In this section we consider a more general form of function that is more complex than the forms of the tables, but which can be reduced to the forms of the tables by means of partial fraction expansion.

To develop the concept of partial fraction expansion, we will first proceed in reverse order. Consider, for example, the function given by

$$F(s) = \frac{4}{s+2} + \frac{5}{s+3} \tag{5-76}$$

This function may be placed over a common denominator if we multiply the first term by $s + 3$ and the second term by $s + 2$. The result obtained is

$$F(s) = \frac{9s + 22}{(s+2)(s+3)} = \frac{9s + 22}{s^2 + 5s + 6} \tag{5-77}$$

However, suppose that we had started with Eq. (5-77) and had wished to express it in the form of Eq. (5-76). This is the problem of partial fraction expansion: to obtain an expansion in terms of the denominator factors.

Why do we desire such an expansion? The reason is that we recognize that both terms of Eq. (5-76) are in the form of transform pair (T-4). However, Eq. (5-77) is not in the form of the table pairs. Yet, if we can decompose the function by partial fraction expansion, we can obtain the inverse transform directly from the tables as a sum of exponential functions.

As it turns out, the Laplace transforms of interest in circuit analysis can, in most cases, be expressed as the ratio of two polynomials. (The reader might refer back to the table and note that all the transforms given are ratios of simple polynomials.) Thus let us assume a transform given by

$$F(s) = \frac{N(s)}{D(s)} \tag{5-78}$$

where $N(s)$ is a numerator polynomial of the form

$$N(s) = a_n s^n + a_{n-1} s^{n-1} + \cdots + a_1 s + a_0 \tag{5-79}$$

and $D(s)$ is a denominator polynomial of the form

$$D(s) = b_m s^m + b_{m-1} s^{m-1} + \cdots + b_1 s + b_0 \tag{5-80}$$

From previous considerations, we know that the numerator polynomial contains n roots, and the denominator polynomial contains m roots. *The roots of the numerator polynomial are called the zeros of $F(s)$, and the roots of the denominator polynomial are called the poles of $F(s)$.* These definitions will be more meaningful later, but at this time, our main interest is primarily that of terminology. It should be clear that if a zero of $F(s)$ coincides with a pole of $F(s)$, the resulting factors will cancel, thus reducing the degrees of both numerator and denominator polynomials by one. Therefore, in the discussion to follow, we assume that such factors, when present, will have been canceled.

We now consider $D(s)$ to be factored, and thus we can express $F(s)$ in the form

$$F(s) = \frac{N(s)}{b_m(s - p_1)(s - p_2) \cdots (s - p_m)} \tag{5-81}$$

where the p's represent the various poles of $F(s)$.

It can be shown that for real physical systems $m \geq n$. For the moment, let us restrict ourselves to the case where $m > n$. Thus, we will assume at this time that the degree of the denominator polynomial is greater than the degree of the numerator polynomial for cases under consideration.

The m poles of $F(s)$ may be arbitrarily classified into four groups:

1. Real poles of first order
2. Complex poles of first order
3. Real poles of multiple order
4. Complex poles of multiple order

In general, $F(s)$ may contain poles belonging to more than one of the foregoing classifications. By means of partial fractions, we will be able to expand a given function in an appropriate series so that the various portions of the function falling in the proper classifications may be identified and inverted individually. We will consider each of the preceding classifications individually. In the remainder of this section, we will consider the first classification, namely, real poles of simple order since this is the case most easily handled by partial fraction techniques.

Assume that the denominator contains r real poles of first order. (Remember that there are m total poles, but some may belong to other classifications. In the case where all poles are real and simple, $r = m$. In all cases, $r \leq m$.) According to the theory of partial fraction expansion, we may expand $F(s)$ as follows:

$$F(s) = \frac{A_1}{s - p_1} + \frac{A_2}{s - p_2} + \cdots + \frac{A_r}{s - p_r} + R(s) \tag{5-82}$$

$R(s)$ is the remaining portion of the expansion due to poles belonging to other classifications. At the moment, we are primarily interested in the first r terms, as we recognize that they may be inverted on a term-by-term basis by means of (T-4).

Our basic problem at present is the determination of the A coefficients. This task may be achieved as follows: Consider an arbitrary coefficient A_k. Let us multiply both sides of Eq. (5-82) by $(s - p_k)$ and rearrange some terms. This results in

$$(s - p_k)F(s) = (s - p_k)\left[\frac{A_1}{s - p_1} + \frac{A_2}{s - p_2} + \cdots + \frac{A_r}{s - p_r} + R(s)\right] + A_k \quad (5\text{-}83)$$

in which the A_k term has been removed from the brackets. Note that the $(s - p_k)$ multiplier has canceled the denominator of the A_k term. It will also cancel the $(s - p_k)$ factor in the denominator of $F(s)$. Since Eq. (5-83) is an equality, it must hold for all values of s. Suppose, then, that we use a "trick" by letting $s = p_k$. As a result of the $(s - p_k)$ factor, all terms on the right-hand side of the equation vanish except the term A_k. However, the left-hand side, in general, is nonzero since the factor will have canceled. Hence a general expression for A_k is deduced as

$$A_k = (s - p_k)F(s)\Big]_{s = p_k} \quad (5\text{-}84)$$

The notation of this formula simply means that we multiply $F(s)$ by $(s - p_k)$, which always cancels the $s - p_k$ term in the denominator, and then we let $s = p_k$. The resulting number is the A_k coefficient. This procedure is repeated at each of the r real poles of simple order.

Once all the A's are determined, the inverse transform of the portion of Eq. (5-82) involving the real poles of simple order can be determined from (T-4). For the moment, let us designate this portion of the time response as $f_1(t)$ to distinguish it from $f(t)$, which would include the effect of other types of poles. We have

$$f_1(t) = A_1 \epsilon^{p_1 t} + A_2 \epsilon^{p_2 t} + \cdots + A_r \epsilon^{p_r t} \quad (5\text{-}85)$$

In most cases of immediate interest, the real poles will be negative in sign, implying that the time function is a sum of decaying exponential terms. Thus *real negative poles of simple order correspond to decaying exponential functions in the time domain.*

So far we have considered only the case where the denominator polynomial is of higher degree than the numerator polynomial (i.e., $m > n$). Practically speaking, a transform describing the voltage or current response of a real physical network excited by real excitations will never have a numerator polynomial of higher degree than the corresponding denominator polynomial. However, under certain conditions, the situation is occasionally encountered in which the numerator and denominator have the same degree (i.e., $m = n$). In this case, the numerator polynomial is first divided by the denominator polynomial, yielding a constant plus a remainder function whose denominator is of higher degree than its numerator. The inverse transform of the constant is an impulse function, and the inverse transform of the remainder function can be determined by the methods of this section or the sections to follow, depending on the nature of the poles. Thus *when the numerator and denominator polynomials have the same degree, an impulse function appears in the time response.* A case of this type will be illustrated in Example 5-15.

Example 5-13

Determine the inverse transform of

$$V(s) = \frac{s + 4}{s^2 + 3s + 2} \qquad (5\text{-}86)$$

Solution   This function was first encountered in Example 5-10, but we were not equipped to solve it at the time. The poles of the function are $p_1 = -1$ and $p_2 = -2$. Thus, in this example, all the poles are real and of simple order, and we are capable of determining the entire time function from the methods of this section. In factored form we have

$$V(s) = \frac{s + 4}{(s + 1)(s + 2)} \qquad (5\text{-}87)$$

The partial fraction expansion reads

$$V(s) = \frac{A_1}{s + 1} + \frac{A_2}{s + 2} \qquad (5\text{-}88)$$

According to Eq. (5-84) the constants are determined as follows:

$$A_1 = \frac{s + 4}{s + 2}\bigg]_{s=-1} = \frac{-1 + 4}{-1 + 2} = 3 \qquad (5\text{-}89)$$

$$A_2 = \frac{s + 4}{s + 1}\bigg]_{s=-2} = \frac{-2 + 4}{-2 + 1} = -2 \qquad (5\text{-}90)$$

Thus

$$V(s) = \frac{3}{s + 1} - \frac{2}{s + 2} \qquad (5\text{-}91)$$

and from (T-4)

$$v(t) = 3\epsilon^{-t} - 2\epsilon^{-2t} \qquad (5\text{-}92)$$

Example 5-14

Determine the exponential portion of the inverse transform of

$$F(s) = \frac{100(s + 3)}{(s + 1)(s + 2)(s^2 + 2s + 5)} \qquad (5\text{-}93)$$

Solution   The reader might verify that the quadratic has complex roots and thus falls into classification 2. We consider this situation in the next section. Instead, we now invert those terms due to the two real poles of simple order, namely $s = -1$ and $s = -2$. Calling this portion of the function $F_1(s)$, and its inverse $f_1(t)$, we have

$$F_1(s) = \frac{A_1}{s+1} + \frac{A_2}{s+2} \tag{5-94}$$

By means of Eq. (5-84), we have

$$A_1 = \frac{100(s+3)}{(s+2)(s^2+2s+5)}\bigg]_{s=-1}$$
$$= \frac{100(2)}{(1)(4)} = 50 \tag{5-95}$$

$$A_2 = \frac{100(s+3)}{(s+1)(s^2+2s+5)}\bigg]_{s=-2}$$
$$= \frac{100(1)}{(-1)(5)} = -20 \tag{5-96}$$

Thus

$$F_1(s) = \frac{50}{s+1} - \frac{20}{s+2} \tag{5-97}$$

$$f_1(t) = 50\epsilon^{-t} - 20\epsilon^{-2t} \tag{5-98}$$

Example 5-15

Determine the inverse transform of

$$F(s) = \frac{2s^2 + 11s + 4}{s(s+1)} \tag{5-99}$$

Solution In this problem $N(s)$ and $D(s)$ have the same degree; hence, we must first divide $N(s)$ by $D(s)$. This is achieved as follows:

$$
\begin{array}{r}
2 \\
s^2 + s \overline{\smash{\big)}\, 2s^2 + 11s + 4} \\
\underline{2s^2 + 2s} \\
9s + 4
\end{array}
\tag{5-100}
$$

Thus

$$F(s) = 2 + \frac{9s+4}{s(s+1)} \tag{5-101}$$

The second quantity may be readily expanded since its denominator is of higher degree than its numerator. The reader can complete the problem and show that

$$f(t) = 2\delta(t) + 4 + 5\epsilon^{-t} \tag{5-102}$$

5-7 COMPLEX POLES OF FIRST ORDER

The determination of a suitable expansion and inversion for complex poles is somewhat more tedious than for real poles. There are several different approaches that
may be employed. We consider three different methods in this section. The first two
methods are approximately equal in complexity, whereas the third method involves
a quick "plug-in" formula.

The reader may wonder why we wish to consider three different procedures.
The reason is that the first two methods, while lengthy, will provide the reader with
a better understanding of the inversion process, whereas the third method, while
much shorter, is somewhat mechanical. Furthermore, the reader who may refer to
other textbooks will undoubtedly need to understand the first two methods as they
are extensively used in the literature.

Let us assume for the moment that a given $F(s)$ contains a single pair of complex
conjugate poles. Let $\bar{p}$ represent one of the poles and $\tilde{\bar{p}}$ represent its complex conjugate. The bar is used to emphasize the fact that the pole is complex, and the "~"
indicates its conjugate. In most cases of interest at present, the real part is negative,
and thus the rectangular form of the poles is

$$\bar{p} = -\alpha + j\omega \tag{5-103}$$

$$\tilde{\bar{p}} = -\alpha - j\omega \tag{5-104}$$

where α is usually a positive number representing a *damping constant* in the time
domain, and ω represents an *angular frequency* in the time domain.

Expansion as a Quadratic

In observing transform pairs (T-5) through (T-8), we note that, to utilize those transform pairs, we must be able to separate quadratic factors containing complex or
imaginary roots from the remainder of the expression. Thus assume that $F(s)$ contains a denominator factor of the form $(s^2 + bs + c)$ whose roots are given by Eqs.
(5-103) and (5-104). According to the theory of partial fraction expansion, we may
expand the function as follows:

$$F(s) = \frac{As + B}{s^2 + bs + c} + R(s) \tag{5-105}$$

where A and B are constants to be determined and $R(s)$ represents everything else
in the expansion. Note that in this case, since the denominator is of second degree,
two constants are required. Furthermore, we are not able to employ a simple "trick"
to determine the constants as was the case for real poles.

Although there are tricks, the most straightforward approach is as follows:
First, determine as much of the expansion for $R(s)$ as possible by other methods.
Next, since Eq. (5-105) expresses an equality, select some convenient values of s to
substitute into the equation and hence produce as many simultaneous equations as
are necessary to solve for the constants. The quadratic term is then inverted by use
of the appropriate transform pairs. In general, we may select any arbitrary values
of s to substitute into the equation except those values of s which are poles of $F(s)$,
since both sides of Eq. (5-105) become indeterminant at these values.

Before considering an example, let us emphasize the following facts: *Complex poles of simple order with negative real parts correspond to a damped sinusoidal response in the time domain. As a special case, purely imaginary poles correspond to an undamped sinusoidal response in the time domain.*

Example 5-16

Determine the entire inverse transform of the function of Example 5-14.
    Solution   Factoring the quadratic results in complex poles

$$\bar{p} = -1 + j2 \tag{5-106}$$

$$\tilde{\bar{p}} = -1 - j2 \tag{5-107}$$

From the results of Example 5-14, a complete partial fraction expansion of $F(s)$ reads

$$F(s) = \frac{100(s + 3)}{(s + 1)(s + 2)(s^2 + 2s + 5)}$$
$$= \frac{50}{s + 1} - \frac{20}{s + 2} + \frac{As + B}{s^2 + 2s + 5} \tag{5-108}$$

We need two simultaneous equations to determine both A and B. First let us pick the simple value $s = 0$ to substitute into the equation. We obtain

$$\frac{100(3)}{(1)(2)(5)} = \frac{50}{1} - \frac{20}{2} + \frac{B}{5} \tag{5-109}$$

resulting in

$$B = -50 \tag{5-110}$$

Next, we choose arbitrarily $s = 1$ and obtain

$$\frac{100(4)}{(2)(3)(8)} = \frac{50}{2} - \frac{20}{3} + \frac{A + B}{8} \tag{5-111}$$

resulting in

$$\frac{A}{8} = -10 - \frac{B}{8} \tag{5-112}$$

$$A = -30 \tag{5-113}$$

Thus a complete partial fraction expansion of $F(s)$ reads

$$F(s) = \frac{50}{s + 1} - \frac{20}{s + 2} + \frac{-30s - 50}{s^2 + 2s + 5} \tag{5-114}$$

The enthusiastic reader might wish to put all these terms together over a common denominator and verify that the original $F(s)$ is obtained. The inversion of the first two terms has already been done in Example 5-14, and the result was designated as $f_1(t)$. Let us designate the portion of the transform of interest at present as $F_2(s)$ and its inverse as $f_2(t)$. Thus

$$F_2(s) = \frac{-30s - 50}{s^2 + 2s + 5} = \frac{-30s - 50}{(s + 1)^2 + (2)^2} \tag{5-115}$$

This quantity may be inverted by means of (T-7) and (T-8) as was demonstrated by a similar function in Example 5-11. The process of placing it in the proper form follows:

$$\frac{-30s - 50}{(s + 1)^2 + (2)^2} = \frac{-30s - 30}{(s + 1)^2 + (2)^2} + \frac{-20}{(s + 1)^2 + (2)^2}$$
$$= \frac{-30(s + 1)}{(s + 1)^2 + (2)^2} - \frac{10(2)}{(s + 1)^2 + (2)^2} \tag{5-116}$$

Inversion yields

$$f_2(t) = -30\epsilon^{-t} \cos 2t - 10\epsilon^{-t} \sin 2t$$
$$= -10\epsilon^{-t}[\sin 2t + 3 \cos 2t] \tag{5-117}$$

The reader can verify that the two sinusoidal terms may be combined to yield either of the following results:

$$f_2(t) = 10\sqrt{10}\epsilon^{-t} \sin (2t - 108.4°) \tag{5-118}$$

or

$$f_2(t) = 10\sqrt{10}\epsilon^{-t} \cos (2t + 161.6°) \tag{5-119}$$

Thus, the entire time function is

$$f(t) = f_1(t) + f_2(t)$$
$$= 50\epsilon^{-t} - 20\epsilon^{-2t} + 10\sqrt{10}\epsilon^{-t} \sin (2t - 108.4°) \tag{5-120}$$

Expansion in Terms of the Poles

The second method is that of expanding the function in terms of the poles directly, as we did for the case of real poles. The main differences are the facts that we have to manipulate complex numbers, and some adjustments must be made on the final time function to put it in the best form. The key to this method is the exponential definition of the cosine function:

$$\cos (\omega t + \phi) = \frac{1}{2} [\epsilon^{j(\omega t + \phi)} + \epsilon^{-j(\omega t + \phi)}] \tag{5-121}$$

In determining the inverse transform by this method, we factor the quadratic into two factors as in the case of real poles, the essential difference being that the factors involve complex numbers. We then expand the function as in the case of real poles, so the procedure of Eq. (5-84) serves equally well for this case. A coefficient is required for both the terms involving $\bar{p}$ and $\tilde{\bar{p}}$. An important point to remember is that *the coefficients of the conjugate poles are themselves complex conjugates.* Once the coefficients are determined, the inverse transform can be manipulated in the form of Eq. (5-121). If we are interested only in the time response due to the complex poles, this procedure allows us to determine that portion alone; whereas the first method required evaluation of other coefficients before we could evaluate the desired coefficients.

Example 5-17

Determine the damped sinusoidal portion of the inverse transform of the function of Examples 5-14 and 5-16 by the method just considered.

Solution   The complex poles are

$$\bar{p} = -1 + j2 \tag{5-122}$$

$$\tilde{p} = -1 - j2 \tag{5-123}$$

In terms of these poles, $F(s)$ may be factored as

$$F(s) = \frac{100(s + 3)}{(s + 1)(s + 2)(s + 1 - j2)(s + 1 + j2)} \tag{5-124}$$

Let us again refer to the portion of the response of interest as $F_2(s)$, and its inverse as $f_2(t)$. We have

$$F_2(s) = \frac{\bar{K}_1}{s + 1 - j2} + \frac{\bar{K}_2}{s + 1 + j2} \tag{5-125}$$

The coefficients are determined as follows:

$$
\begin{aligned}
\bar{K}_1 &= \frac{100(s + 3)}{(s + 1)(s + 2)(s + 1 + j2)}\bigg]_{s = -1 + j2} \\
&= \frac{100(2 + j2)}{(j2)(1 + j2)(j4)} = \frac{100(2\sqrt{23}\underline{/45°})}{(-8)(\sqrt{5}\underline{/63.4°})} \\
&= 5\sqrt{10}\underline{/161.6°} = 5\sqrt{10}\,\epsilon^{j161.6°}
\end{aligned} \tag{5-126}
$$

(Strictly speaking, we should convert the angle to radians when expressing $\bar{K}_1$ in exponential form, but we have retained the angle in degrees for convenience.)
    The reader might verify that

$$\bar{K}_2 = 5\sqrt{10}\underline{/-161.6°} = 5\sqrt{10}\,\epsilon^{-j161.6°} \tag{5-127}$$

and thus

$$\bar{K}_2 = \tilde{\bar{K}}_1 \tag{5-128}$$

as predicted. Therefore, in theory we need only calculate one of the coefficients, and the other follows by inspection. However, it is often a good check to actually calculate both. The partial fraction expansion of $F_2(s)$ is

$$F_2(s) = \frac{5\sqrt{10}\,\epsilon^{j161.6°}}{s + 1 - j2} + \frac{5\sqrt{10}\,\epsilon^{-j161.6°}}{s + 1 + j2} \tag{5-129}$$

Inversion of $F_2(s)$ by means of (T-4) yields

$$f_2(t) = 5\sqrt{10}\,\epsilon^{j161.6°}\epsilon^{(-1+j2)t} + 5\sqrt{10}\,\epsilon^{-j161.6°}\epsilon^{(-1-j2)t}$$
$$= 5\sqrt{10}\,\epsilon^{-t}[\epsilon^{j(2t+161.6°)} + \epsilon^{-j(2t+161.6°)}] \tag{5-130}$$

Comparing the quantity in the brackets with Eq. (5-121), we have

$$f_2(t) = 10\sqrt{10}\,\epsilon^{-t}\cos(2t + 161.6°) \tag{5-131}$$

which certainly agrees with Example 5-16.

A "Trick" Formula

We now wish to turn our attention to a reasonably simple formula for determining the portion of the inverse transform of a given function that corresponds to a pair of complex poles of simple order. Like the second method, it allows us to determine the damped or undamped sinusoidal part of a time function without considering the remainder of the function. The method will be stated here without proof. The interested reader can refer to Appendix C for a proof.

Assume that $F(s)$ is written in the form

$$F(s) = \frac{Q(s)}{s^2 + bs + c} = \frac{Q(s)}{(s + \alpha)^2 + \omega^2} \tag{5-132}$$

where $Q(s)$ represents everything else in $F(s)$ besides the quadratic factor. If desired, we may also write

$$Q(s) = (s^2 + bs + c)F(s) \tag{5-133}$$

The complex poles are $\bar{p} = -\alpha + j\omega$ and $\tilde{\bar{p}} = -\alpha - j\omega$, of course. We now evaluate $Q(s)$ for the particular value of $s = \bar{p} = -\alpha + j\omega$. The resulting complex number is expressed in polar form as a magnitude and an angle. Thus let

$$Q(-\alpha + j\omega) = M\underline{/\theta} \tag{5-134}$$

where M is the magnitude and θ is the angle.

Let us designate the portion of the time response of interest as $f_1(t)$. (Do not confuse the subscript 1 here with that of $f_1(t)$ in the previous few examples. In any problem involving several types of time functions, the notation must be modified anyway.) It can be shown that $f_1(t)$ is given by

$$f_1(t) = \frac{M}{\omega}\,\epsilon^{-\alpha t}\sin(\omega t + \theta) \tag{5-135}$$

where M and θ are defined by Eq. (5-134).

An important special case: suppose the pair of poles are purely imaginary, (i.e., $\alpha = 0$). Then

$$F(s) = \frac{Q(s)}{s^2 + \omega^2} \tag{5-136}$$

or

$$Q(s) = (s^2 + \omega^2)F(s) \tag{5-137}$$

We then substitute $s = j\omega$ into $Q(s)$ and obtain

$$Q(j\omega) = M\underline{/\theta} \tag{5-138}$$

The time function is then

$$f_1(t) = \frac{M}{\omega} \sin(\omega t + \theta) \tag{5-139}$$

Now that we have developed such a simple procedure, we will employ it in most of our subsequent work. Again, however, the wise reader should keep in mind the other two approaches for future reference. In particular, the techniques of network synthesis and design frequently require expansions of the forms of the first two methods.

Example 5-18

Determine the damped sinusoidal response of $F(s)$ in Examples 5-16 and 5-17 by means of the formula just discussed.

Solution   The function is

$$F(s) = \frac{100(s + 3)}{(s + 1)(s + 2)(s^2 + 2s + 5)} \tag{5-140}$$

The quantity $Q(s)$ is

$$Q(s) = \frac{100(s + 3)}{(s + 1)(s + 2)} \tag{5-141}$$

We must now substitute $s = -1 + j2$. Thus

$$Q(-1 + j2) = \frac{100(2 + j2)}{(j2)(1 + j2)} = \frac{100(2\sqrt{2}\underline{/45°})}{(2\underline{/90°})(\sqrt{5}\underline{/63.4°})} \tag{5-142}$$
$$= 20\sqrt{10}\underline{/-108.4°}$$

Thus

$$M = 20\sqrt{10} \tag{5-143}$$

$$\theta = -108.4° \tag{5-144}$$

To conform with Examples 5-16 and 5-17, we will again call the time function $f_2(t)$. By means of Eq. (5-135), we have

$$f_2(t) = 10\sqrt{10}\epsilon^{-t} \sin(2t - 108.4°) \tag{5-145}$$

No one will question the simplicity of this method as compared with the previous two methods.

Example 5-19

Determine the sinusoidal steady-state (undamped) portion of the voltage time response corresponding to the transform

$$V(s) = \frac{500(s + 5)}{(2s^2 + 200)(s^2 + 20s + 200)} \qquad (5\text{-}146)$$

Solution   The sinusoidal steady-state response is due to the quadratic term $(2s^2 + 200)$. The other quadratic has poles at $-10 \pm j10$ which corresponds to a damped sinusoidal respose whose steady-state value would be zero. Thus the response of interest is that due to the $(2s^2 + 200)$ factor. Factoring the coefficient of the highest degree results in $2(s^2 + 100)$, and thus the poles of interest are $\bar{p} = j10$ and $\tilde{\bar{p}} = -j10$. We first cancel the 2 factor and write $V(s)$ as

$$V(s) = \frac{250(s + 5)}{(s^2 + 100)(s^2 + 20s + 200)} \qquad (5\text{-}147)$$

which is in the required form of Eq. (5-136).   Next, $Q(s)$ is determined to be

$$Q(s) = \frac{250(s + 5)}{s^2 + 20s + 200} \qquad (5\text{-}148)$$

Substituting $s = j10$ results in

$$Q(j10) = \frac{250(5 + j10)}{100 + j200} = 12.5 \qquad (5\text{-}149)$$

Thus

$$M = 12.5 \qquad (5\text{-}150)$$

$$\theta = 0° \qquad (5\text{-}151)$$

The sinusoidal portion of the time response is

$$v_1(t) = \frac{12.5}{10} \sin 10t$$
$$= 1.25 \sin 10t \qquad (5\text{-}152)$$

5-8 MULTIPLE-ORDER POLES

The most common types of poles occurring in the transforms of circuit responses are the first-order real, complex, and imaginary poles considered in the preceding two sections. Occasionally, however, multiple-order poles do occur, and it is necessary to know how to handle them. Although a number of special tricks have been devised, we feel that the relatively small frequency of occurrence of such poles warrants only the thorough consideration of one reasonably straightforward expansion method. We will present without formal proof the procedure involved in this section.

First, let us consider the case of multiple-order real poles. Assume that the denominator of $F(s)$ contains a factor of the form $(s - p)^r$, indicating that $s = p$ is a real pole of rth order. We may thus write $F(s)$ in the form

$$F(s) = \frac{Q(s)}{(s - p)^r} \tag{5-153}$$

or

$$Q(s) = (s - p)^r F(s) \tag{5-154}$$

where $Q(s)$ represents everything else in $F(s)$ besides the $(s - p)^r$ factor. The partial fraction expansion of $F(s)$ requires the following form:

$$F(s) = \frac{A_1}{(s - p)^r} + \frac{A_2}{(s - p)^{r - 1}} + \cdots + \frac{A_k}{(s - p)^{r - k + 1}}$$
$$+ \cdots + \frac{A_r}{(s - p)} + R(s) \tag{5-155}$$

where $R(s)$ is the expansion due to all other poles. Let $F_1(s)$ represent the expansion of interest at present.

$$F_1(s) = \frac{A_1}{(s - p)^r} + \frac{A_2}{(s - p)^{r - 1}} + \cdots + \frac{A_k}{(s - p)^{r - k + 1}} + \cdots + \frac{A_r}{(s - p)} \tag{5-156}$$

It can be shown that the general kth coefficient is given by the formula

$$A_k = \frac{1}{(k - 1)!} \frac{d^{k - 1}}{ds^{k - 1}} Q(s) \bigg]_{s = p} \tag{5-157}$$

According to this formula, we successively differentiate $Q(s)$ with respect to s a total of $r - 1$ times and evaluate $Q(s)$ and each derivative for $s = p$. With the proper constant multipliers, the sequence of numbers generated yields the A coefficients. A total of r coefficients is required.

Once the coefficients are known, the inverse transform can be determined by means of (T-10). The general form is

$$f_1(t) = \left[\frac{A_1 t^{r - 1}}{(r - 1)!} + \frac{A_2 t^{r - 2}}{(r - 2)!} + \cdots + \frac{A_k t^{r - k}}{(r - k)!} + \cdots + A_r \right] \epsilon^{pt} \tag{5-158}$$

We might note that for a simple-order pole, there is only one coefficient, which by Eq. (5-157) is

$$A_1 = Q(s) \bigg]_{s = p} \tag{5-159}$$

This is in perfect agreement with Eq. (5-84) since $Q(s) = (s - p)F(s)$. Furthermore, Eq. (5-158) reduces to

$$f_1(t) = A_1 \epsilon^{pt} \tag{5-160}$$

as expected.

The most important case of a multiple-order real pole is the second-order real pole. We will tabulate the equations for this case separately. Thus if $F(s)$ contains a denominator factor of the form $(s - p)^2$, the expansion for $F_1(s)$ reads

$$F_1(s) = \frac{A_1}{(s - p)^2} + \frac{A_2}{(s - p)} \tag{5-161}$$

From Eq. (5-157), the coefficients are

$$A_1 = Q(s)\Big]_{s = p} \tag{5-162}$$

$$A_2 = \frac{dQ(s)}{ds}\Big]_{s = p} \tag{5-163}$$

The resulting time function is

$$f_1(t) = (A_1 t + A_2)\epsilon^{pt} \tag{5-164}$$

Multiple-order complex poles are best handled by combining the method of this section with the second technique described in the section on complex poles of simple order. In order words, the quadratic is factored as in the case of real poles, and the procedure of this section is used to obtain an expression in the time domain. Finally, the exponentials with imaginary arguments are manipulated to obtain an appropriate time expression.

Example 5-20

Determine the inverse transform of

$$F(s) = \frac{s^2 + 4}{s(s + 1)(s + 2)^3} \tag{5-165}$$

Solution   First we will determine the response due to the third-order pole, $s = -2$. Letting $F_1(s)$ represent this portion of the transform and $f_1(t)$ its inverse, we have

$$F_1(s) = \frac{A_1}{(s + 2)^3} + \frac{A_2}{(s + 2)^2} + \frac{A_3}{(s + 2)} \tag{5-166}$$

Furthermore,

$$Q(s) = \frac{s^2 + 4}{s(s + 1)} \tag{5-167}$$

The required derivatives are

$$\frac{dQ(s)}{ds} = \frac{2s(s^2 + s) - (2s + 1)(s^2 + 4)}{(s^2 + s)^2} = \frac{s^2 - 8s - 4}{s^2(s + 1)^2} \tag{5-168}$$

$$\frac{d^2Q(s)}{ds^2} = \frac{s^2(s + 1)^2(2s - 8) - (s^2 - 8s - 4)[2s^2(s + 1) + 2s(s + 1)^2]}{s^4(s + 1)^4} \tag{5-169}$$

There is not much point in simplifying the last expression since no further differentiation is required, and when the proper value is inserted shortly, it will reduce fairly quickly.

By means of Eq. (5-157) the coefficients are

$$A_1 = \frac{4 + 4}{(-2)(-1)} = 4 \tag{5-170}$$

$$A_2 = \frac{4 + 16 - 4}{(4)(1)} = 4 \tag{5-171}$$

$$A_3 = \frac{1}{2} \left[\frac{(4)(1)(-12) - (4 + 16 - 4)(2)[(4)(-1) + (-2)(1)]}{(16)(1)} \right]$$
$$= \frac{1}{2} \left[\frac{-48 - (16)(-12)}{16} \right] = \frac{9}{2} = 4.5 \tag{5-172}$$

Thus

$$F_1(s) = \frac{4}{(s + 2)^3} + \frac{4}{(s + 2)^2} + \frac{4.5}{(s + 2)} \tag{5-173}$$

and from Eq. (5-158)

$$f_1(t) = [2t^2 + 4t + 4.5]\epsilon^{-2t} \tag{5-174}$$

Let us designate the remainder of the time function as $f_2(t)$. It is left as an exercise for the reader to show that

$$f_2(t) = 0.5 - 5\epsilon^{-t} \tag{5-175}$$

The total time function is, of course, given by

$$f(t) = f_1(t) + f_2(t) \tag{5-176}$$

Example 5-21

Determine the inverse transform of

$$I(s) = \frac{100}{(s^2 + 2s + 5)^2(s + 1)} \tag{5-177}$$

Solution   The roots of the quadratic are $s = -1 \pm j2$. Furthermore, the roots are second-order complex roots. Proceeding in the manner that we discussed in this section, we factor the quadratic as follows:

$$I(s) = \frac{100}{(s + 1 - j2)^2(s + 1 + j2)^2(s + 1)} \tag{5-178}$$

Thus there are actually two pairs of second-order poles. In principle, the procedure of the preceding section should be applied at both distinct complex poles.

However, as in the case of complex poles of simple order, the coefficients of the conjugate poles are themselves conjugates, and we will employ the procedure only once.

First, let $I_a(s)$ represent the expansion due to the second-order pole $s = -1 + j2$. The expansion reads

$$I_a(s) = \frac{\bar{A}_1}{(s + 1 - j2)^2} + \frac{\bar{A}_2}{(s + 1 - j2)} \tag{5-179}$$

where the bars emphasize that the coefficients are complex numbers. Let

$$Q(s) = \frac{100}{(s + 1 + j2)^2(s + 1)} \tag{5-180}$$

Its first derivative is

$$\frac{dQ(s)}{ds} = -100 \left[\frac{(s + 1 + j2)^2 + 2(s + 1)(s + 1 + j2)}{(s + 1 + j2)^4(s + 1)^2} \right] \tag{5-181}$$

By means of Eqs. (5-162) and (5-163) extended to this case, we have

$$\bar{A}_1 = Q(s)\Big]_{s = -1 + j2}$$
$$= \frac{100}{(j4)^2(j2)} = 3.125 \underline{/90^\circ} = 3.125\epsilon^{j90^\circ} \tag{5-182}$$

$$\bar{A}_2 = -100 \left[\frac{(j4)^2 + 2(j2)(j4)}{(j4)^4(j2)^2} \right] = -100 \left[\frac{-32}{-1024} \right] \tag{5-183}$$
$$= -3.125$$

Now let $I_b(s)$ represent the expansion due to the second-order pole $s = -1 - j2$. The expansion reads

$$I_b(s) = \frac{\tilde{A}_1}{(s + 1 + j2)^2} + \frac{\tilde{A}_2}{(s + 1 + j2)} \tag{5-184}$$

where

$$\tilde{A}_1 = 3.125\epsilon^{-j90^\circ} \tag{5-185}$$

and

$$\tilde{A}_2 = -3.125 \tag{5-186}$$

Let us designate the expansion due to both pairs of second-order poles as $I_1(s)$. We have

$$I_1(s) = I_a(s) + I_b(s)$$
$$= 3.125 \left[\frac{\epsilon^{j90^\circ}}{(s + 1 - j2)^2} + \frac{\epsilon^{-j90^\circ}}{(s + 1 + j2)^2} \right]$$
$$- 3.125 \left[\frac{1}{(s + 1 - j2)} + \frac{1}{(s + 1 + j2)} \right] \tag{5-187}$$

By means of Eq. (5-164) the time function is

$$i_1(t) = 3.125t[\epsilon^{j90°}\epsilon^{(-1+j2)t} + \epsilon^{-j90°}\epsilon^{(-1-j2)t}] - 3.125[\epsilon^{(-1+j2)t} + \epsilon^{(-1-j2)t}]$$

$$= 3.125t\epsilon^{-t}[\epsilon^{j(2t+90°)} + \epsilon^{-j(2t+90°)}] - 3.125\epsilon^{-t}[\epsilon^{j2t} + \epsilon^{-j2t}] \qquad (5\text{-}188)$$

By definition of the cosine function, this result is expressed as

$$i_1(t) = 6.25t\epsilon^{-t}\cos(2t + 90°) - 6.25\epsilon^{-t}\cos 2t$$

$$= -6.25\epsilon^{-t}[t\sin 2t + \cos 2t] \qquad (5\text{-}189)$$

The reader might verify that the time response due to the real pole of simple order, $s = -1$, is

$$i_2(t) = 6.25\epsilon^{-t} \qquad (5\text{-}190)$$

Of course, the complete response is

$$i(t) = i_1(t) + i_2(t) \qquad (5\text{-}191)$$

GENERAL PROBLEMS

In Problems 5-1 through 5-18, determine the Laplace transforms of the functions given using Table 5-1 (and Table 5-2 if appropriate).

5-1. $v(t) = 8$

5-2. $i(t) = 4\epsilon^{-2t}$

5-3. $e(t) = 5t$

5-4. $f(t) = 2t^2$

5-5. $v(t) = 20\sin 3t$

5-6. $i(t) = 0.1\cos 4t$

5-7. $v(t) = 4\epsilon^{-2t}\sin 3t$

5-8. $f(t) = 8\epsilon^{-3t}\cos 5t$

5-9. $v(t) = 6\sin(3t + 60°)$

5-10. $i(t) = 4\sin(3t - 30°)$

5-11. $v(t) = 6\epsilon^{-2t}\sin(3t + 60°)$

5-12. $i(t) = 4\epsilon^{-t}\sin(3t - 30°)$

5-13. $e(t) = 12\epsilon^{-5t}t^3$

5-14. $i(t) = \begin{cases} 200 \text{ mA} & \text{for } 0 < t < 10 \text{ ms} \\ 0 & \text{otherwise} \end{cases}$

(The pulse width is short compared with the applicable circuit time constant.)

5-15. $v(t) = 5(t-2)u(t-2)$

5-16. $f(t) = 20\epsilon^{-2(t-3)}u(t-3)$

5-17. See Fig. P5-17.

5-18. See Fig. P5-18.

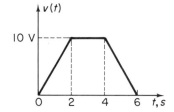

Figure P5-17

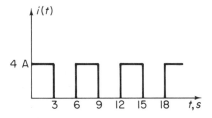

Figure P5-18

5-19. Determine the roots of the following polynomials and classify them according to real, complex, or imaginary. Write each polynomial in factored form.
  (a) $2s + 8$ **(b)** $4s^2 + 20s + 16$ **(c)** $s^2 + 4s + 29$
  (d) $s^2 + 6s + 9$ **(e)** $s^2 + 25$

5-20. Determine the roots of the following polynomials and classify them according to real, complex, or imaginary. Write each polynomial in factored form.
  (a) $s^2 + 200s + 50,000$
  (b) $s^2 + 10,000$
  (c) $s^3 + 6s^2 + 33s + 50$ ($s = -2$ is a root.)
  (d) $s^4 + 4s^3 + 14s^2 + 20s + 25$ ($s = -1 + j2$ is a root.)
  (e) $s^4 + 13s^2 + 36$

In Problems 5-21 through 5-30, determine the inverse Laplace transforms of the functions given using Table 5-1 (and Table 5-2 if appropriate).

5-21. $V(s) = \dfrac{8}{s + 3}$

5-22. $I(s) = 2 + \dfrac{3}{s} + \dfrac{4}{s^2}$

5-23. $F(s) = \dfrac{32}{s^2 + 64}$

5-24. $I(s) = \dfrac{10s}{s^2 + 25}$

5-25. $V(s) = \dfrac{6s + 15}{s^2 + 9}$

5-26. $I(s) = \dfrac{8 - 3s}{s^2 + 4}$

5-27. $F(s) = \dfrac{4s - 22}{s^2 + 4s + 29}$

5-28. $V(s) = \dfrac{100s}{s^2 + 200s + 50,000}$

5-29. $V(s) = \dfrac{s + 2}{s + 1}$

5-30. $F(s) = \dfrac{\pi}{s^2 + \pi^2}(1 + \epsilon^{-s} + \epsilon^{-2s} + \epsilon^{-3s} + \cdots)$ (Sketch the time function.)

In Problems 5-31 through 5-48, determine the inverse transforms of the functions given using partial fraction expansion and the related procedures of Sections 5-6 through 5-8.

5-31. $F(s) = \dfrac{9s + 23}{(s + 2)(s + 3)}$

5-32. $I(s) = \dfrac{7s + 23}{(s + 3)(s + 4)}$

5-33. $V(s) = \dfrac{3s^2 + 13s + 8}{s(s^2 + 3s + 2)}$

5-34. $F(s) = \dfrac{4s^2 + 6s + 3}{s(s^2 + 4s + 3)}$

5-35. $V(s) = \dfrac{20s + 56}{4s^2 + 24s + 32}$

5-36. $F(s) = \dfrac{50s + 15,000}{s^2 + 300s + 2 \times 10^4}$

5-37. $I(s) = \dfrac{2(s + 2)}{(s + 1)(s^2 + 4s + 13)}$

5-38. $F(s) = \dfrac{50s + 100}{(s + 1)(2s^2 + 8s + 26)}$

5-39. $V(s) = \dfrac{120(s + 1)}{(4s + 8)(2s^2 + 50)}$

5-40. $I(s) = \dfrac{10}{(s^2 + 4)(s^2 + 2s + 5)}$

5-41. $V(s) = \dfrac{20(s^2 + 4)}{s(s + 1)(s^2 + 2s + 2)}$

5-42. $V(s) = \dfrac{100s^2}{(s^2 + 3s + 2)(s^2 + 2s + 2)}$

5-43. $I(s) = \dfrac{2s + 5}{(s^2 + 3s + 2)(s^2 + 6s + 25)}$

5-44. $I(s) = \dfrac{20s + 10}{s^4 + 5s^2 + 4}$

5-45. $F(s) = \dfrac{30}{(s + 2)^4}$

5-46. $F(s) = \dfrac{6s}{(s + 2)^4}$

5-47. $V(s) = \dfrac{20}{(s + 1)(s + 2)^2}$

5-48. $I(s) = \dfrac{4s}{(s^2 + 1)(s + 1)^3}$

DERIVATION PROBLEMS

5-49. Derive the Laplace transform of $f(t) = t$ [i.e., (T-3) of Table 5-1].

5-50. Derive the Laplace transform of $f(t) = \epsilon^{-at}$ [i.e., (T-4) of Table 5-1].

5-51. Derive the Laplace transform of $\cos \omega t$ by starting with the Laplace transform of $\sin \omega t$ and employing operation (O-1) of Table 5-2.

5-52. Derive the Laplace transform of $u(t)$ by starting with the Laplace transform of the impulse function and employing operation (O-2) of Table 5-2.

5-53. Derive the Laplace transform of the damped sine function [i.e., (T-7)] by starting with the Laplace transform of the undamped sine function [i.e., (T-5)] and employing operation (O-4) from Table 5-2.

5-54. Derive the Laplace transform of $f(t) = t^2$ by starting with the Laplace transform of t and employing operation (O-2) of Table 5-2.

5-55. Show that the Laplace transform of $f(t) = \sin (\omega t + \theta)$ is

$$F(s) = \frac{s \sin \theta + \omega \cos \theta}{s^2 + \omega^2}$$

5-56. Show that the Laplace transform of $f(t) = \cos (\omega t + \theta)$ is

$$F(s) = \frac{s \cos \theta - \omega \sin \theta}{s^2 + \omega^2}$$

COMPUTER PROBLEMS

One area in which the computer can be readily used in support of Laplace transform analysis is for factoring higher-order polynomials. The libraries of most scientific computer centers and an increasing number of software packages for microcomputers contain polynomial factoring programs. The development of such programs is a topic in numerical analysis and not within the objective of this book. The reader is encouraged to investigate the availability of such programs, and the following problems can be used to test the programs available.

Utilize an available computer program to determine the roots of the polynomials in Problems 5-57 through 5-60.

5-57. $s^3 + 2s^2 + 2s + 1$

5-58. $s^4 + 2.6131s^3 + 3.4142s^2 + 2.6131s + 1$

5-59. $s^5 + 3.2361s^4 + 5.2361s^3 + 5.2361s^2 + 3.2361s + 1$

5-60. $s^6 + 3.8637s^5 + 7.4641s^4 + 9.1416s^3 + 7.4641s^2 + 3.8637s + 1$

6

CIRCUIT ANALYSIS
BY LAPLACE TRANSFORMS

OVERVIEW

In Chapter 5 the concept of the Laplace transform was developed, and many of its basic mathematical properties were introduced. Consideration was made both of determining the transform of a given time function and determining the time function corresponding to a given transform function. It should be recalled that these transformations represent the initial and final steps of a complete network solution. There remains the problem of solving the problem in the transform domain.

The major purpose of this chapter is to develop the procedure for solving a complete circuit problem by transform techniques. Transform-domain equivalent circuits are developed for representing the voltage–current relationships of all circuit components. The use of these equivalent circuits permits the application of basic algebraic circuit analysis schemes to be applied directly to complex circuits.

The properties of certain common circuit forms are investigated in some detail. This includes first- and second-order circuits with arbitrary excitations.

OBJECTIVES

After completing this chapter, the reader should be able to

1. Draw the transform equivalent circuit of a *capacitor* and represent the effects of any initial energy storage by a suitable *s*-domain source.
2. Draw the transform equivalent circuit of an *inductor* and represent the effect of any initial energy storage by a suitable *s*-domain source.
3. Draw the transform equivalent circuit of a *resistor*.

171

4. Transform a complete circuit, including the effects of sources, and represent the circuit in the best form for either mesh or node analysis.

5. Apply basic circuit analysis methods (e.g., mesh analysis, node analysis, Thévenin's theorem, etc.) to transform-domain circuits to obtain desired response functions.

6. Discuss the concepts of *natural* and *forced* response and how these terms relate to *transient* and *steady-state* response.

7. Define the terms *impulse response* and *step response*.

8. Determine complete responses of first-order circuits with arbitrary excitations using transform methods.

9. Determine complete responses of second-order circuits with arbitrary excitations using transform methods.

10. For a second-order response, discuss the mathematical and physical nature of the following types of responses: (a) *overdamped*, (b) *critically damped*, and (c) *underdamped*.

11. Define the following terms with respect to a second-order response: (a) *damping constant*, (b) *undamped natural frequency*, and (c) *damped natural frequency*.

12. Analyze the response of a series *RLC* circuit excited by a step function of voltage.

13. Analyze the response of a parallel *RLC* circuit excited by a step function of current.

14. Discuss the relationship between the number of energy-storage elements and the order of the circuit.

15. Solve an ordinary constant-coefficient linear differential equation using transform methods.

16. For a given circuit, write the differential equations directly in the time domain, and solve for a desired variable using transform methods.

6-1 TRANSFORM EQUIVALENT OF CAPACITANCE

In Chapter 3 we studied the voltage–current relationships for a charged capacitor. We found that a charged capacitor could be represented as an uncharged capacitor in series with a dc (or step) voltage source as long as the voltage source was considered to be included in the effective terminals of the capacitor. Since the effect of the initial voltage is treated like other sources in the network, let us consider for the moment an uncharged capacitor as shown in Fig. 6-1. Using $t = 0$ as an initial reference, the

Figure 6-1 Time-domain representation of uncharged capacitor.

time-domain relationship is

$$v(t) = \frac{1}{C} \int_0^t i(t)\, dt \tag{6-1}$$

Now let us "operate" on Eq. (6-1) by determining the Laplace transforms of both sides. Since $v(t)$ and $i(t)$ are not specified, we can use the definitions

$$V(s) = \mathscr{L}[v(t)] \tag{6-2}$$

$$I(s) = \mathscr{L}[i(t)] \tag{6-3}$$

Transformation of Eq. (6-1) yields

$$\mathscr{L}[v(t)] = \mathscr{L}\left[\frac{1}{C} \int_0^t i(t)\, dt \right] \tag{6-4}$$

or

$$V(s) = \frac{1}{C}\, \mathscr{L}\left[\int_0^t i(t)\, dt \right] \tag{6-5}$$

By means of operation (O-2) of Chapter 5, the Laplace transform of the integral term is

$$\mathscr{L}\left[\int_0^t i(t)\, dt \right] = \frac{I(s)}{s} \tag{6-6}$$

Substitution of Eq. (6-6) into Eq. (6-5) results in

$$V(s) = \frac{1}{sC}\, I(s) \tag{6-7}$$

A most important characteristic of Eq. (6-7) is that, while the time-domain equation involves an integration, the transform-domain equation is an algebraic equation. Of course, the quantity s appears in the equation, but it may be manipulated by any normal algebraic procedure.

In a dc resistive circuit, the voltage is a constant times the current. Although it is not correct to speak of s as a constant (it is actually a type of *operator*), we can compare Eq. (6-7) to a dc circuit in the sense that the voltage is expressed as a function times the current.

Since we cannot use the term "resistance," we will borrow the term *impedance* from steady-state ac circuit theory and attach the adjective *transform* for descriptive reasons. We define the *transform impedance of a capacitor* as

$$Z(s) = \frac{1}{sC} \tag{6-8}$$

The quantity impedance has the same dimensions as resistance, namely ohms. Impedance in the transform domain may be treated, from an algebraic point of view, in the same manner as resistance is treated in dc circuits. The essential difference is that the s operator is ever present and must not be misplaced in manipulations.

Although we have only looked at a capacitor so far, it will help to slightly generalize the concept at this point. We can define *Ohm's law in the transform domain*

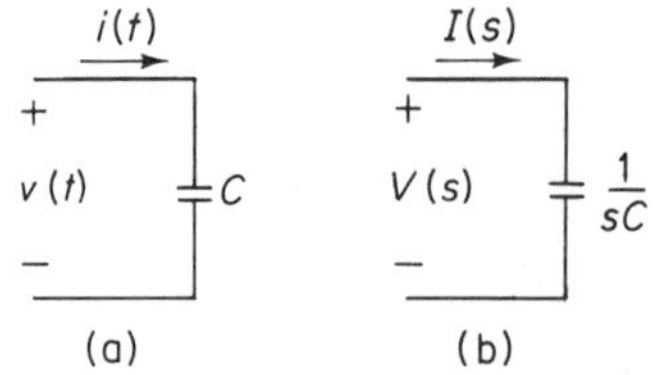

Figure 6-2 Time-domain and transform-domain representations of uncharged capacitor.

by

$$V(s) = Z(s)I(s) \tag{6-9}$$

or

$$I(s) = \frac{V(s)}{Z(s)} \tag{6-10}$$

or

$$Z(s) = \frac{V(s)}{I(s)} \tag{6-11}$$

We can also speak of the reciprocal of impedance. This quantity will be called the *transform admittance* and will be denoted by $Y(s)$. Thus

$$Y(s) = \frac{1}{Z(s)} \tag{6-12}$$

For the capacitor, the transform admittance is

$$Y(s) = sC \tag{6-13}$$

Returning to the capacitor and considering Fig. 6-2a, we can transform the capacitor by expressing it as an impedance $1/sC$ as shown in (b). The circuit is now in a form suitable for transform analysis. Strictly speaking, there is no reason why we could not designate the transformed circuit in terms of admittance. However, for the same reason that resistors are usually designated in ohms instead of siemens, namely consistency, we will normally designate the transforms of circuits in terms of impedances.

6-2 TRANSFORM EQUIVALENT OF INDUCTANCE

In Chapter 3 we found that an inductor with an initial current could be represented as an unfluxed inductor in parallel with a current source as long as the current source was considered to be included in the effective terminals of the inductor. Since the effect of the initial current is treated like other sources in the network, let us consider for the moment an unfluxed inductor as shown in Fig. 6-3. Using $t = 0$ as an initial reference, the time-domain relationship is

$$v(t) = L\frac{di(t)}{dt} \tag{6-14}$$

Figure 6-3   Time-domain representation of unfluxed inductor.

Now let us "operate" on Eq. (6-14) by determining the Laplace transforms of both sides. As in the case of the capacitor, and in essentially all future work, we employ the transform definitions.

$$V(s) = \mathscr{L}[v(t)] \tag{6-15}$$

$$I(s) = \mathscr{L}[i(t)] \tag{6-16}$$

Transformation of Eq. (6-14) yields

$$\mathscr{L}[v(t)] = \mathscr{L}\left[L\frac{di}{dt}\right] \tag{6-17}$$

or

$$V(s) = L\mathscr{L}\left[\frac{di}{dt}\right] \tag{6-18}$$

By means of (O-1), with the assumption that the initial current is zero, we obtain

$$\mathscr{L}\left[\frac{di}{dt}\right] = sI(s) \tag{6-19}$$

Substitution of Eq. (6-19) into Eq. (6-18) yields

$$V(s) = sLI(s) \tag{6-20}$$

As in the case of the capacitor, the transform-domain equation for an inductor is an algebraic equation. Thus *the transform impedance of an inductor is*

$$Z(s) = sL \tag{6-21}$$

If the transform admittance of an inductor is desired, it can be expressed as

$$Y(s) = \frac{1}{sL} \tag{6-22}$$

Referring to Fig. 6-4a, we can transform an unfluxed inductor by expressing it as an impedance sL as shown in (b).

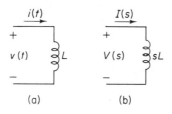

Figure 6-4   Time-domain and transform-domain representations of unfluxed inductor.

(a)   (b)

6-3 TRANSFORM EQUIVALENT OF RESISTANCE

The transform impedance of a resistor is very easily obtained and needs very little explanation. The time-domain relationship is

$$v(t) = Ri(t) \tag{6-23}$$

Transformation of both sides of Eq. (6-23) yields

$$V(s) = RI(s) \tag{6-24}$$

Thus the *transform impedance of a resistor is simply*

$$Z(s) = R \tag{6-25}$$

6-4 TRANSFORMING COMPLETE CIRCUITS

A complete linear circuit under excitation consists of one or more sources (both voltage and current sources, in general) and any arbitrary combination of circuit components (resistance, capacitance, and inductance, in general). The sources in the network are either actual excitations or hypothetical sources due to initial conditions. *As far as transform manipulations are concerned, initial condition sources are treated exactly like external sources.* The essential difference lies in determining the effective terminals of reactive components.

To transform sources, we apply the techniques developed in Chapter 5 for determining the transforms of time functions. To transform components, we employ the results of the preceding three sections of this chapter.

Example 6-1

The switch in the circuit of Fig. 6-5a is closed at $t = 0$. The initial values of inductive currents and capacitive voltages are shown. Draw the transformed circuit in a form most suitable for mesh current analysis.

   Solution Although the circuit may or may not be in a steady-state condition at $t = 0$, so long as we know the initial voltages on capacitors and initial currents through inductors, it is immaterial. We first draw the time-domain circuit, expressing initial condition generators as shown in Fig. 6-5b. Since we are primarily interested in mesh current analysis, we have used the equivalent circuits involving voltage sources. Thus, the two inductor initial condition sources are impulse sources. Recall that in Chapter 4 we stated that, for time-domain differential equation solutions, it was best to avoid impulse functions in general. However, *in transform analysis, impulse sources should be retained in the circuit as they provide the required initial conditions.* Note also that we have converted the right-hand dc current source to a dc voltage source.

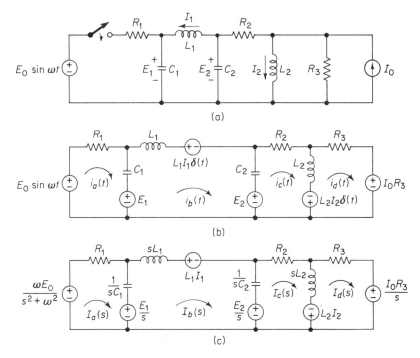

Figure 6-5   Circuit of Ex. 6-1.

Recalling the appropriate transform pairs from Chapter 5, the transformed circuit is shown in Fig. 6-5c. It should be observed again that the transforms of the impulse sources have the simplest possible form, since they are constants.

This circuit could now be solved or simplified by any standard algebraic or circuit analysis procedure for the desired transform variable. Finally, the inverse transform of the quantity desired could be determined. We consider these ideas gradually as we go along; so we will not attempt to solve this circuit at the moment.

6-5 SOLUTIONS OF COMPLETE CIRCUITS IN THE TRANSFORM DOMAIN

After a circuit is completely transformed according to the procedure discussed in Section 6-4, it may then be manipulated by any standard algebraic or circuit analysis technique. Among the possible methods of solution are mesh current analysis, node voltage analysis, Thévenin's theorem, Norton's theorem, successive reduction techniques, and many others. *In general, any dc circuit analysis scheme may be employed as long as we remember that both sources and impedances which appear in the circuit are functions of the variable s.*

After the desired voltage or current is obtained in the transform domain, its inverse transform can be determined to yield the final time function. In many problems such as are encountered in network and control system design studies, the entire analysis and design may be carried out in the transform domain, and no inversion may be necessary. In other words, engineers and technologists have learned to "think in the transform domain." Let us now illustrate the approach of transform solution methods by solving some examples.

Example 6-2

Referring back to the circuit of Example 6-1, write a set of mesh current equations that characterize the network.

    Solution   The transformed circuit is shown again in Fig. 6-6 with mesh currents assigned. Remember, the transform impedances and transform sources are treated exactly like dc quantities in writing equations. The mesh equations are

$$\left[R_1 + \frac{1}{sC_1}\right]I_a(s) - \left[\frac{1}{sC_1}\right]I_b(s) = \frac{\omega E_0}{s^2 + \omega^2} - \frac{E_1}{s} \quad (6\text{-}26)$$

$$\left[\frac{-1}{sC_1}\right]I_a(s) + \left[\frac{1}{sC_1} + sL_1 + \frac{1}{sC_2}\right]I_b(s) - \left[\frac{1}{sC_2}\right]I_c(s) = \frac{E_1}{s} - L_1I_1 - \frac{E_2}{s} \quad (6\text{-}27)$$

$$-\left[\frac{1}{sC_2}\right]I_b(s) + \left[\frac{1}{sC_2} + R_2 + sL_2\right]I_c(s) - [sL_2]I_d(s) = \frac{E_2}{s} + L_2I_2 \quad (6\text{-}28)$$

$$-[sL_2]I_c(s) + [sL_2 + R_3]I_d(s) = -L_2I_2 - \frac{I_0R_3}{s} \quad (6\text{-}29)$$

    The reader should be careful not to be confused by the use of capital letters for both the transform quantities and initial conditions, since the quantities are quite different. We have carefully kept the argument (s) to ensure that the transform functions are recognized. The reader is urged to do the same when working problems.

    Theoretically, the equations could be solved simultaneously to yield any desired current, and of course, branch voltages and currents could be determined from the

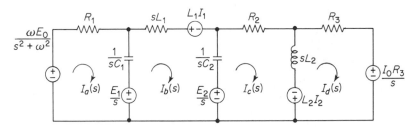

Figure 6-6   Circuit of Ex. 6-2.

mesh currents. Practically speaking, however, this could be quite a chore from a computational viewpoint for this circuit. Also, to obtain the inverse transform of a particular current would be most unwieldy even if values were specified for the components. The completion of a problem as complex as this would be enhanced by the availability of computer facilities. Thus we will leave this problem without further consideration.

Example 6-3

The switch in Fig. 6-7a is opened at $t = 0$, thus applying the current source to the network. Write the transform node voltage equations, determine $V_2(s)$, and invert to determine $v_2(t)$. The capacitor is initially uncharged.

    Solution   The transformed circuit applicable for $t > 0$ is shown in Fig. 6-7b. Although in node voltage analysis we will work with *admittances*, we have still followed the consistent convention of labeling the schematic in terms of impedances. The equations are

$$V_1(s)[2 + 8s] - V_2(s)[8s] = \frac{10}{s} \tag{6-30}$$

$$-V_1(s)[8s] + V_2(s)[4 + 8s] = 0 \tag{6-31}$$

These equations must be solved simultaneously for $V_2(s)$. This may be achieved by either determinant methods or by substitution. The result is

$$V_2(s) = \frac{10}{6s + 1} = \frac{5/3}{s + 1/6} \tag{6-32}$$

Inversion of $V_2(s)$ follows directly from (T-4). Thus

$$v_2(t) = \frac{5}{3}\epsilon^{-t/6}$$

This circuit could have easily been solved by means of the single time constant circuit concept of Chapter 4. We have solved it by transform methods for illustrative purposes.

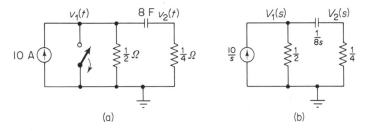

(a)   (b)

Figure 6-7 Circuit of Ex. 6-3.

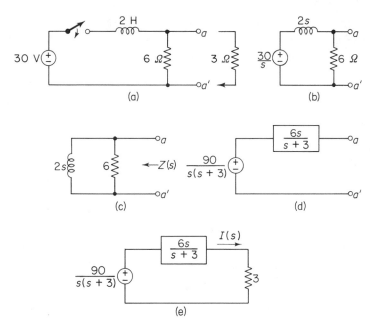

Figure 6-8   Circuit of Ex. 6-4.

Example 6-4

The switch in the circuit of Fig. 6-8a is closed at $t = 0$. The inductor is initially unfluxed.

    (a) Determine a transform Thévenin equivalent circuit at the terminals a-a'.

    (b) Using the result of (a), determine the current through a 3-Ω res stor which is assumed connected to the circuit at the time the switch is closed.

    Solution   (a) The transformed circuit is shown in Fig. 6-8b. To obtain the Thévenin equivalent circuit, we first measure $V_{oc}(s)$ across the terminals a-a'. By means of the voltage-divider rule, this voltage is

$$V_{oc}(s) = \frac{30}{s} \times \frac{6}{2s + 6} = \frac{90}{s(s + 3)} \tag{6-33}$$

We determine the Thévenin equivalent impedance $Z(s)$ by shorting the source and combining the impedances, $3s$ and 6, in parallel. Thus, as shown in (c),

$$Z(s) = \frac{2s \times 6}{2s + 6} = \frac{6s}{s + 3} \tag{6-34}$$

The Thévenin equivalent circuit is shown in (d).

    (b) Knowing the Thévenin equivalent circuit, we can determine the current through the 3-Ω resistor from the series circuit shown in Fig. 6-8e. We have

$$I(s) = \frac{V_{oc}(s)}{Z(s) + 3} = \frac{90/s(s + 3)}{(6s)/(s + 3) + 3}$$

$$= \frac{10}{s(s + 1)}$$

(6-35)

We may readily expand $I(s)$ into partial fractions and obtain

$$I(s) = \frac{10}{s} - \frac{10}{s + 1}$$

(6-36)

The time function is

$$i(t) = 10(1 - \epsilon^{-t})$$

(6-37)

Example 6-5

Consider the circuit of Fig. 6-9a, which is identical to the circuit of Example 6-4, except that the circuit to the left of a-a' has been connected for a sufficiently long time so that a steady-state situation exists at $t = 0^-$. The 3-Ω resistor is to be connected at $t = 0$.

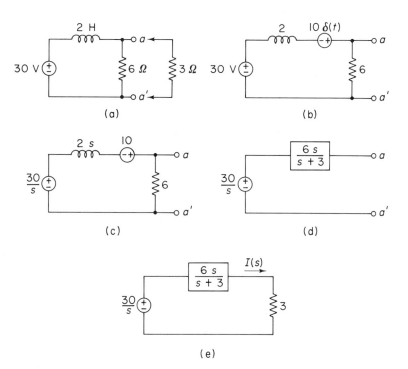

(a)

(b)

(c)

(d)

(e)

Figure 6-9   Circuit of Example 6-5.

(a) Determine the transform Thévenin equivalent circuit applicable for $t > 0$.

(b) Determine the current through the 3-Ω resistor.

Solution (a) First, we must determine the initial current through the inductor. Since the inductor acts as a short circuit at $t = 0^-$, we have

$$i_L(0^-) = \frac{30}{6} = 5 \text{ A} \tag{6-38}$$

Replacing the fluxed inductor by its Thévenin equivalent circuit, we obtain the circuit shown in Fig. 6-9b. The transformed circuit is shown in (c). The open-circuit voltage in this case is

$$V_{oc}(s) = \left[\frac{30}{s} + 10\right] \times \frac{6}{2s + 6} = \frac{30}{s} \tag{6-39}$$

The transform impedance is the same as in Example 6-4, namely,

$$Z(s) = \frac{6s}{s + 3} \tag{6-40}$$

The equivalent circuit is shown in (d). Notice that the presence of the initial current through the inductor modifies the Thévenin voltage.

(b) With the 3-Ω resistor connected as shown in Fig. 6-9e, the current is

$$I(s) = \frac{30/s}{(6s)/(s + 3) + 3} = \frac{(10/3)(s + 3)}{s(s + 1)} \tag{6-41a}$$

Partial fraction expansion of $I(s)$ yields

$$I(s) = \frac{10}{s} - \frac{20/3}{s + 1} \tag{6-41b}$$

The time function is

$$i(t) = 10 - \frac{20}{3}\epsilon^{-t} \tag{6-42}$$

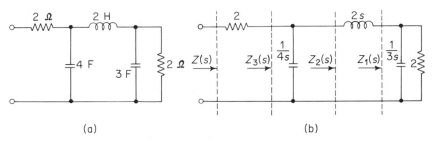

(a)   (b)

Figure 6-10 Circuit of Ex. 6-6.

Example 6-6

(a) Determine the equivalent transform impedance at the terminals of the relaxed network shown in Fig. 6-10a.

(b) Determine the transform of the current that would flow from a sinusoidal voltage generator, $e(t) = 10 \sin 4t$, connected to the terminals.

Solution   (a) Due to the ladder configuration of this circuit, it is best solved by means of successive series and parallel reductions. First we draw the transformed circuit as shown in Fig. 6-10b. The various Z's refer to the impedances looking to the right at various points *if* the circuits to the left, in each case, were disconnected. We have

$$Z_1(s) = \frac{2 \times (1/3s)}{2 + (1/3s)} = \frac{2}{6s + 1} \tag{6-43}$$

$$Z_2(s) = 2s + \frac{2}{6s + 1} = \frac{12s^2 + 2s + 2}{6s + 1} \tag{6-44}$$

At this point in determining $Z_3(s)$, it is probably best to switch to admittance momentarily. Thus

$$Y_3(s) = 4s + \frac{6s + 1}{12s^2 + 2s + 2} = \frac{48s^3 + 8s^2 + 14s + 1}{12s^2 + 2s + 2} \tag{6-45}$$

or

$$Z_3(s) = \frac{12s^2 + 2s + 2}{48s^3 + 8s^2 + 14s + 1} \tag{6-46}$$

and

$$Z(s) = 2 + \frac{12s^2 + 2s + 2}{48s^3 + 8s^2 + 14s + 1}$$
$$= \frac{96s^3 + 28s^2 + 30s + 4}{48s^3 + 8s^2 + 14s + 1} \tag{6-47}$$

(b) The sinusoidal excitation applied to the network is given by

$$e(t) = 10 \sin 4t \tag{6-48}$$

The transform is

$$E(s) = \frac{40}{s^2 + 16} \tag{6-49}$$

The transform current is

$$I(s) = \frac{E(s)}{Z(s)} = \frac{40(48s^3 + 8s^2 + 14s + 1)}{(s^2 + 16)(96s^3 + 28s^2 + 30s + 4)} \tag{6-50}$$

6-6 FORMS FOR INITIAL CONDITIONS

The initial conditions that must be considered in transform analysis are initial currents through inductors and initial voltages on capacitors. In the past few sections, we have dealt with such conditions by first drawing the equivalent time-domain circuits and then transforming the hypothetical sources along with other sources in the network. This approach is certainly correct and adequate. However, the reader who may have occasion to work many problems involving initial conditions may wish to avoid the intermediate step of writing down the time-domain equivalent circuit. Instead, it is possible, with a little practice, to go directly to the transform representations for initial conditions. Also, it is easier to manipulate between Thévenin and Norton forms in the transform domain.

At the beginning, we recognize that since only step and impulse sources are involved, the transforms will be either of the form $1/s$ or a constant. First, let us consider a charged capacitor as shown in Fig. 6-11a. If we mentally visualize replacing the capacitor by its Thévenin equivalent circuit and transforming, we obtain the representation shown in (b). Thus, the transform Thévenin equivalent circuit follows naturally from basic considerations. To obtain the Norton equivalent circuit shown in (c), we apply Norton's theorem directly in the transform domain. The short-circuit current would be

$$I_{sc}(s) = \frac{V_0/s}{1/sC} = CV_0 \qquad (6\text{-}51)$$

and the impedance is simply $1/sC$. Thus, the Norton equivalent circuit is easily obtained directly in the transform domain, without dealing specifically with an impulse function. If desired, one can begin each problem with the easily remembered Thévenin form.

The opposite situation exists in the case of the fluxed inductor shown in Fig. 6-12a. The most natural representation is the Norton equivalent circuit whose transform is shown in (b). To obtain the Thévenin equivalent circuit shown in (c), we could measure V_{oc} and would obtain

$$V_{oc}(s) = \frac{I_0}{s} sL = LI_0 \qquad (6\text{-}52)$$

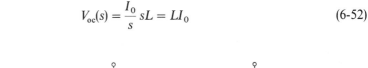

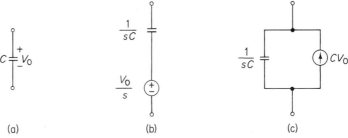

(a)   (b)   (c)

Figure 6-11   Transform representations for charged capacitor.

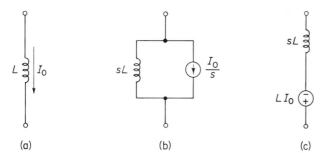

Figure 6-12 Transform representation for fluxed inductor.

(a)   (b)   (c)

and the impedance is sL. Thus, in the case of the inductor, the easily remembered form is the Norton form, and the Thévenin form may be easily obtained in the transform domain. As in the time domain, the sources must be considered to be within the effective terminals of the component.

6-7 TRANSIENT AND STEADY-STATE PHENOMENA

The concepts of transient and steady-state phenomena were introduced in Chapter 4. It was stated then that, immediately after a circuit is excited by one or more arbitrary excitations, there is an initial "adjustment" response called the *transient response*. Eventually, in most cases, the circuit reaches an equilibrium or semiequilibrium state in which the response follows a more uniform behavior. The latter type of response is called the *steady-state response*.

Before we pursue any more transform problems, it is wise to pause here and look at this concept again in view of having solved a few transform problems. Such considerations will help us to "see the forest through the trees" as we go along.

From the problems already considered, we note that the nature of the time function is dependent on the type of poles (denominator roots) of the desired transform variable. Corresponding to each pole (or pair of poles for complex poles), there is a corresponding term in the time response. If a given pole is real and simple, the time response is an exponential term. If a pair of poles are complex, the time response is a damped sinusoidal term, etc. Thus *a knowledge of the poles of a transform response is sufficient to predict the type of time response.* Of course, inverse transformation is necessary to determine the actual magnitudes of the various terms, but a quick inspection to determine the type of response is often helpful.

As we work through many problems in the remainder of this chapter, the reader should observe that the poles of a given transform response arise from two distinct sources: (a) poles due to the network structure, and (b) poles due to the excitation. Thus, any time response, in general, will consist of (a) terms due to the circuit itself, and (b) terms due to the excitation.

The poles due to the network structure may be considered to produce the portion of the total response known as the natural response. In essentially all practical circuits not containing electronic feedback or negative resistance, the natural response vanishes after a sufficiently long time. In such cases, this natural response may be considered to be the *transient response*, although the remainder of the total response is simultaneously present and affects the entire shape of the response during this transient phase.

The poles due to the excitation may be considered to produce the portion of the total response known as the forced response. If the excitation and network structure are such that the natural response eventually vanishes, the remaining forced response may be considered to be the *steady-state response.*

In summary:

1. The form of the *natural response* is determined by the nature of the circuit, and this function may be considered to be the *transient response* whenever it is transient in nature.

2. The form of the *forced response* is determined by the nature of the excitation, and this function may be considered to be the *steady-state* response whenever the natural response is transient in nature.

In studying the natural transient behavior of circuits, it is often necessary to assume some specific type of excitation. The two most common types of excitations for this purpose are the impulse function and the step function. The impulse function is particularly simple for this purpose since $\mathscr{L}[\delta(t)] = 1$. Thus since the transform of the impulse function contains no poles at all, the forms of all terms of the response resulting from such an excitation are due to the network alone and are thus *natural response terms.*

Although the transform of the step function is not quite as simple since it contains a pole, $s = 0$, in some cases the response resulting from a step excitation is simpler in form than the response resulting from an impulse excitation. This is particularly true in cases where the pole of the transform of the step function cancels a numerator factor, resulting only in poles due to the network structure. Furthermore, the response resulting from a step excitation may be readily measured in the laboratory.

A response resulting from an impulse excitation is called an *impulse response*, and a response resulting from a step excitation is called a *step response.* Both the impulse response and step response concepts are widely used in systems analysis.

6-8 FIRST-ORDER CIRCUITS

In this section and in the next few sections, we consider some specific classes of circuits that are readily solvable by transform methods. The purpose is twofold: (a) we can certainly gain more practice in transform analysis by solving more problems, and (b) we can learn to associate certain types of responses with certain special classes of networks that occur frequently in practical applications.

In this section we consider a few problems involving circuits whose time-domain differential equations are of first order. Such circuits were considered in Chapter 4 and were appropriately designated as single time constant circuits. It was demonstrated that such circuits consisted of either resistance and capacitance (*RC*) or resistance and inductance (*RL*). Such circuits have appeared in a few examples in this chapter.

When single time constant circuits were considered in the time domain in Chapter 4, we restricted the excitations to include only dc sources and initial conditions. For those cases it was indicated that the solutions could be obtained most easily by inspection. However, there is certainly no reason why such circuits couldn't be solved by transform methods. When the general framework of analysis is in the transform domain, it may be more convenient to do so. Furthermore, and most important, with transform methods, we can easily generalize our excitations to include such waveforms as impulse functions, sinusoidal functions, and many others.

An important point to remember is that with arbitrary excitations, the responses may involve more than just exponential terms and a dc term even though the circuit itself is of the single time constant form. Such general terms are due to the nature of the excitation terms. Thus if a sinusoidal source excites a single time constant circuit, the response may consist of an exponential term and a sinusoidal term. We have preferred to define a different term for describing such circuits with general excitations. We will refer to a circuit that can be described by a first-order differential equation with arbitrary excitations as a *first-order circuit*. Let us consider some examples.

Example 6-7

The circuit shown in Fig. 6-13a is excited at $t = 0$ by a voltage source $e(t)$. The capacitor is initially uncharged. By transform methods, determine the current $i(t)$ and the voltage across the capacitor $v(t)$, for $t > 0$, for each of the following excitations.

    (a) $e(t) = 20$ V (dc source)
    (b) $e(t) = 20 \sin 2t$
    (c) $e(t) = 20\epsilon^{-t}$
    (d) $e(t) = 20\epsilon^{-2t}$

    Solution   First, we transform the circuit as shown in Fig. 6-13b. We have used a general $E(s)$ at this point since we ultimately must consider four different cases. Since two variables, $i(t)$ and $v(t)$, must be determined in each case, we could solve for one of the desired variables in the transform domain, invert it, and determine the other variable quite easily in the time domain, since the voltage and current associated with a capacitor are easily related. However, to enhance our understanding of transform manipulations, we will choose to

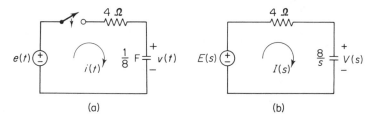

(a)   (b)

Figure 6-13   Circuit of Ex. 6-7.

solve for all variables in the transform domain and invert each separate quantity, except for the case of excitation (b).

First, let us obtain a general expression for $I(s)$ and $V(s)$ in terms of $E(s)$ so that we may "plug in" the different values of $E(s)$. From the circuit of Fig. 6-13b, we have

$$I(s) = \frac{E(s)}{4 + (8/s)} = \frac{sE(s)}{4(s + 2)} \tag{6-53}$$

The quantity $V(s)$ is given by

$$V(s) = \frac{8}{s} I(s) = \frac{2E(s)}{s + 2} \tag{6-54}$$

Now let us consider each of the excitations. The reader is encouraged to check the case (a) by use of the single time constant concept of Chapter 4.

(a) $e(t) = 20\ V$. The transform of $e(t)$ is

$$E(s) = \frac{20}{s} \tag{6-55}$$

Substitution of $E(s)$ into Eq. (6-53) yields

$$I(s) = \frac{5}{s + 2} \tag{6-56}$$

Thus, by means of pair (T-4), we have

$$i(t) = 5\epsilon^{-2\tau} \tag{6-57}$$

Substitution of $E(s)$ into Eq. (6-54) yields for $V(s)$

$$V(s) = \frac{40}{s(s + 2)} = \frac{20}{s} - \frac{20}{s + 2} \tag{6-58}$$

Thus

$$v(t) = 20(1 - \epsilon^{-2t}) \tag{6-59}$$

Sketches of $i(t)$ and $v(t)$ are shown in Fig. 6-14a. It is observed that the steady-state response of $i(t)$ is zero, whereas the steady-state response of $v(t)$ is a constant 20 V.

(b) $e(t) = 20 \sin 2t$.

$$E(s) = \frac{40}{s^2 + 4} \tag{6-60}$$

$$I(s) = \frac{10s}{(s + 2)(s^2 + 4)} \tag{6-61}$$

We may readily determine the portion of the response due to the real pole of simple order by partial-fraction expansion. Calling this portion of the

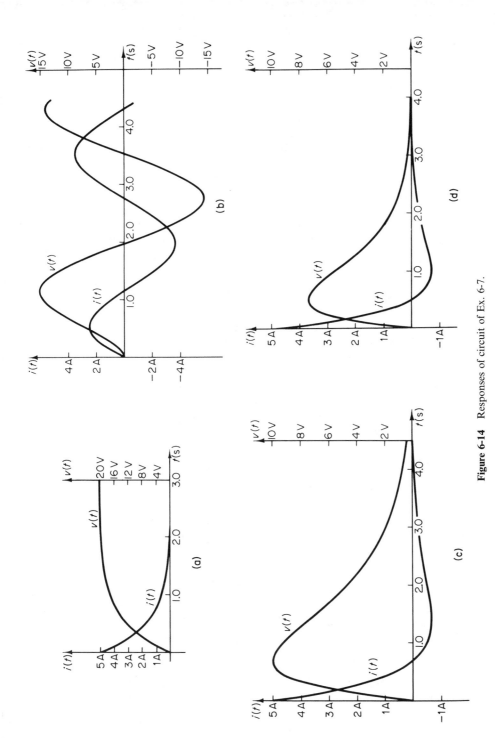

Figure 6-14 Responses of circuit of Ex. 6-7.

189

expression $I_1(s)$, we have

$$I_1(s) = \frac{A}{s+2} \tag{6-62}$$

$$A = \frac{10s}{s^2+4}\bigg]_{s=-2} = -\frac{5}{2} \tag{6-63}$$

$$i_1(t) = -\frac{5}{2}\epsilon^{-2t} \tag{6-64}$$

The sinusoidal component of the time response, which is due to the pair of imaginary roots, can be determined by the "trick" formula of Section 5-7. Letting $i_2(t)$ represent this function and employing the notation of that section, we have

$$Q(s) = \frac{10s}{s+2} \tag{6-65}$$

$$Q(j2) = \frac{20j}{2+j2} = \frac{10}{\sqrt{2}}\;\underline{/45°} \tag{6-66}$$

$$i_2(t) = \frac{5}{\sqrt{2}}\sin(2t+45°) \tag{6-67}$$

Thus,

$$i(t) = i_1(t) + i_2(t) = -\frac{5}{2}\epsilon^{-2t} + \frac{5}{\sqrt{2}}\sin(2t+45°) \tag{6-68}$$

In this case, we will determine $v(t)$ in the time domain

$$\begin{aligned}
v(t) &= 8\int_0^t i(t)\,dt \\
&= 8\int_0^t \left[-\frac{5}{2}\epsilon^{-2t} + \frac{5}{\sqrt{2}}\sin(2t+45°)\right]dt \\
&= \left[10\epsilon^{-2t} - \frac{20}{\sqrt{2}}\cos(2t+45°)\right]_0^t \\
&= 10[\epsilon^{-2t} - \sqrt{2}\cos(2t+45°)]
\end{aligned} \tag{6-69}$$

Sketches of $i(t)$ and $v(t)$ are shown in Fig. 6-14b.

The transient and steady-state portions of the responses are readily determined. After the exponential term becomes negligible, the remaining steady-state response in each case is a sinusoidal function. Thus

$$i_{ss}(t) = \frac{5}{\sqrt{2}}\sin(2t+45°) \tag{6-70}$$

and

$$v_{ss}(t) = -10\sqrt{2}\cos(2t+45°) \tag{6-71}$$

(c) $e(t) = 20\epsilon^{-t}$.

$$E(s) = \frac{20}{s + 1} \tag{6-72}$$

$$I(s) = \frac{5s}{(s + 1)(s + 2)} = \frac{-5}{s + 1} + \frac{10}{s + 2} \tag{6-73}$$

$$i(t) = 10\epsilon^{-2t} - 5\epsilon^{-t} = 5\epsilon^{-t}(2\epsilon^{-t} - 1) \tag{6-74}$$

$$V(s) = \frac{40}{(s + 1)(s + 2)} = \frac{40}{s + 1} - \frac{40}{s + 2} \tag{6-75}$$

$$v(t) = 40(\epsilon^{-t} - \epsilon^{-2t}) = 40\epsilon^{-t}(1 - \epsilon^{-t}) \tag{6-76}$$

Sketches of $i(t)$ and $v(t)$ are shown in Fig. 6-14c.

In this case, the entire responses are transient in nature, and the steady-state solution is zero as a result of the nature of the excitation.

(d) $e(t) = 20\epsilon^{-2t}$. The reader may wonder why we wish to consider another exponential function since, at first glance, this might appear to be a repeat of the type of solution in (c). However, further calculation will reveal the difference. In this case,

$$E(s) = \frac{20}{s + 2} \tag{6-77}$$

Substitution of $E(s)$ into Eq. (6-53) yields

$$I(s) = \frac{5s}{(s + 2)^2} \tag{6-78}$$

We note that in this case the time constant of the excitation coincides with the time constant of the circuit, thus producing poles of second order in the response. This is a special case of forced resonance which will be discussed more fully later. Referring to the procedure of Section 5-8, we have

$$I(s) = \frac{A_1}{(s + 2)^2} + \frac{A_2}{(s + 2)} \tag{6-79}$$

$$Q(s) = 5s \tag{6-80}$$

$$A_1 = 5s]_{s=-2} = -10 \tag{6-81}$$

$$A_2 = 5 \tag{6-82}$$

Thus

$$i(t) = -10t\epsilon^{-2t} + 5\epsilon^{-2t} = 5\epsilon^{-2t}(1 - 2t) \tag{6-83}$$

Substituting $E(s)$ into Eq. (6-54) we have for $V(s)$

$$V(s) = \frac{40}{(s + 2)^2} \tag{6-84}$$

In this case the expression is recognized to be of the form of (T-10). Thus

$$v(t) = 40t\epsilon^{-2t} \tag{6-85}$$

Sketches of $i(t)$ and $v(t)$ are shown in Fig. 6-14d. As in the case of (c), the entire responses are transient in nature.

Example 6-8

Consider the circuit of Example 6-7, which is shown again in Fig. 6-15a. However, now the capacitor is initially charged to 5 V in the direction shown. The circuit is excited at $t = 0$ by the exponential source shown. Solve for $i(t)$ and $v(t)$.

  Solution First, we transform the circuit as shown in Fig. 6-15b. The current in this case is given by

$$I(s) = \frac{[20/(s + 1)] + (5/s)}{4 + (8/s)} = \frac{5s}{(s + 1)(s + 2)} + \frac{1.25}{(s + 2)} \tag{6-86}$$

Although we could readily expand both terms in partial fraction expansions, if we refer back to part (c) of Example 6-7, we note that the first term is identical with the entire expression for $I(s)$ obtained in that case. Thus the second term in this case must be due to the initial voltage on the capacitor. The reader is invited to use the superposition principle on this circuit and verify that the second term is, in fact, equal to the current that would be produced by the 5-V source acting alone.

  In any event, a complete partial fraction expansion yields

$$I(s) = \frac{-5}{s + 1} + \frac{11.25}{s + 2} \tag{6-87}$$

Thus

$$i(t) = 11.25\epsilon^{-2t} - 5\epsilon^{-t} \tag{6-88}$$

Again, we will choose to find $V(s)$ and invert. We have from the transform circuit

$$V(s) = \frac{8}{s} I(s) - \frac{5}{s}$$

$$= \frac{40}{(s + 1)(s + 2)} + \frac{10}{s(s + 2)} - \frac{5}{s} \tag{6-89}$$

A partial fraction expansion reads

$$V(s) = \frac{40}{s + 1} - \frac{45}{s + 2} \tag{6-90}$$

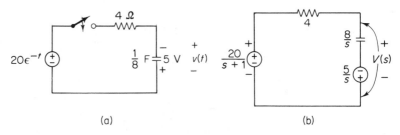

(a)   (b)

Figure 6-15 Circuit of Ex. 6-8.

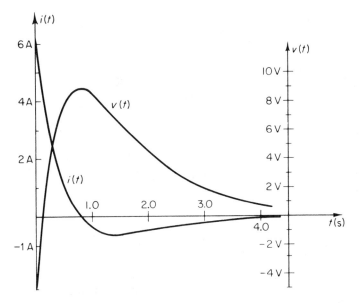

Figure 6-16   Responses of circuit of Ex. 6-8.

in which the net coefficient of the $1/s$ term turns out to be zero. Inversion yields

$$v(t) = 40\epsilon^{-t} - 45\epsilon^{-2t}$$
$$= 5\epsilon^{-t}(8 - 9\epsilon^{-t})$$

(6-91)

Sketches of $i(t)$ and $v(t)$ are shown in Fig. 6-16.

Example 6-9

The *RL* circuit of Fig. 6-17a is excited at $t = 0$ by a narrow voltage pulse whose magnitude is 1000 V and whose width is 10 ms as shown in (b).

(a) Using transform analysis, solve for the exact current response $i(t)$.

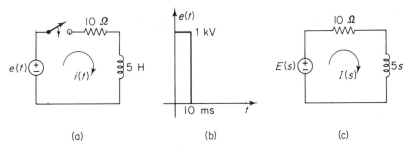

(a)   (b)   (c)

Figure 6-17   Circuit and excitation of Ex. 6-9.

(b) Justify using the impulse approximation and solve for the current response due to an impulse approximation of the excitation. Compare the two results.

Solution   (a) The transformed circuit is shown in Fig. 6-17c. To obtain $E(s)$, we first write $e(t)$ as the sum of two step functions.

$$e(t) = 1000u(t) - 1000u(t - 0.01)$$
$$= 1000[u(t) - u(t - 0.01)] \tag{6-92}$$

Using pair (T-2) and operation (O-3), we have

$$E(s) = \frac{1000}{s}[1 - \epsilon^{-0.01s}] \tag{6-93}$$

In terms of $E(s)$, the current is

$$I(s) = \frac{E(s)}{5s + 10} = \frac{200}{s(s + 2)}[1 - \epsilon^{-0.01s}] \tag{6-94}$$

The time response is recognized as being composed of a set of terms beginning at $t = 0$ and a set of terms beginning at $t = 0.01$ s.

Let us designate the first set of terms as $i_0(t)$ and the second set of terms as $i_1(t)$. Referring to their transforms, we have

$$I_0(s) = \frac{200}{s(s + 2)} = \frac{100}{s} - \frac{100}{s + 2} \tag{6-95}$$

and

$$i_0(t) = 100(1 - \epsilon^{-2t})u(t) \tag{6-96}$$

where the $u(t)$ has been used for clarity. Next we have

$$I_1(s) = \frac{-200\epsilon^{-0.01s}}{s(s + 1)} \tag{6-97}$$

To expand $I_1(s)$, we mentally remove the delay term until after the expansion has been carried out, and then place it as a factor for all terms. In other words, it should *not* be kept in the expression when we are determining the partial fraction coefficients, as this may lead to an erroneous interpretation. Thus

$$I_1(s) = -\frac{200}{s(s + 1)}\epsilon^{-0.01s} = -\left[\frac{100}{s} - \frac{100}{s + 2}\right]\epsilon^{-0.01s} \tag{6-98}$$

By means of (O-3) used in conjunction with the appropriate pairs, we have

$$i_1(t) = -100[1 - \epsilon^{-2(t - 0.01)}]u(t - 0.01) \tag{6-99}$$

Finally,

$$i(t) = i_0(t) + i_1(t)$$
$$= 100(1 - \epsilon^{-2t})u(t) - 100[1 - \epsilon^{-2(t - 0.01)}]u(t - 0.01) \tag{6-100}$$

If desired, we may write $i(t)$ in two different intervals as follows:

$0 < t < 0.01$ s:

$$i(t) = 100(1 - \epsilon^{-2t}) \tag{6-101}$$

$t > 0.01$ s:

$$i(t) = 100 - 100\epsilon^{-2t} - 100 + 100\epsilon^{-2t}\epsilon^{0.02}$$
$$= 100\epsilon^{-2t}(\epsilon^{0.02} - 1) \approx 2\epsilon^{-2t} \tag{6-102}$$

since

$$\epsilon^{0.02} - 1 \approx 0.02 \tag{6-103}$$

A sketch of $i(t)$ is shown in Fig. 6-18.

(b) Let us now investigate the possibility of approximating the narrow pulse as an impulse excitation. In previous situations of this sort, we have assumed that the width of pulses used were short in comparison to the circuit time constants. In this case, we observe that the time constant of the circuit is 0.5 s and that the pulse width is 0.01 s. Thus the time constant is 50 times the width of the pulse, and we may proceed to use the approximation. The area of the pulse is

$$\text{area} = 1000 \text{ V} \times 0.01 \text{ s} = 10 \text{ V·s} \tag{6-104}$$

Let us designate the impulse approximation to the source as $e_a(t)$ and the response to this impulse as $i_a(t)$. We have

$$e_a(t) = 10\delta(t) \tag{6-105}$$

$$E_a(s) = 10 \tag{6-106}$$

Referring to the circuit, we have

$$I_a(s) = \frac{10}{5s + 10} = \frac{2}{s + 2} \tag{6-107}$$

and

$$i_a(t) = 2\epsilon^{-2t} \tag{6-108}$$

The impulse response is shown on the same scale as the actual response in Fig. 6-18. From the figure and the results of part (a) of this problem, it is seen that the results coincide almost exactly for $t > 0.01$ s, whereas for $t < 0.01$ s, the two responses differ. However, we recognize that the time scale relative to

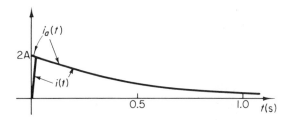

Figure 6-18 Responses of Ex. 6-9.

the time constant of the circuit would probably be the scale of interest in this problem, and hence, the deviation in the initial interval may not be important at all. The savings in computation by using the impulse approximation is obvious. Thus, along with other uses of the impulse function, it may be used as approximation to a very narrow pulse. From the point of view of reasonable accuracy, the shortest time constant of a circuit should be at least 10 to 20 times longer than the pulse width.

6-9 SERIES RLC CIRCUIT

In Section 6-8 our consideration was directed toward first-order circuits, in which case the configuration could be an RL or an RC form, but could never be a RLC or an LC form. In this section we begin considering second-order circuits, that is, circuits whose describing differential equation is of second order.

A second-order system may still be an RL or an RC form, or it may be an RLC form. Specifically in this section, we introduce the second-order circuit by means of a special case of importance, the series RLC circuit.

Consider the circuit shown in Fig. 6-19a, with no initial energy storage assumed, and its transform shown in (b). Depending on the desired quantity, we can solve for a transform response by writing a mesh current equation or by means of the impedance concept. The latter interpretation results in

$$I(s) = \frac{E(s)}{Z(s)} \tag{6-109}$$

where

$$
\begin{aligned}
Z(s) &= sL + R + \frac{1}{sC} \\
&= \frac{s^2 LC + sRC + 1}{sC} \\
&= \frac{s^2 + sR/L + 1/LC}{s/L}
\end{aligned}
\tag{6-110}
$$

Substitution of Eq. (6-110) into Eq. (6-109) yields

$$I(s) = \frac{sE(s)/L}{s^2 + sR/L + 1/LC} \tag{6-111}$$

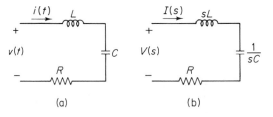

Figure 6-19 Series RLC circuit and its transform.

The poles of $I(s)$, which determine the form of the time response, are determined by the poles of $E(s)$ and the roots of the quadratic $s^2 + sR/L + 1/RC$. The latter roots are the zeros (numerator roots) of the impedance $Z(s)$. Since $E(s)$ may be almost anything in general, let us turn our attention at the moment to the form of the transient response of the network due primarily to the poles of the circuit. In this case we assume a step function excitation because of its practical significance and because the simplest possible expression for $I(s)$ will result.

Therefore, let us assume that $e(t)$ is a dc voltage of E volts applied at $t = 0$. The transform of the excitation is $E(s) = E/s$. Substituting $E(s)$ into Eq. (6-111) results in

$$I(s) = \frac{E/L}{s^2 + s(R/L) + 1/LC} \tag{6-112}$$

Notice that the pole, $s = 0$, of $E(s)$ canceled the s term in the numerator of Eq. (6-111). The poles due to the network are determined from the equation

$$s^2 + s\frac{R}{L} + \frac{1}{LC} = 0 \tag{6-113}$$

Letting s_1 and s_2 represent these poles, we have

$$\begin{cases} s_1 \\ s_2 \end{cases} = \frac{-R}{2L} \pm \sqrt{\left(\frac{R}{2L}\right)^2 - \frac{1}{LC}} \tag{6-114}$$

We need to consider three separate cases:

1. *Overdamped.* If $R/2L > 1/\sqrt{LC}$, the roots are both real, negative in sign, and of simple order. The circuit is said to be *overdamped*. In this case, we may write

$$I(s) = \frac{E/L}{(s + \alpha_1)(s + \alpha_2)} = \frac{A_1}{s + \alpha_1} + \frac{A_2}{s + \alpha_2} \tag{6-115}$$

where $\alpha_1 = -s_1$ and $\alpha_2 = -s_2$. The coefficients A_1 and A_2 can be readily determined by partial fraction expansion. The time response is then of the form

$$i(t) = A_1 \epsilon^{-\alpha_1 t} + A_2 \epsilon^{-\alpha_2 t} \tag{6-116}$$

Thus, the overdamped second-order system produces two separate time constants in the time response. A typical response is shown in Fig. 6-20a.

2. *Critically damped.* If $R/2L = 1/\sqrt{LC}$, the roots are both real, negative in sign, and equal. The circuit is said to be *critically damped*. In this case, we may write

$$I(s) = \frac{E/L}{(s + \alpha)^2} \tag{6-117}$$

where $\alpha = -R/2L$. By means of (T-10), we determine the time response to be

$$i(t) = \frac{Et}{L} \epsilon^{-\alpha t} \tag{6-118}$$

A typical response is shown in Fig. 6-20b.

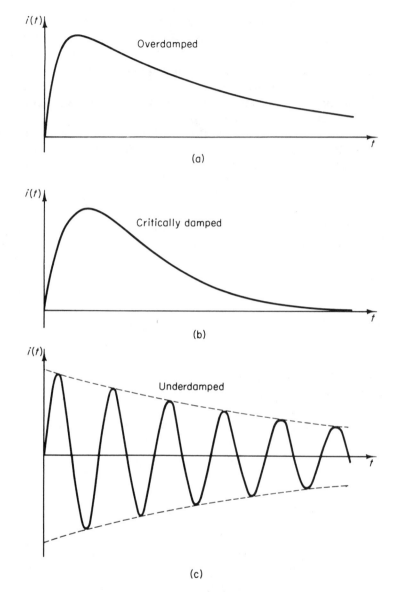

Figure 6-20   Typical step responses of series RLC circuit.

3. *Underdamped.* If $R/2L < 1/\sqrt{LC}$, the roots are complex and of first order with negative real parts. In this case, the roots may be written in the form

$$\begin{cases} s_1 \\ s_2 \end{cases} = -\frac{R}{2L} \pm j\,\sqrt{\frac{1}{LC} - \left(\frac{R}{2L}\right)^2} \tag{6-119}$$

Since the real part corresponds to a damping constant and the imaginary part corresponds to an oscillatory response, let us define some useful terms. Let

$$\alpha = \frac{R}{2L} = \text{damping constant} \tag{6-120}$$

$$\omega_0 = \frac{1}{\sqrt{LC}} = \text{undamped natural resonant frequency} \tag{6-121}$$

$$\omega_d = \sqrt{\omega_0^2 - \alpha^2} = \text{damped natural resonant frequency} \tag{6-122}$$

The quantity ω_0 is the angular frequency of oscillation if there were no resistance in the circuit (i.e., $R = 0$). However, the damped frequency ω_d is always less than the undamped frequency, as can be seen from Eq. (6-122).

For the underdamped case, we may write

$$I(s) = \frac{E/L}{(s + \alpha)^2 + \omega_d^2} \tag{6-123}$$

By means of (T-7), the time response is readily determined to be

$$i(t) = \frac{E\epsilon^{-\alpha t}}{\omega_d L} \sin \omega_d t \tag{6-124}$$

A typical response is shown in Fig. 6-20c.

It is an interesting problem to investigate how the response changes as the damping factor is increased. Consider the circuit of Fig. 6-21 in which L and C are fixed but R is adjustable. Thus ω_0 is fixed, but α varies directly with R and ω_d decreases with an increase of R, according to Eq. (6-122). The results are shown in Fig. 6-22 for some different ratios of α to ω_0. The circuit becomes overdamped for $\alpha/\omega_0 = \zeta = 1$.

When an *RLC* circuit is excited by a more general excitation, the response will consist of two parts. The *natural* part will be due to the circuit itself and will always be similar to one of the forms discussed in this section, depending on whether the circuit is underdamped, critically damped, or overdamped. As long as there is any resistance at all in the circuit, this response will be transient in nature and will disappear after a sufficiently long time, as can be seen from the previous figures. The *forced* part of the response will be due to the nature of the source, and if the source is such as to maintain a response after the transient disappears, such response is, of course, the steady-state response as previously discussed.

In this section we have considered the behavior of *natural resonance* in a typical *RLC* circuit. The reader should not confuse this concept with the closely related concept of *forced resonance* in the sinusoidal steady-state analysis of ac circuits. In

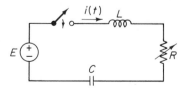

Figure 6-21   Circuit in which the damping is to varied.

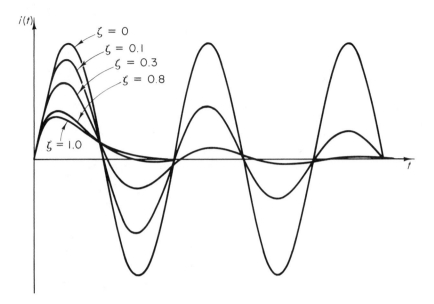

Figure 6-22   Step responses of series *RLC* circuit for different values of damping.

natural resonance, we deal with an oscillation whose form is produced by the circuit itself without regard to the type of excitation. In forced resonance, we deal with circuits having specific steady-state impedance properties, such as minimum or maximum impedances, unity power factor, etc. In fact, the forced steady-state resonant angular frequency of a series *RLC* circuit is usually defined by Eq. (6-121) irrespective of the amount of resistance present. At this frequency, the steady-state reactances cancel, and the impedance is minimum. However, the natural resonance concepts under consideration have a different meaning. The adjectives *natural* and *forced* should help to clarify the meanings.

Example 6-10

The relaxed series *RLC* circuit of Fig. 6-23a is excited at $t = 0$ by the sinusoidal source shown. Solve for the current $i(t)$ for $t > 0$.

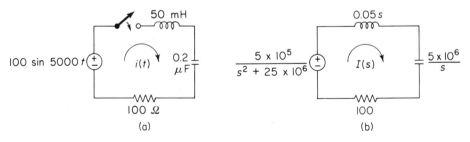

Figure 6-23   Circuit of Ex. 6-10.

Solution   Although the mathematics will eventually reveal the type of response, a preliminary calculation should prove interesting. We will first calculate $R/2L$ and $1/\sqrt{LC}$.

$$\frac{R}{2L} = \frac{100}{2 \times 0.05} = 10^3 \tag{6-125a}$$

$$\frac{1}{\sqrt{LC}} = \frac{1}{\sqrt{0.05 \times 0.2 \times 10^{-6}}} = 10^4 \tag{6-125b}$$

Since $R/2L < 1/\sqrt{LC}$, the circuit is underdamped and oscillatory. We have

$$\alpha = 10^3 \text{ nepers}^\dagger \tag{6-126}$$

$$\omega_0 = 10^4 \text{ rad/s} \tag{6-127}$$

$$\omega_d = \sqrt{\omega_0^2 - \alpha^2} = 9.95 \times 10^3 \text{ rad/s} \tag{6-128}$$

As a result of the relatively small amount of damping, the damped resonant frequency differs from the undamped resonant frequency by only 0.5%. As a matter of interest, the damped repetition frequency is $f_d = \omega_d/2\pi = 1548$ Hz. Notice that the natural damped frequency is about twice the frequency of the excitation. Again, we point out that these preliminary calculations are not absolutely necessary as the results will "fall out" of the math that follows.

The transformed circuit is shown in Fig. 6-23b. Using the impedance concept, we have

$$Z(s) = 0.05s + 100 + \frac{5 \times 10^6}{s}$$

$$= \frac{0.05s^2 + 100s + 5 \times 10^6}{s} \tag{6-129}$$

$$= \frac{s^2 + 2000s + 10^8}{20s}$$

The current is

$$I(s) = \frac{E(s)}{Z(s)} = \frac{10^7 s}{(s^2 + 25 \times 10^6)(s^2 + 2000s + 10^8)} \tag{6-130}$$

The poles due to the quadratic with three terms are

$$\begin{cases} s_1 \\ s_2 \end{cases} = -10^3 \pm j9.95 \times 10^3 \tag{6-131}$$

which agrees with our preliminary calculations.

We obtain the final desired result by finding the inverse transform of $I(s)$. Since one quadratic has imaginary roots and the other has complex roots, we

---

†Strictly speaking, both ω_0 and α have dimensions of time^{-1}, since radians are dimensionless. However, since ω_0 and α define different phenomena, it has been common practice to use a different unit for a damping constant. The common convention is to employ the unit *neper*.

may invert the function by applying the special formula of Section 5-7 individually to the two quadratic factors. The reader is invited to show that the result is

$$i(t) = 0.133\epsilon^{-1000t} \sin(9.95 \times 10^3 t - 99.51°)$$
$$+ 0.132 \sin(5000t + 82.41°) \tag{6-132}$$

The response is seen to consist of a damped sinusoidal term whose frequency is the natural damped resonant frequency of the circuit, and an undamped sinusoid whose frequency is that of the excitation. The former term is transient in nature, whereas the latter term is the steady-state response. After the transient disappears, the steady-state or forced response is

$$i_{ss}(t) = 0.132 \sin(5000t + 82.41°) \tag{6-133}$$

6-10 PARALLEL RLC CIRCUIT

In this section we turn our attention to another special case of a second-order circuit, the parallel *RLC* configuration. The reader familiar with the duality principle may recognize that the parallel *RLC* circuit is the dual of the series *RLC* circuit, and many of its properties could be deduced directly from the work of the preceding section. However, we develop the properties independently for the benefit of readers not familiar with this principle.

Consider the circuit shown in Fig. 6-24a, with no initial energy storage assumed, and its transform shown in (b). In analyzing parallel circuits such as this, it is better to assume current source excitation. Thus if the circuit is excited by a voltage source in series with a resistance, the source can be converted to an equivalent current source for analysis purposes. If the network is excited by a source that approximates an ideal voltage source, we could analyze the circuit by determining the individual currents by means of the basic voltage–current relationships for the elements.

The admittance of the network is given by

$$Y(s) = sC + \frac{1}{R} + \frac{1}{sL} = \frac{s^2LC + sL/R + 1}{sL}$$
$$= \frac{s^2 + s/RC + 1/LC}{s/C} \tag{6-134}$$

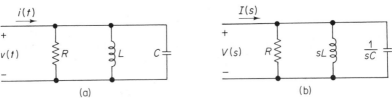

(a)   (b)

Figure 6-24 Parallel *RLC* circuit and its transform.

The impedance is

$$Z(s) = \frac{s/C}{s^2 + s/RC + 1/LC} \tag{6-135}$$

If a current $I(s)$ excites the network, the resulting voltage is

$$V(s) = Z(s)I(s) = \frac{sI(s)/C}{s^2 + s/RC + 1/LC} \tag{6-136}$$

As in the preceding section, let us direct our attention to the step response. Letting $I(s) = I/s$, we have

$$V(s) = \frac{I/C}{s^2 + s/RC + 1/LC} \tag{6-137}$$

We determine the poles by factoring the denominator polynomial.

$$\begin{cases} s_1 \\ s_2 \end{cases} = -\frac{1}{2RC} \pm \sqrt{\left(\frac{1}{2RC}\right)^2 - \frac{1}{LC}} \tag{6-138}$$

As in the case of the series RLC circuit, we have three possibilities:

1. *Overdamped*

$$\frac{1}{2RC} > \frac{1}{\sqrt{LC}} \tag{6-139}$$

2. *Critically damped*

$$\frac{1}{2RC} = \frac{1}{\sqrt{LC}} \tag{6-140}$$

3. *Underdamped*

$$\frac{1}{2RC} < \frac{1}{\sqrt{LC}} \tag{6-141}$$

In the underdamped case, we may define the undamped resonant frequency as

$$\omega_0 = \frac{1}{\sqrt{LC}} \tag{6-142}$$

Note that ω_0 is the same as in the series case. The damping factor in this case, however, is different and is

$$\alpha = \frac{1}{2RC} \tag{6-143}$$

As in the series case, the damped resonant frequency is

$$\omega_d = \sqrt{\omega_0^2 - \alpha^2} \tag{6-144}$$

The relative shapes and forms of the three possible types of voltage responses in this case are analogous to the current responses of the series RLC circuit of the preceding section.

Example 6-11

The relaxed circuit of Fig. 6-25a is excited at $t = 0$ by a pulse which approximates an impulse of area 100 V·s. Determine the voltage across the tuned circuit, $v(t)$, for $t > 0$.

    Solution   The first step in solving the problem is to rearrange the circuit in the simplest form for analysis. We do this by converting the impulse-voltage source to an impulse-current source and combining the two resistors. The result is shown in Fig. 6-25b and its transform is shown in (c). The admittance is

$$Y(s) = \frac{1}{20}s + \frac{1}{2} + \frac{4}{5s} = \frac{s^2 + 10s + 16}{20s} \tag{6-145}$$

$$Z(s) = \frac{20s}{s^2 + 10s + 16} \tag{6-146}$$

The voltage $V(s)$ is

$$V(s) = Z(s)I(s) = \frac{500s}{s^2 + 10s + 16} \tag{6-147}$$

The poles are

$$\begin{cases} s_1 \\ s_2 \end{cases} = -5 \pm \sqrt{25 - 16} = -5 \pm 3 = -2 \quad \text{and} \quad -8 \tag{6-148}$$

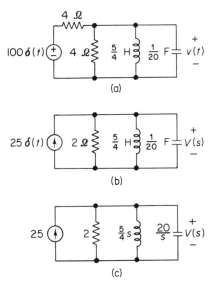

(a)

(b)

(c)

Figure 6-25   Circuits of Ex. 6-11.

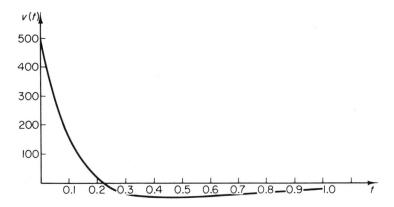

Figure 6-26   Responses of the circuit of Example 6-11.

Thus the response in this case is overdamped. We may write $V(s)$ as

$$V(s) = \frac{500s}{(s + 2)(s + 8)} = \frac{-500/3}{s + 2} + \frac{2000/3}{s + 8} \tag{6-149}$$

The time response is thus

$$v(t) = \frac{500}{3} \left[4\epsilon^{-8t} - \epsilon^{-2t} \right] \tag{6-150}$$

A sketch of the response is shown in Fig. 6-26.

6-11 REDUNDANCY AND HIGHER ORDER

In the preceding two sections we have considered two special common cases of second-order circuits, the series RLC and parallel RLC circuits. We saw that the response of such circuits could be either underdamped and oscillatory, overdamped, or critically damped. Practically speaking, component values are always subject to deviation, and thus the exact critically damped case probably occurs very rarely in practice. However, the relative shape of the time response curve varies so little as the system crosses the critically damped point that the assumption of such a response can be quite accurate.

There are many other forms for second-order RLC circuits, some of which are shown in Fig. 6-27. No generators are shown, as they could be connected in almost

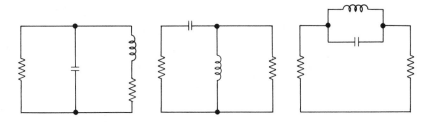

Figure 6-27   Some possible forms for second-order circuits.

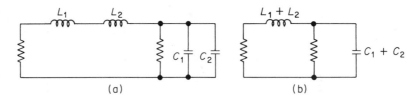

Figure 6-28 All illustration of redundancy in a circuit.

any imaginable manner. The reader might observe one important feature of all of these circuits which is identical with the series and parallel *RLC* circuits previously considered. All of these circuits contain one capacitor and one inductor. The question then arises as to whether or not there is a relationship between the number of energy-storage elements and the order of the system. The reader will recall that all first-order systems considered either contained only one reactive element or could be reduced to a system with only one reactive element. This reduction idea can be explained by the concept of redundancy. *A circuit containing two or more circuit components of the same type is said to be redundant if these components can be reduced to a single equivalent component under all conditions external to their terminals.*

Referring to Fig. 6-28a, we see that one of the two inductors, L_1 and L_2, is redundant since they can be reduced to a single inductor of value $L_1 + L_2$, as shown in (b). Similarly, the capacitors C_1 and C_2 can be reduced to a single capacitor of value $C_1 + C_2$, as shown in (b). Thus although this circuit might have appeared at first glance to be a higher-order circuit, it is really a second-order circuit after redundancy is considered.

However, the circuit of Fig. 6-29 is one in which no redundancy exists. Nowhere in the circuit are there two or more components that can be replaced by a single component.

The most common form of redundancy is either simple series or parellel connections of like components; thus such redundancy may be easily detected.

Some formal rules may now be stated: *The order of a circuit is the number of energy-storage elements after redundancy is removed.* Furthermore, *the order of the circuit is the order of the denominator polynomial of a given response when the only excitations in the circuit are impulse functions.* In effect, this latter statement says that poles due to excitations must not be counted in determining the order of a circuit from a transform function.

We may thus realize second- or higher-order circuits using only *RC* or *RL* forms. In other words, it is not necessary to have an *RLC* form to achieve a second- or higher-order circuit. Thus a circuit containing two nonredundant capacitors will be a second-order form. However, there is a basic limitation on the pole values of such a circuit containing only one type of energy-storage element. It can be shown that

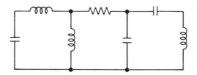

Figure 6-29 A circuit without redundancy.

the impulse response of a circuit without electronic feedback and containing only one type of energy-storage component (i.e., either RC or RL circuits) will always be either overdamped or critically damped. This is equivalent to saying that *the poles of the impulse response of such a circuit will always be real numbers and are never complex numbers.* In other words, to achieve an underdamped or oscillatory response in a circuit without feedback, it is necessary to employ both inductance and capacitance in the circuit. Since resistance is usually present, it is always realistic to refer to the latter form as an *RLC* circuit.

The reader might observe that we have excluded the possibility of electronic feedback. With feedback, it is possible to achieve an oscillatory response even with an *RC* form. Such considerations will be pursued somewhat in the next chapter.

In this chapter we have studied the general approaches for transforming a complete circuit so that it could be solved by transform methods. Furthermore, we have investigated in some detail first-order and second-order circuits and have obtained some complete solutions for such cases. In theory, we should be prepared to solve a more complex higher-order circuit by extending the basic network equations to such a problem.

In practice, the principal limitation to such higher-order solutions is the complexity of the computations. One of the principal problems that arises is the determination of the roots of the denominator polynomial for certain higher-order systems.

Modern technology provides the engineer and technologist with many computing facilities such as microcomputers and mainframes. The author feels that, beyond a certain order of complexity, it is best to depend, at least partially, on such aids. For example, most digital computers have standard numerical subroutines for determining the roots of higher-degree polynomials. Thus, during this phase of a problem solution, one could rely on a digital computer to actually determine the roots of a particular denominator polynomial.

6-12 SOLUTION FROM THE DIFFERENTIAL EQUATION

In many cases, the circuit analyst will have occasion to solve a given differential equation by transform methods. In this case, the circuit itself is not given, only the resulting differential equation and some initial conditions. In fact, the differential equation may not even represent a circuit but, perhaps, some other physical system entirely. In this section we consider how the transform process may be applied to such a representation.

The Laplace transform approach is best suited to ordinary linear differential equations of the constant-coefficient type. Such an equation of order m appears in the form

$$b_m \frac{d^m y}{dt^m} + b_{m-1} \frac{d^{m-1} y}{dt^{m-1}} + \cdots + b_0 y = f(t) \qquad (6\text{-}151)$$

The equation is linear since there are no products of y and itself, or products of various derivatives, and because the coefficients are constant. The quantity $f(t)$ represents a functional combination of all excitations in the system. In Chapter 7 this quantity will be expanded to yield a different form. But for the purpose at hand, this form is adequate.

In working directly with circuit components in this chapter, we have never had to transform a derivative of any order higher than the first. The transform of the first derivative was stated in Chapter 5 to be

$$\mathscr{L}\left[\frac{dy}{dt}\right] = sY(s) - y(0) \tag{6-152}$$

To transform Eq. (6-151) we need an expression for the transform of higher-order derivatives. This can be readily deduced if we note that

$$\frac{d^2y}{dt^2} = \frac{d}{dt}\left[\frac{dy}{dt}\right] \tag{6-153}$$

In other words, the second derivative is the first derivative of the first derivative; hence operation (O-1) can be interpreted for the second derivative if $y(t)$ is replaced by dy/dt. Thus

$$\mathscr{L}\left[\frac{d^2y}{dt^2}\right] = s\mathscr{L}\left[\frac{dy}{dt}\right] - y'(0)$$
$$= s^2 Y(s) - sy(0) - y'(0) \tag{6-154}$$

where the prime signifies a derivative. A similar procedure for the third derivative yields

$$\mathscr{L}\left[\frac{d^3y}{dt^3}\right] = s\mathscr{L}\left[\frac{d^2y}{dt^2}\right] - y''(0)$$
$$= s^3 Y(s) - s^2 y(0) - sy'(0) - y''(0) \tag{6-155}$$

In general,

$$\mathscr{L}\left[\frac{d^k y}{dt^k}\right] = s^k Y(s) - s^{k-1}y(0) - s^{k-2}y'(0) - \cdots - y^{k-1}(0) \tag{6-156}$$

From the relationships above, it is clear that the transform of a kth-order derivative requires specification of the function and its first $k - 1$ derivatives at $t = 0$, or a total of k quantities. Since the highest-order derivative requires the most information, it is deduced that a differential equation of order m can be uniquely solved by a knowledge of the desired function and its first $m - 1$ derivatives evaluated at $t = 0$. The point $t = 0$ is usually interpreted to mean $t = 0^+$ in the sense that initial conditions are specified after the excitation is applied.

After the differential equation is transformed, $Y(s)$ may be determined by algebraic manipulation using the methods of Chapter 5. This phase of the problem is identical with the approach we have already considered.

In many cases, a system of simultaneous differential equations in terms of several variables will be given rather than a single differential equation. The same rule on the number of equations required holds as for the case of simultaneous algebraic equations. In other words, if there are N unknowns, it is necessary to have N simultaneous equations. Furthermore, each equation requires the specification of a number of initial conditions equal to its order.

Example 6-12

The response of a given physical system is described for $t > 0$ by the differential equation

$$4\frac{d^2y}{dt^2} + 24\frac{dy}{dt} + 32y = 100 \tag{6-157}$$

The initial values of y and dy/dt are

$$y(0) = +10 \tag{6-158}$$

and

$$y'(0) = -20 \tag{6-159}$$

Solve for $y(t)$, for $t > 0$, using Laplace transforms.

Solution   Transforming the equation according to Eqs. (6-154) and (6-152), we have

$$4[s^2Y(s) - s(10) - (-20)] + 24[sY(s) - (10)] + 32Y(s) = \frac{100}{s} \tag{6-160}$$

After some rearrangement, we have

$$Y(s)[s^2 + 6s + 8] = \frac{25}{s} + 10s + 40 \tag{6-161}$$

or

$$Y(s) = \frac{25}{s(s^2 + 6s + 8)} + \frac{10s + 40}{s^2 + 6s + 8} \tag{6-162}$$

The reader is invited to complete the problem and obtain the solution:

$$y(t) = \frac{5}{8}(5 + 6\epsilon^{-2t} + 5\epsilon^{-4t}) \tag{6-163}$$

The reader might also check that the two initial conditions are satisfied by the solution.

Example 6-13

The initially relaxed circuit of Fig. 6-30a is excited at $t = 0$ by the dc voltage shown. Rather than transforming the circuit directly according to the early work of this chapter, write a set of simultaneous mesh differential equations in the time domain, deduce the necessary initial conditions, and solve for $i_2(t)$ by transform methods.

Solution   The simultaneous mesh equations are

$$2i_1 + 8\int_0^t i_1\, dt + 2\left[\frac{di_1}{dt} - \frac{di_2}{dt}\right] = 10 \tag{6-164}$$

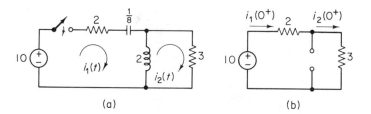

Figure 6-30   Circuit of Ex. 6-13.

and

$$2\left[\frac{di_2}{dt} - \frac{di_1}{dt}\right] + 3i_2 = 0 \tag{6-165}$$

Since the highest-order derivative in both equations is the first derivative, it is only necessary to know $i_1(0^+)$ and $i_2(0^+)$. These quantities are readily deduced from the initial circuit shown in Fig. 6-30b. We have

$$i_1(0^+) = i_2(0^+) = 2 \text{ A} \tag{6-166}$$

Using operation (O-2) and transform (T-2), we have

$$2I_1(s) + \frac{8}{s}I_1 + 2sI_1(s) - 4 - 2sI_2(s) + 4 = \frac{10}{s} \tag{6-167}$$

and

$$2sI_2(s) - 4 - 2sI_1(s) + 4 + 3I_2(s) = 0 \tag{6-168}$$

Simultaneous algebraic solution of these equations for $I_2(s)$ yields

$$I_2(s) = \frac{20s}{10s^2 + 22s + 24}$$
$$= \frac{2s}{s^2 + 2.2s + 2.4} \tag{6-169}$$

Inversion of $I_2(s)$ is accomplished by means of the procedure of Section 5-7 since the poles are complex. The reader is invited to verify that

$$i_2(t) = 2.84\epsilon^{-1.1t} \sin (1.09t + 135.2°) \tag{6-170}$$

GENERAL PROBLEMS

6-1. An uncharged capacitance of 0.2 F is given. Write expressions for both the transform impedance and admittance. Draw the transform-domain equivalent circuit and label with the impedance value.

6-2. An uncharged capacitance of 0.5 μF is given. Write expressions for both the transform impedance and admittance. Draw the transform-domain equivalent circuit and label with the impedance value.

6-3. An unfluxed inductance of 4 H is given. Write expressions for both the transform impedance and admittance. Draw the transform-domain equivalent circuit and label with the impedance value.

6-4. An unfluxed inductance of 80 mH is given. Write expressions for both the transform impedance and admittance. Draw the transform-domain equivalent circuit and label with the impedance value.

6-5. For the charged capacitor of Fig. P6-5, draw both the Thévenin and Norton equivalent circuits in the transform domain.

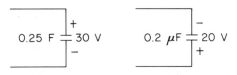

Figure P6-5   **Figure P6-6**

6-6. For the charged capacitor of Fig. P6-6, draw both the Thévenin and Norton equivalent circuits in the transform domain.

6-7. For the fluxed inductor of Fig. P6-7, draw both the Thévenin and Norton equivalent circuits in the transform domain.

Figure P6-7   **Figure P6-8**

6-8. For the fluxed inductor of Fig. P6-8, draw both the Thévenin and Norton equivalent circuits in the transform domain.

6-9. For the circuit of Fig. P6-9, draw the transform-domain equivalent circuit in a form suitable for writing mesh current equations. Write the two transform-domain mesh current equations.

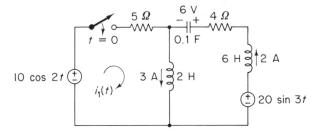

Figure P6-9

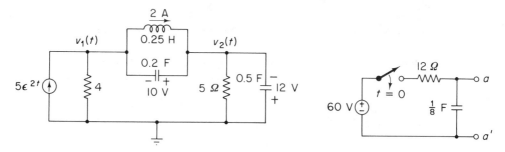

Figure P6-10

Figure P6-11

6-10. For the circuit of Fig. P6-10, draw the transform-domain equivalent circuit in a form suitable for writing node voltage equations. Write the two transform-domain node voltage equations.

6-11. (a) Obtain a transform Thévenin equivalent circuit at terminals a-a' for the circuit of Fig. P6-11 if the switch is closed at $t = 0$.

(b) Using the result of (a), determine the time-domain current $i(t)$ that would flow through a 6-Ω resistance connected to the terminals a-a' at $t = 0$.

6-12. Assume in Problem 6-11 that the switch has been closed a long time before $t = 0$ so that the circuit to the left of a-a' is in a steady-state condition. Repeat the analysis of Problem 6-11. It is assumed again that the 6-Ω resistance is connected at $t = 0$.

6-13. Determine the transform impedance $Z(s)$ seen at the terminals of the circuit of Fig. P6-13. Arrange as a ratio of polynomials.

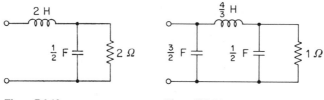

Figure P6-13

Figure P6-14

6-14. Determine the transform impedance $Z(s)$ seen at the terminals of the circuit of Fig. P6-14. Arrange as a ratio of polynomials

6-15. Assume that the circuit of Problem 6-13 is excited at $t = 0$ by a dc voltage source of value 10 V. Determine an expression for the transform current $I(s)$ that would flow from the source.

6-16. Assume that the circuit of Problem 16-14 is excited at $t = 0$ by a dc current source of value 5 A. Determine an expression for the transform voltage that would appear across the source.

In Problems 6-17 through 6-20, (a) write the forms of the inverse transforms of the functions with arbitrary coefficients assumed. In each case, $N(s)$ is some arbitrary numerator polynomial whose roots are assumed not to coincide with any roots of the denominator. (b) In each case, identify the transient and steady-state response terms.

6-17. $V(s) = \dfrac{N(s)}{s(s^2 + 4s + 3)}$

6-18. $I(s) = \dfrac{N(s)}{s(s^2 + 4s + 13)}$

6-19. $V(s) = \dfrac{N(s)}{(s^2 + 6s + 25)(s^2 + 4)}$ **6-20.** $I(s) = \dfrac{N(s)}{(s^2 + 2000s + 5 \times 10^6)(s^2 + 10^6)}$

6-21. For the circuit of Fig. P6-21, determine $i(t)$ and $v_C(t)$ if $e(t) = 10$ V. Identify transient and steady-state portions of the response.

6-22. Repeat the analysis of Problem 6-21 if $e(t) = 10 \sin 2t$.

6-23. Repeat the analysis of Problem 6-21 if $e(t) = 10 \cos 5t$.

6-24. Repeat the anaylsis of Problem 6-21 if $e(t) = 10\epsilon^{-5t}$.

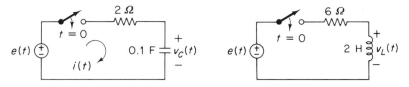

Figure P6-21   Figure P6-25

6-25. For the circuit of Fig. P6-25, determine $i(t)$ and $v_L(t)$ if $e(t) = 30$ V. Identify transient and steady-state portions of the response.

6-26. Repeat the analysis of Problem 6-25 if $e(t) = 30 \sin 2t$.

6-27. Repeat the analysis of Problem 6-25 if $e(t) = 30 \cos 4t$.

6-28. Repeat the analysis of Problem 6-25 if $e(t) = 30\delta(t)$.

6-29. The circuit of Fig. P6-29 is in a steady-state condition at $t = 0^-$. At $t = 0$, the switch is thrown to position 2. Solve for $v(t)$ for $t > 0$.

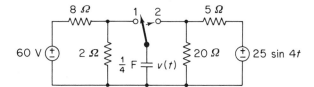

Figure P6-29

6-30. The circuit shown in Fig. P6-30a is excited at $t = 0$ by the narrow current pulse shown in (b).

    (a) Solve for $v(t)$ by an exact analysis.

    (b) Justify that the pulse may be approximated by an impulse source, and solve for the response due to this impulse. Compare the results.

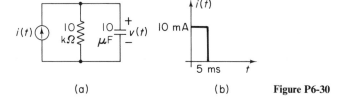

(a)   (b)   Figure P6-30

6-31. A series RLC circuit has $R = 2.4\ \Omega$, $L = 0.4$ H, and $C = 0.1$ F. Indicate whether the circuit is overdamped, critically damped, or underdamped. If the circuit is overdamped or critically damped, determine the damping factor(s). If underdamped, determine the damping factor, undamped natural frequency, and damped natural frequency.

6-32. Repeat the analysis of Problem 6-31 if $R = 5$ kΩ, $L = 2.5$ H, and $C = 0.4$ μF.

6-33. Repeat the analysis of Problem 6-31 if $R = 1.5$ kΩ, $L = 0.5$ H, and $C = 1$ μF.

6-34. The current in the series RLC circuit shown in Fig. P6-34 is know to be $i(t) = 2t$. Solve for $v(t)$ by any method. (*Hint:* This has been presented to refresh the reader's thinking on a basic point.)

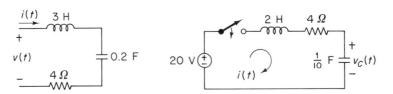

Figure P6-34   Figure P6-35

6-35. The switch in the circuit of Fig. P6-35 is closed at $t = 0$, and the circuit is initially relaxed. Determine the current $i(t)$ and the voltage $v_C(t)$.

6-36. The switch in the circuit of Fig. P6-36 is closed at $t = 0$, and the circuit is initially relaxed. Determine the current $i(t)$ and the voltage $v_C(t)$.

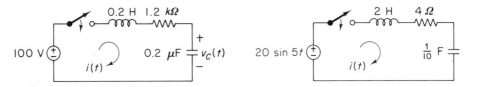

Figure P6-36   Figure P6-37

6-37. The switch in the circuit of Fig. P6-37 is closed at $t = 0$, and the circuit is initially relaxed. Determine the current $i(t)$.

6-38. The switch in the circuit of Fig. P6-38 is closed at $t = 0$, and the circuit is initially relaxed. Determine the current $i(t)$.

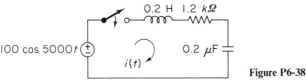

Figure P6-38

6-39. A parallel RLC circuit has $R = 1$ kΩ, $L = 2.5$ H, and $C = 0.1$ μF. Indicate whether the circuit is overdamped, critically damped, or underdamped. If the circuit is overdamped or critically damped, determine the damping factor(s). If underdamped, determine damping factor, undamped natural frequency, and damped natural frequency.

6-40. Repeat the analysis of Problem 6-39 if $R = 500$ Ω, $L = 0.1$ H, and $C = 2$ μF.

6-41. Repeat the analysis of Problem 6-39 if $R = 20$ kΩ, $L = 0.1$ H, and $C = 0.001$ μF.

6-42. The voltage across the parallel *RLC* circuit shown in Fig. P6-42 is known to be $v(t) = 40t$. Solve for $i(t)$ by any method. (*Hint:* This has been presented to refresh the reader's thinking on a basic point.)

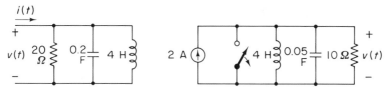

Figure P6-42   Figure P6-43

6-43. The switch in the circuit of Fig. P6-43 is opened at $t = 0$, and the circuit is initially relaxed. Determine the voltage $v(t)$.

6-44. The switch in the circuit of Fig. P6-44 is opened at $t = 0$, and the circuit is initially relaxed. Determine the voltage $v(t)$.

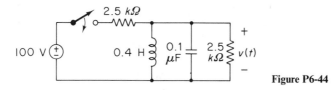

Figure P6-44

6-45. Determine the order of the circuit shown in Fig. P6-45.

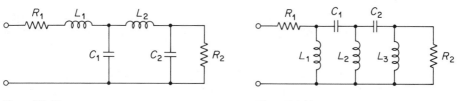

Figure P6-45   Figure P6-46

6-46. Determine the order of the circuit shown in Fig. P6-46.

6-47. Determine the order of the circuit shown in Fig. P6-47.

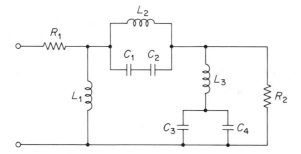

Figure P6-47

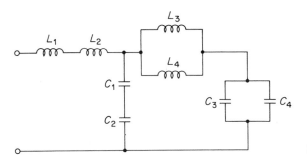

Figure P6-48

6-48. Determine the order of the circuit shown in Fig. P6-48.

6-49. The circuit of Fig. P6-49 is one form of a low-pass second-order passive Butterworth filter normalized to a cutoff frequency of 1 rad/s. The circuit is excited at $t = 0$ by the dc voltage shown. Determine the output voltage $v(t)$.

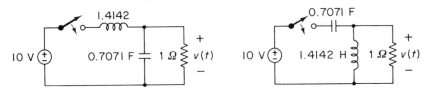

Figure P6-49   **Figure P6-50**

6-50. The circuit of Fig. P6-50 is one form of a high-pass second-order passive Butterworth filter normalized to a cutoff frequency of 1 rad/s. The circuit is excited at $t = 0$ by the dc voltage shown. Determine the output voltage $v(t)$.

6-51. The circuit of Fig. P6-51 is one form of a low-pass third-order passive Butterworth filter normalized to a cutoff frequency of 1 rad/s. The circuit is excited at $t = 0$ by the dc voltage shown. Determine the output voltage $v(t)$. (*Hint:* One pole is $s = -1$.)

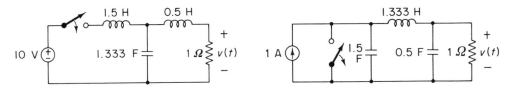

Figure P6-51   **Figure P6-52**

6-52. The circuit of Fig. P6-52 is one form of a low-pass third-order passive Butterworth filter normalized to a cutoff frequency of 1 rad/s. The circuit is excited at $t = 0$ by the dc current shown. Determine the output voltage $v(t)$. (*Hint:* One pole is $s = -1$.)

In Problems 6-53 through 6-56, solve the differential equations given using Laplace transform methods. Note the initial conditions stated.

6-53. $2\dfrac{d^2y}{dt^2} + 6\dfrac{dy}{dt} + 4y = 24$

$y(0) = 12, \; y'(0) = -4$

6-54. $4\dfrac{dy}{dt} + 12y = 20 \sin 4t$

$y(0) = -4$

6-55. $2\dfrac{dy}{dt} + 8y + 26\displaystyle\int_0^t y\,dt = 10\cos t$

$y(0) = 0$

6-56. $\dfrac{d^2y}{dt^2} + 3000\dfrac{dy}{dt} + 2\times 10^6 y = 10^8$

$y(0) = 75,\ y'(10) = 10^4$

6-57. For the circuit of Problem 6-35 (Fig. P6-35), solve for the current $i(t)$ by first writing a time-domain mesh current integrodifferential equation, specifying appropriate initial conditions, and solving the equation using Laplace transforms.

6-58. For the circuit of Problem 6-44 (Fig. P6-44), solve for the voltage $v(t)$ by first writing a time-domain node voltage integrodifferential equation, specifying appropriate initial conditions, and solving the equation using Laplace transforms.

DERIVATION PROBLEMS

6-59. The derivation of transform capacitive impedance was achieved in the text with operation (O-2) of Table 5-2. Provide an alternate derivation by first starting with the derivative relationship for current in terms of voltage and applying operation (O-1) of Table 5-2.

6-60. The derivation of transform inductive impedance was achieved in the text with operation (O-1) of Table 5-2. Provide an alternate derivation by first starting with the integral relationship for current in terms of voltage and applying operation (O-2) of Table 5-2.

6-61. Under certain conditions, it is possible for a given initial condition to result in a total cancellation of the natural or transient response. Consider the RC circuit of Fig. P6-61 with the capacitor initially charged to a voltage V_0 as shown. Assume a sinusoidal excitation at $t = 0$. Show that the transient response completely vanishes if the following condition is met:

$$V_0 = \frac{\omega RCV_p}{1 + \omega^2 R^2 C^2}$$

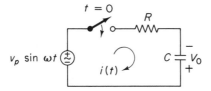

Figure P6-61

6-62. For the parallel RLC circuit of Section 6-10 excited by a step current of magnitude I, show that the voltage for the underdamped case is

$$v(t) = \frac{I}{\omega_d C}\,\epsilon^{-\alpha t}\sin\omega_d t$$

6-63. Consider the current response of the critically damped series RLC circuit as given by Eq. (6-118).

(a) Show that the maximum value of the current occurs at a time t_m given by

$$t_m = \frac{2L}{R}$$

(b) Show that the maximum value of the current I_m is given by

$$I_m = 0.7358 \frac{E}{R}$$

6-64. Consider the current response of the underdamped series RLC circuit as given by Eq. (6-124). Show that the first maximum of the current occurs at a time t_m given by

$$t_m = \frac{1}{\omega_d} \tan^{-1} \frac{\omega_d}{\alpha}$$

COMPUTER PROBLEMS

6-65. Any response $y(t)$ of a first-order circuit with a sinusoidal input may be expressed as

$$y(t) = A\epsilon^{-t/\tau} + B \sin(\omega t + \theta)$$

Write a computer program to evaluate $y(t)$ at intervals of Δt in the range from t_1 to t_2.

Input data: A, B, τ, ω, θ, t_1, t_2, Δt
Output data: t, $y(t)$

(*Note:* The sinusoidal function should be programmed to accept angles in radians, and θ should be expressed in radians.)

6-66. Write a computer program to evaluate the current $i(t)$ for the step response of a series RLC circuit at intervals of Δt in the range from t_1 to t_2. Means should be provided to test whether the response is overdamped, critically damped, or underdamped and to branch to appropriate segments as required.

Input data: E, R, L, C, t_1, t_2, Δt
Output data: t, $i(t)$, a statement to indicate form of response.

(*Note:* Underdamped response should be programmed to manipulate angle in radians.)

7

SYSTEM CONSIDERATIONS

OVERVIEW
The techniques for solving general linear circuit analysis problems using Laplace transform methods were developed in Chapter 6. In all cases considered, the actual circuit forms were given, and desired responses were determined by utilizing circuit analysis theorems in conjunction with transform manipulations.

In this chapter the approach is broadened to the systems level. The actual details of circuits will not be as relevant as the mathematical input–output relationships. It will be seen that two or more different circuits can display the same input-output characteristics. Mastery of the techniques of this chapter will allow the reader to deal with many of the basic concepts of control and communications systems.

OBJECTIVES
After completing this chapter, the reader should be able to

1. Define the requirements for a system to be linear and discuss its properties.
2. Define the concept of a transfer function.
3. Determine the transfer function for a given circuit.
4. Apply the transfer function concept to determine the output from the input for a given system.
5. Determine the transfer function from the input–output differential equation, and vice versa.
6. Determine the step response of a system from the impulse response, and vice versa.
7. Compare the similarities and differences between a transfer function and an impedance function.

219

8. **Determine the poles and zeros for a given transfer function and plot them in the *s*-plane.**

9. **Determine a transfer function from an *s*-plane pole–zero plot.**

10. **Define *stable system*, *unstable system*, and *marginally stable system*.**

11. **Show how the pole locations are related to the relative stability of a system.**

12. **Apply block diagram simplification methods to determine the net transfer function of (a) a cascade connection, (b) a parallel connection, and (c) a feedback loop.**

13. **State the convolution integral.**

14. **Apply the convolution integral to determine the response of a system from the impulse response and the input excitation.**

15. **Discuss the general properties of driving-point impedance functions.**

7-1 TRANSFER FUNCTION OF LINEAR SYSTEMS

In essentially all circuit analysis problems considered thus far in the text, the basic theme has been as follows: Given one or more voltage or current sources exciting a network, determine one or more voltage or current responses in the circuit. Thus, if we are given one or more *excitations*, we must determine the resulting *responses*.

At this point, let us restrict our consideration to the case where there is one particular response of interest and a particular point of excitation which produces the given response. This single-input, single-output type of system is one of the most important types of systems occurring in practice. Furthermore, a multi-input system can often be treated as a combination of several single-input systems.

The input–output system concept is illustrated in *block diagram* form in Fig. 7-1. It is assumed that there are no sources inside the equivalent block.

The quantity $x(t)$ represents the excitation, and $y(t)$ represents the response. We have used these general symbols to signify that the input could be either a voltage or a current, and the output could be either a voltage or a current. For that matter, many of the principles we develop here will be applicable to many linear systems other than circuits. Thus, in a mechanical system the excitation might be a force, and the response might be a movement of a machine component.

The problem we wish to consider is to determine some convenient mathematical description of a system from an overall external point of view. The description should permit us to determine the response from a knowledge of the excitation, without necessarily knowing the details of the hardware inside the box. This, of course, is a basic requirement of systems analysis, since a systems designer must work with "black boxes" whose exact contents are not known, but whose input–ouput relationship can be determined.

Figure 7-1 Input–output concept of a circuit or system.

Throughout the text, we have considered only linear circuits. In the framework of our systems approach, let us define more rigorously the concept of a linear system. In the definition that follows, we must assume an *initially relaxed system* so that the total output is produced only by the input and not by initial energy storage.

Definition of linear system. A system is said to be linear with respect to an excitation $x(t)$ and a response $y(t)$ if the following two properties are satisfied:

Property 1 (amplitude linearity). If an excitation $x(t)$ produces a response $y(t)$, then an excitation $Kx(t)$ should produce a response $Ky(t)$ for any value of K, where K is a *constant.*

As an example, if an input voltage of 10 V produces an output current of 2 A, then an input voltage of 25 V should produce an output current of 5 A if the system is linear.

Property 2 (superposition principle). If an excitation $x_1(t)$ produces a response $y_1(t)$, and an excitation $x_2(t)$ produces a response $y_2(t)$, then an excitation $x_1(t) + x_2(t)$ should produce a response $y_1(t) + y_2(t)$ for any arbitrary waveforms $x_1(t)$ and $x_2(t)$.

For example, suppose that an input step of 10 units produces an output ramp $20t$, and an input sinusoid $5 \cos t$ produces an output sinusoid $10 \sin t$. Then if the two input signals are summed together to yield an input $10 + 5 \cos t$, the output should be $20t + 10 \sin t$ if the system is linear.

There is a large body of elegant mathematical theory that has been built around linear system concepts, much of which is beyond the scope of this book. As far as circuit analysis is concerned, a circuit is linear as long as the parameters of the circuit (R, L, and C) are either constants or functions only of the independent variable time. Thus a circuit is nonlinear if any of the parameters are dependent on the amplitudes of the excitation (note Property 1) or the nature of the excitation (note Property 2). A good example of a nonlinear element is a resistor whose resistance changes as the current through it increases.

Actually, nothing in the real world is completely linear over an infinite range, as all real systems are limited by such effects as saturation and heat dissipation. When we imply that a system is linear, we mean that over its normal working range the linear model is sufficiently accurate to describe its behavior.

What do we mean by a circuit whose parameters are functions of time? A rather crude example of this phenomenon would be a motor-driven variable resistor in which the variation of resistance vs. time could be predicted. In general, such phenomena frequently result from using a nonlinear device such as a voltage-variable capacitor in conjunction with a special source such that the nonlinearity is controlled in a specific manner, resulting in the effect of a time-variable parameter. This concept is the basis for such devices as *parametric amplifiers* and *parametric frequency converters.* As long as the parameters can be considered to be functions of time only, the resulting system can be shown to be linear.

In this book we have restricted our consideration to lumped linear circuits and systems in which all parameters are constant. *A system in which all parameters are constant is said to be a time-invariant or stationary system.* Thus our sole attention is directed toward *lumped, linear, time-invariant* circuits and systems.

Now let us turn our attention to the problem of finding a means of representing a lumped, linear, time-invariant system from an external point of view. We saw in Chapter 6 how the use of transform methods permits us to replace the processes of differentiation and integration by simple algebra. As it turns out, the transform method is equally suited for providing a description of a circuit or system satisfying our stated restrictions.

Assume, then, that a circuit or system is initially relaxed and is excited at $t = 0$ by an arbitrary excitation $x(t)$. Let

$$X(s) = \mathscr{L}[x(t)] \tag{7-1}$$

$$Y(s) = \mathscr{L}[y(t)] \tag{7-2}$$

As a result of the excitation, a series of transform equations may be written describing the circuit or system. All dependent variables but the desired variable $Y(s)$ may be eliminated, resulting in an algebraic equation expressing $Y(s)$ and $X(s)$. As a result of the algebraic characteristics of the Laplace transform, the output transform will always turn out to be proportional to the input transform. Hence, in general,

$$Y(s) = G(s)X(s) \tag{7-3}$$

where $G(s)$ is an algebraic function of s. Solving for $G(s)$, we obtain

$$G(s) = \frac{Y(s)}{X(s)} \tag{7-4}$$

The quantity $G(s)$ is called the *transfer function* of the circuit or system, and it is the ratio of the output transform to the input transform. As simple as Eqs. (7-3) and (7-4) are, they form the basis for a considerable body of circuit and system analysis and are very important relationships.

It should be emphasized that the quantity $G(s)$ *is fixed by the nature of the system and is not dependent on the type of excitation. The poles and zeros of $G(s)$ are due only to the circuit or system, and the order of the denominator polynomial of $G(s)$ is the order of the circuit or system.* Thus, a circuit containing m nonredundant energy-storage elements will, in general, have a transfer function whose denominator polynomial is of order m.

In determining the transfer function of an arbitrary system, it is usually convenient simply to carry along an arbitrary $X(s)$ in the calculations, and then divide through according to Eq. (7-4) when $Y(s)$ has been determined. Also, the choice of an impulse excitation is often convenient, as will be seen shortly.

Referring to the basic defining relationship of Eq. (7-3), suppose we assume as an excitation a unit impulse function. Thus let

$$x(t) = \delta(t) \tag{7-5}$$

$$X(s) = 1 \tag{7-6}$$

Substitution of Eq. (7-6) into Eq. (7-3) yields for this special case

$$Y(s) = G(s) \tag{7-7}$$

The *time-domain impulse response* is

$$y(t) = \mathscr{L}^{-1}[G(s)] \tag{7-8}$$

Let us define this inverse transform as follows: Let

$$g(t) = \mathscr{L}^{-1}[G(s)] \tag{7-9}$$

The quantity $g(t)$ is called the *impulse response* of the system.

Thus *the inverse transform of the transfer function is the impulse response.* In effect, then, we may describe a system in one of two ways: (1) In the transform domain, we may specify the *transfer function* $G(s)$, (2) In the time domain, we may specify the *impulse response* $g(t)$. Of course, if either one is known, the other may be determined by transform manipulation. These ideas are compared in block diagram fashion in Fig. 7-2.

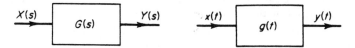

Figure 7-2   Transfer function and impulse response descriptions of linear circuit or system.

At this point, let us consider a common error made by beginners using transform analysis. We know that if $G(s)$ and $X(s)$ are given, we may readily determine $Y(s)$ by multiplying $G(s)$ and $X(s)$. On the other hand, if $g(t)$ is given, it is *not correct* to multiply $x(t)$ by $g(t)$ in an attempt to obtain $y(t)$. In fact, if this concept were true, there would be no need for transform methods at all! The simplicity of the transform approach is the fact that algebra may be used in the transform domain. The output may be determined directly in the time domain using a concept called *convolution*, which is the topic of Section 7-8.

Example 7-1

(a) Determine the transfer function and the impulse response of the circuit of Fig. 7-3a. The excitation is $v_1(t)$ and the response is $v_2(t)$.

(b) Using the transfer function concept, determine the response due to an excitation $v_1(t) = 5 \sin 2t$.

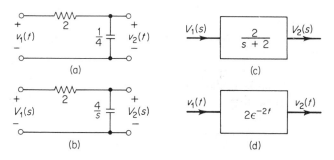

(a)

(c)

(b)

(d)

Figure 7-3   Circuits of Example 7-1.

Solution (a) We first transform the circuit as shown in Fig. 7-3b with arbitrary $V_1(s)$ assumed. In this case, the transfer function is readily obtained by means of the voltage-divider rule. Thus

$$G(s) = \frac{V_2(s)}{V_1(s)} = \frac{4/s}{2 + 4/s} = \frac{2}{s + 2} \tag{7-10}$$

We observe that the denominator is a first-order polynomial resulting from the fact that the circuit is a first-order circuit.

The impulse response $g(t)$ is

$$g(t) = \mathcal{L}^{-1}\left[\frac{2}{s + 2}\right] = 2\epsilon^{-2t} \tag{7-11}$$

Block diagrams representing the transform-domain and time-domain representations are shown in (c) and (d) of Fig. 7-3.

(b) For the sinusoidal excitation, we have

$$V_1(s) = \frac{10}{s^2 + 4} \tag{7-12}$$

and since

$$V_2(s) = G(s)V_1(s) \tag{7-13}$$

we have

$$V_2(s) = \frac{20}{(s + 2)(s^2 + 4)} \tag{7-14}$$

The reader is invited to show that

$$v_2(t) = \frac{5}{2}\epsilon^{-2t} + \frac{5}{\sqrt{2}}\sin(2t - 45°) \tag{7-15}$$

We note that $v_2(t)$ contains an exponential term expressing the natural behavior of the circuit as predicted by the impulse response, and a sinusoidal term expressing the form of the excitation. The exponential term is *transient* in nature and is due to the circuit, whereas the sinusoidal term is a *steady-state* term and is due to the excitation.

Example 7-2

Determine the transfer function of the circuit shown in Fig. 7-4a. The input is $v_1(t)$ and the desired output is $i_2(t)$.

Solution We first draw the transformed circuit as shown in Fig. 7-4b. In this case, a relationship between $I_2(s)$ and $V_1(s)$ can be obtained if we write a pair of simultaneous mesh equations and eliminate $I_1(s)$. We have

$$\left(2s + \frac{4}{s}\right)I_1(s) - \frac{4}{s}I_2(s) = V_1(s) \tag{7-16}$$

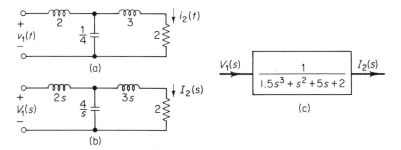

Figure 7-4   Circuits of Example 7-2.

and

$$-\frac{4}{s} I_1(s) + \left(\frac{4}{s} + 3s + 2\right) I_2(s) = 0 \tag{7-17}$$

The quantity $I_1(s)$ may be eliminated by substitution, or $I_2(s)$ may be solved for by determinants. The reader may verify that the result obtained is

$$I_2(s) = \frac{V_1(s)}{1.5s^3 + s^2 + 5s + 2} \tag{7-18}$$

The transfer function is

$$G(s) = \frac{I_2(s)}{V_1(s)} = \frac{1}{1.5s^3 + s^2 + 5s + 2} \tag{7-19}$$

A block diagram representation is shown in Fig. 7-4c. Note that since the circuit is a third-order circuit, the transfer function is also of third order.

Example 7-3

The circuit of Fig. 7-5a is one particular active realization of a second-order Butterworth low-pass filter using an operational amplifier, and it is normalized to a cutoff frequency of 1 rad/s with 1-Ω resistors. With $v_1(t)$ as the input and $v_2(t)$ as the output, determine the transfer function. (An actual realistic circuit is derived from the normalized circuit by scaling the frequency and resistance levels.)

      Solution   The operational amplifier is connected in what is known as a "voltage-follower" configuration in which the output voltage acts as a VCVS with a voltage gain of one. The controlling voltage is the voltage from the noninverting input terminal (labeled as +) to ground. No current is assumed to flow into this input terminal.

      The s-domain circuit is shown in Fig. 7-5b. The input and output node voltages are labeled as $V_1(s)$ and $V_2(s)$, respectively. However, two additional voltages $V_3(s)$ and $V_4(s)$ are introduced at the two intermediate nodes. Note

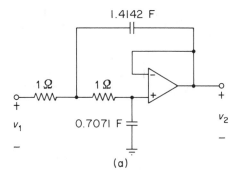

(a)

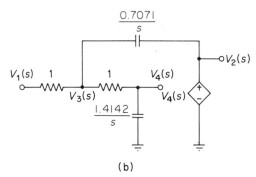

(b)

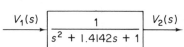

Figure 7-5 Normalized second-order active low-pass Butterworth filter, s-domain model with VCVS, and transfer function as determined by Example 7-3.

that the dependent source has a voltage value equal to the voltage of the controlling node since the gain is unity.

If it is desired to eliminate n variables in a set of simultaneous equations, it is necessary to have $n + 1$ independent equations. A relationship between $V_1(s)$ and $V_2(s)$ is desired so it is necessary to eliminate $V_3(s)$ and $V_4(s)$. Thus three independent equations are required.

First, a node equation will be written at the $V_3(s)$ node. Applying KCL at this node, we have

$$\frac{V_3(s) - V_1(s)}{1} + \frac{s}{0.7071}[V_3(s) - V_2(s)] + \frac{V_3(s) - V_4(s)}{1} = 0 \tag{7-20}$$

Next, a node equation will be written at the $V_4(s)$ node. We have

$$\frac{V_4(s) - V_3(s)}{1} + \frac{s}{1.4142}V_4(s) = 0 \tag{7-21}$$

Finally, the dependent source voltage and the output voltage are exactly the same, that is,

$$V_2(s) = V_4(s) \tag{7-22}$$

The voltages $V_3(s)$ and $V_4(s)$ are eliminated by the procedure that follows. First, the relationship of Eq. (7-22) is used to eliminate $V_4(s)$ in Eqs. (7-20) and (7-21). This leaves two equations with only $V_3(s)$ remaining as a superfluous variable. Next, Eq. (7-21) is solved for $V_3(s)$ in terms of $V_2(s)$ following the earlier substitution. This result is substituted in Eq. (7-20). After several manipulations, we obtain

$$V_2(s) = \frac{V_1(s)}{s^2 + 1.4142s + 1} \tag{7-23}$$

The transfer function is then determined as

$$G(s) = \frac{V_2(s)}{V_1(s)} = \frac{1}{s^2 + 1.4142s + 1} \tag{7-24}$$

Example 7-4

The input to a certain system is $x(t)$ and the output is $y(t)$. The impulse response of the system is given by

$$g(t) = 10\epsilon^{-t} \sin 2t \tag{7-25}$$

Determine the response resulting from an excitation $x(t) = 5 \sin t$

   Solution   First we must determine the transfer function $G(s)$. We have

$$G(s) = \frac{20}{(s + 1)^2 + 4} \tag{7-26}$$

The transform of the input is

$$X(s) = \frac{5}{s^2 + 1} \tag{7-27}$$

We then have

$$Y(s) = G(s)X(s) = \frac{100}{[(s + 1)^2 + 4](s^2 + 1)} \tag{7-28}$$

The reader is invited to show that the time response is

$$y(t) = 11.18\epsilon^{-t} \sin (2t + 116.57°) + 22.36 \sin (t - 26.57°) \tag{7-29}$$

Example 7-5

A certain circuit has an input $x(t)$ and an output $y(t)$. When the input is a constant of 5 units [i.e., $x(t) = 5$], the output is observed to be

$$y(t) = 10\epsilon^{-2t} + 5\epsilon^{-t} \sin 2t \tag{7-30}$$

Determine the transfer function $G(s)$.

Solution First we determine the transform of $y(t)$:

$$Y(s) = \frac{10}{s+2} + \frac{5(2)}{(s+1)^2 + 4}$$

$$= \frac{10(s^2 + 2s + 5) + 10(s+2)}{(s+2)(s^2 + 2s + 5)} \tag{7-31}$$

$$= \frac{10(s^2 + 3s + 7)}{(s+2)(s^2 + 2s + 5)}$$

However, since the input was not an impulse, a further manipulation is required to determine $G(s)$. Since

$$G(s) = \frac{Y(s)}{X(s)} \tag{7-32}$$

and

$$X(s) = \frac{5}{s} \tag{7-33}$$

then

$$G(s) = \frac{s}{5} \cdot Y(s)$$

$$= \frac{2s(s^2 + 3s + 7)}{(s+2)(s^2 + 2s + 5)} \tag{7-34}$$

We could now determine the response due to any other input if desired.

7-2 DIFFERENTIAL EQUATION AND TRANSFER FUNCTION

We saw in Section 7-1 how it was possible to describe the input-output relationship of a lumped, linear, time-invariant circuit or system by means of a transfer function. In developing the transfer functions of circuits considered, we worked directly in the transform domain, as has been our practice through most of the last few chapters. In some cases, however, we will encounter a circuit or system whose mathematical description is given by a differential equation rather than a transfer function or an impulse response. Furthermore, there is a definite relationship between a given transfer function and the corresponding differential equation of the system. In this section, we will study this relationship and learn to determine a transfer function from a differential equation, and vice versa.

First, let us assume our standard lumped, linear, time-invariant system with excitation $x(t)$ and response $y(t)$. The most general differential-equation description of the system is of the form

$$b_m \frac{d^m y}{dt^m} + b_{m-1} \frac{d^{m-1} y}{dt^{m-1}} + \cdots + b_0 y = a_n \frac{d^n x}{dt^n} + a_{n-1} \frac{d^{n-1} x}{dt^{n-1}} + \cdots + a_0 x \tag{7-35}$$

The highest derivative of y determines the order of the system. Therefore, Eq. (7-35) describes a system of order m. Furthermore, we will require that $m \geq n$.

Now as our next step, we wish to transform both sides of Eq. (7-35) to obtain a transform-domain description. The transform of a higher-order derivative was developed in Chapter 6 and was shown to be

$$\mathscr{L}\left[\frac{d^k y}{dt^k}\right] = s^k Y(s) - s^{k-1} y(0) - \cdots - y^{(k-1)}(0) \tag{7-36}$$

It appears that if Eq. (7-36) is applied to each term of Eq. (7-35), a rather monstrous number of terms will result on both sides of the equation. This is certainly true, but let us proceed to transform Eq. (7-35), writing explicitly only the terms containing $Y(s)$ and $X(s)$. Hence, an expression is obtained as follows:

$$(b_m s^m + b_{m-1} s^{m-1} + \cdots + b_0) Y(s) + I_y \tag{7-37}$$
$$= (a_n s^n + a_{n-1} s^{n-1} + \cdots + a_0) X(s) + I_x$$

where I_y is defined as the "mass collection" of all possible initial-condition terms for y, and I_x is the same for x. Solving for $Y(s)$, we obtain

$$Y(s) = \left[\frac{a_n s^n + a_{n-1} s^{n-1} + \cdots + a_0}{b_m s^m + b_{m-1} s^{m-1} + \cdots + b_0}\right] X(s) \tag{7-38}$$
$$+ \frac{I_y - I_x}{b_m s^m + b_{m-1} s^{m-1} + \cdots + b_0}$$

Now let us assume that the system is *initially relaxed*. This assumption has been paramount throughout the basic development of the transfer function. If there is no initial energy present in the system, a serious look at Eq. (7-38) reveals that the quantity $(I_y - I_x)$ must be identically zero. This is true since $Y(s)$ must be linearly proportional to $X(s)$, and furthermore, if $X(s) = 0$, we know that $Y(s)$ must be zero, which would not be true except for $I_y - I_x = 0$.

Thus in the case of the initially relaxed system, the initial condition terms between x and y cancel, leaving the basic relationship

$$Y(s) = \left[\frac{a_n s^n + a_{n-1} s^{n-1} + \cdots + a_0}{b_m s^m + b_{m-1} s^{m-1} + \cdots + b_0}\right] X(s) \tag{7-39}$$

or

$$\frac{Y(s)}{X(s)} = G(s) = \frac{a_n s^n + a_{n-1} s^{n-1} + \cdots + a_0}{b_m s^m + b_{m-1} s^{m-1} + \cdots + b_0} \tag{7-40}$$

For convenience, let

$$N(s) = a_n s^n + a_{n-1} s^{n-1} + \cdots + a_0 \tag{7-41}$$

and

$$D(s) = b_m s^m + b_{m-1} b^{m-1} + \cdots + b_0 \tag{7-42}$$

Thus,

$$G(s) = \frac{N(s)}{D(s)} \tag{7-43}$$

and of course,

$$Y(s) = G(s)X(s) \tag{7-44}$$

Thus, $G(s)$ is recognized as the transfer function which has been derived directly from the differential equation in this case.

Comparing Eqs. (7-35) and (7-40), we see the definite relationships between the various terms of the differential equation and the corresponding terms of the transfer function. The reader is encouraged to learn to convert back and forth between these forms. The problems that follow should clarify this procedure.

Example 7-6

Determine a differential equation expressing the input–output relationship for the circuit of Fig. 7-6. Determine the transfer function for the circuit.

Solution   Assuming an arbitrary input voltage, $v_1(t)$, we can best describe the circuit by writing a node voltage equation at the output node. Summing currents leaving the node, we have

$$4\frac{dv_2}{dt} + 6v_2 + 8\frac{d}{dt}(v_2 - v_1) + d\int_0^t (v_2 - v_1)\,dt = 0 \tag{7-45}$$

To remove the integral sign, we will differentiate all terms. Furthermore, v_1 terms will be moved to the right-hand side of the equation. This results in

$$12\frac{d^2v_2}{dt^2} + 6\frac{dv_2}{dt} + 2v_2 = 8\frac{d^2v_1}{dt^2} + 2v_1 \tag{7-46}$$

which is recognized to be of the form of Eq. (7-35). Transformation of this equation, with the recognition of the fact that initial conditions cancel, yields

$$(12s^2 + 6s + 2)V_2(s) = (8s^2 + 2)V_1(s) \tag{7-47}$$

resulting in the transfer function

$$G(s) = \frac{V_2(s)}{V_1(s)} = \frac{8s^2 + 2}{12s^2 + 6s + 2} = \frac{4s^2 + 1}{6s^2 + 3s + 1} \tag{7-48}$$

The reader is invited to verify the result by transforming the circuit and working in the transform domain.

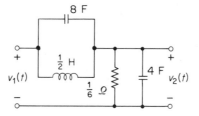

Figure 7-6 Circuit of Example 7-6.

7-3 STEP AND IMPULSE RESPONSES

In the preceding work of this chapter, we have seen that the transfer function or its inverse transform, the impulse response, provides a mathematical description of the input–output relationship of a linear system. While the impulse response is probably the easiest type of response to calculate mathematically, it is not always the most convenient to determine experimentally. From an experimental standpoint, the response due to a step-function excitation is usually somewhat easier to determine. In this section we will investigate the relationships between the impulse and step responses of a system.

Continuing the notation of the past few sections, assume that a given system has a transfer function $G(s)$ and an impulse response $g(t)$. Assume that the system is excited by a unit step function [i.e., $x(t) = 1$]. Let $h(t)$ represent the step response and $H(s)$ represent the transform of the step response. According to the basic theory of transfer functions, we have

$$H(s) = \frac{1}{s} \cdot G(s) = \frac{G(s)}{s} \tag{7-49}$$

The step response is

$$h(t) = \mathscr{L}^{-1}\left[\frac{G(s)}{s}\right] \tag{7-50}$$

By means of operation (O-2), Eq. (7-50) can be expressed as

$$h(t) = \int_0^t g(t)\, dt \tag{7-51}$$

Thus *in the same manner that the step function is the integral of the impulse function, the step response is the integral of the impulse response.*

Equation (7-51) tells us that if we know the impulse response of a system, we may determine the step response directly in the time domain by integrating the impulse response. This approach is often easier than transform manipulations whenever $g(t)$ is known.

A relationship for determining the impulse response from the step response may also be obtained. If we solve for $G(s)$ in Eq. (7-49), we have

$$G(s) = sH(s) \tag{7-52}$$

Immediately, this relationship is recognized to be of the form of operation (O-1). However, a constant term representing the initial value is required. Using $t = 0^+$ as a reference, we can modify Eq. (7-52) as follows:

$$G(s) = sH(s) - h(0^+) + h(0^+) \tag{7-53}$$

The first two terms on the right of Eq. (7-53) are recognized as the exact transform of the derivative of $h(t)$. The third term is a constant, so its inverse transform is an impulse function. Thus

$$g(t) = \frac{dh(t)}{dt} + h(0^+)\delta(t) \tag{7-54}$$

Thus *the impulse response is the derivative of the step response with a possible additional impulse function term.*

Equation (7-54) tells us that if we know the step response, we may readily determine the impulse response directly in the time domain by differentiating the step response and, in some cases, adding an impulse term. Note that the impulse term does not appear if $h(0^+) = 0$. Actually, in systems where the denominator polynomial of the transfer function is of higher degree than the numerator polynomial, it can be shown that the step response is always zero at $t = 0^+$. It is only in those cases where the numerator and denominator polynomials are of the same degree that the step response is nonzero at $t = 0^+$. In such cases, an impulse term appears in the impulse response, providing the required jump in the step response at $t = 0^+$.

Example 7-7

Verify the relationships of this section for the circuit of Fig. 7-7.

Solution   Mental transformation of this simple circuit and application of the voltage-divider rule yields

$$G(s) = \frac{V_2(s)}{V_1(s)} = \frac{2}{2 + 4/s} = \frac{s}{s + 2} \tag{7-55}$$

Since the numerator and denominator are of the same degree, we must divide the numerator by the denominator to separate the constant before we can determine the impulse response. This yields

$$G(s) = 1 - \frac{2}{s + 2} \tag{7-56}$$

Ths impulse response is

$$g(t) = \delta(t) - 2\epsilon^{-2t} \tag{7-57}$$

Therefore, since the numerator and denominator are of the same degree, an impulse term appears in the impulse response.

By means of Eq. (7-51) the response due to a unit step function $(v_1(t) = 1)$ is

$$
\begin{aligned}
h(t) &= \int_0^t \left[\delta(t) - 2\epsilon^{-2t}\right] dt = u(t) + \epsilon^{-2t}\Big]_0^t \\
&= u(t) + \epsilon^{-2t} - 1 = 1 + \epsilon^{-2t} - 1 \qquad \text{for } t > 0 \tag{7-58} \\
&= \epsilon^{-2t} \qquad \text{for } t > 0
\end{aligned}
$$

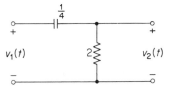

Figure 7-7   Circuit of Example 7-7.

Now let us use this result to verify Eq. (7-54). We have

$$g(t) = \frac{d}{dt} \left[\epsilon^{-2t} \right] + \epsilon^0 \delta(t)$$

$$= -2\epsilon^{-2t} + \delta(t)$$

$$(7\text{-}59)$$

agreeing with our previous result.

Example 7-8

The impulse response of a certain system is

$$g(t) = 10 + 18\epsilon^{-2t} + 25 \sin 5t \qquad (7\text{-}60)$$

Determine the response due to a step excitation of magnitude 8.

      Solution   The desired response $y(t)$ is 8 times $h(t)$, the response due to a unit step function. By means of Eq. (7-51) we have

$$y(t) = 8h(t) = 8 \int_0^t \left[10 + 18\epsilon^{-2t} + 25 \sin 5t \right] dt$$

$$= 8 \left[10t - 9\epsilon^{-2t} - 5 \cos 5t \right]_0^t$$

$$= 8 \left[10t - 9\epsilon^{-2t} - 5 \cos 5t + 9 + 5 \right]$$

$$= 8 \left[14 + 10t - 9\epsilon^{-2t} - 5 \cos 5t \right]$$

$$(7\text{-}61)$$

      The reader is invited to verify this result by transorm analysis. It will be seen that the direct use of time-domain analysis was easier in this case.

7-4 COMPARISON BETWEEN TRANSFER AND IMPEDANCE FUNCTIONS

In considering the concept of the transfer function in electric circuits, we have seen that the excitation may be either a voltage or a current, and the response may be either a voltage or a current. In most cases when the term "transfer" is used, there seems to be an implication that the excitation terminals are physically displaced from the response terminals as has been the case in most problems considered.

      On the other hand, this physical separation may not always be true. For example, consider the circuit shown in Fig. 7-8. Suppose that the excitation is considered

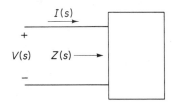

Figure 7-8 Driving-point impedance function.

to be the current $I(s)$, and the response is considered to be the voltage $V(s)$. In this case the excitation and response are actually not physically separated as in the case of a network with input and output terminals. However, we may certainly develop a relationship between $V(s)$ and $I(s)$. In fact, the "transfer function" is

$$G(s) = \frac{V(s)}{I(s)} = Z(s) \qquad (7\text{-}62)$$

In other words the "transfer function" in this case is simply the input impedance $Z(s)$. An impedance or admittance function relating the voltage and current at the same terminals is called a *driving-point impedance or admittance function*.

Thus a driving-point impedance function may often be considered as a special type of transfer function whenever the excitation is a voltage or current and the corresponding response is the other variable associated with the same terminals. In such cases the "input" and "output" are not really physically separated in the same sense as most transfer functions. In discussing the properties of the transfer function, we will imply that many of these properties also apply to a driving-point impedance or admittance function whenever it is considered as a transfer function.

On the other hand, the reader should not expect all the properties of the true transfer function to be true for impedance functions, or vice versa. There are certain special properties of each type of function that will be made clearer in the sections that follow.

One basic difference in the two types of functions lies in the fact that, when a driving-point impedance is considered as a transfer function, the role of excitation and response may be interchanged, and the new transfer function is the reciprocal of the original. For example, referring back to Fig. 7-8 we note that if $V(s)$ is considered as the excitation and $I(s)$ as the response, then

$$\frac{I(s)}{V(s)} = \frac{1}{Z(s)} = Y(s) \qquad (7\text{-}63)$$

which is the reciprocal of the transfer function obtained earlier.

However, transfer functions in general do not obey this inverse property. In fact, if one interchanges the roles of excitation and response, an entirely new transfer function, which may not be simply related to the original, is obtained. For example, consider the simple resistive circuit of Fig. 7-9. First let v_1 be the excitation, v_2 be the response, and G_1 be the transfer function. Since no current flows in the 5-Ω resistor, we have

$$G_1 = \frac{v_2}{v_1} = \frac{10}{90 + 10} = \frac{1}{10} \qquad (7\text{-}64)$$

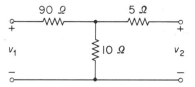

Figure 7-9 Simple resistive circuit.

Now let us assume that v_2 is the excitation, v_1 is the response, and G_2 is the appropriate transfer function. We have

$$G_2 = \frac{v_1}{v_2} = \frac{10}{10 + 5} = \frac{2}{3} \tag{7-65}$$

As can be seen, there is no simple relationship between G_1 and G_2.

Certain types of circuit transfer functions fit the reciprocity theorem exactly, and in such cases, the network may be turned completely around without affecting the transfer function. This type of situation exists whenever the transfer function is the ratio of an open-circuit voltage to an input current such as shown in Fig. 7-10a or

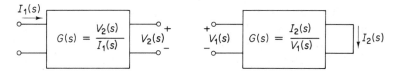

Figure 7-10   Transfer functions fitting the reciprocity theorem.

the ratio of a short-circuit current to an input voltage as shown in Fig. 7-10b. These points will be demonstrated in Problems 7-25 and 7-26.

7-5 POLES AND ZEROS

Let us return our consideration to the transfer function concept. Let

$$G(s) = \frac{N(s)}{D(s)} \tag{7-66}$$

where $N(s)$ is the numerator polynomial and $D(s)$ is the denominator polynomial. Let

$$N(s) = a_n s^n + a_{n-1} s^{n-1} + \cdots + a_0 \tag{7-67}$$

and

$$D(s) = b_m s^m + b_{m-1} s^{m-1} + \cdots + b_0 \tag{7-68}$$

Thus we are assuming a general circuit or system of order m.

In previous work we have used the term *poles* to refer to the roots of $D(s)$ and, in a few instances, the term *zeros* has been used to refer to the roots of $N(s)$. We now wish to review and strengthen this terminology, as it will be used extensively in the remainder of this chapter.

The following definitions are in order:

1. *Poles (finite).* The m roots of $D(s)$ are called the *finite poles* of the transfer function $G(s)$.
2. *Zeros (finite).* The n roots of $N(s)$ are called the *finite zeros* of the transfer function $G(s)$.
3. *Critical frequencies.* All the poles and zeros of a transfer function are said to be the *critical frequencies* of the function.

The use of the adjective "finite" will be explained shortly. First, let us develop some reasoning as to why we use the terms "pole" and "zero." Suppose we consider that s is, in some sense, a variable that can assume many values. If s is made to equal the value of one of the finite zeros, then $N(s)$ is zero since, by definition, if s assumes a value of a root of $N(s)$, the polynomial is zero. Inspection of Eq. (7-66) reveals that $G(s)$ is zero when $N(s)$ is zero and hence the term "zero" has a direct meaning in this sense.

However, suppose that s assumes the value of a pole. In this case $D(s) = 0$, and inspection of Eq. (7-66) reveals that $G(s)$ "blows up" or approaches infinity under this condition. Thus since a pole represents something "long or high," there is a direct meaning associated with this term.

Thus the zeros of a transfer function tell us what values of s will cause the transfer function to be zero, and the poles of a transfer function tell us what values of s will cause the transfer function to be infinite.

If the quantity s is considered as a variable, it must be considered as a *complex variable*, since we know that the poles and zeros of a transform function may be complex in general. To that end, let us represent s as

$$s = \sigma + j\omega \qquad (7\text{-}69)$$

where σ is the real part of s representing a variable *damping constant*, and ω is the imaginary part representing a variable angular *frequency*. The quantity s is considered as a *complex frequency* in a general sense.

Since the quantity s is complex, in order to display information associated with the possible values of s that the poles and zeros may assume, we will define a two-dimensional *complex s-plane* or *complex-frequency plane* as shown in Fig. 7-11a. The horizontal (real) axis is called the *σ-axis* or *damping axis*, and the vertical (imaginary) axis is called the *$j\omega$-axis* or *frequency axis*.

Referring to Fig. 7-11b in the s-plane we may display points whose complex coordinates represent the values of poles and zeros. Such a representation is called a *pole–zero plot*. The $\times$ symbol is used to designate a pole, and the $\circ$ symbol is used

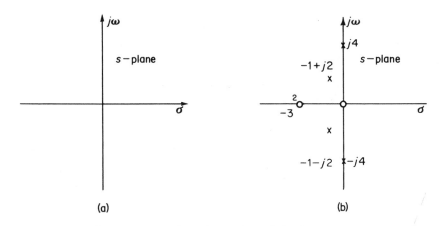

Figure 7-11 Complex s-plane and a typical pole–zero plot.

to designate a zero. A higher-order pole or zero is designated by means of an exponent representing the order. In referring to a pole or zero, it is often convenient to say *a pole at* (*value*) or *a zero at* (*value*), where the value is the complex representation of the given quantity.

For the *s*-plane plot of Fig. 7-11b, there are four finite poles and three finite zeros. The poles are located at $-1 + j2$, $-1 - j2, j4$, and $-j4$. The zeros are located at 0, -3, and -3 (second-order zero).

A property of most true transfer functions is that the degree of the denominator polynomial is greater than or equal to the degree of the numerator polynomial, i.e., $m \geq n$. This means that the transfer function has m finite poles and n finite zeros. If $m = n$, the number of finite poles is equal to the number of finite zeros. On the other hand, if $m > n$, the number of finite poles is greater than the number of finite zeros. For certain descriptive reasons, it is desirable to employ a definition whereby the number of zeros is equal to the number of poles. This is achieved for the case where $m > n$ by defining *zeros at infinity*.

A transfer function in which $m > n$ *is said to have* r *zeros at infinity, where*

$$r = m - n \tag{7-70}$$

$$= \text{(degree of denominator polynomial)} - \text{(degree of numerator polynomial)}$$

What is the physical meaning of this concept? To answer this question, let us look at the asymptotic nature of $N(s)$, $D(s)$, and $G(s)$ for large s. A basic property of a polynomial is that, for a sufficiently large value of the argument, the polynomial becomes "dominated" by its highest-degree term. That is, the percentage of error caused by approximating a polynomial by its highest-degree term becomes increasingly smaller as the argument increases. In mathematics, the polynomial is said to be *asymptotic* to its highest-degree term for large argument. For engineering purposes, the polynomial may be approximated by this term. Using the conventions for $N(s)$ and $D(s)$ established by Eqs. (7-67) and (7-68) we have

$$N(s) \approx a_n s^n \qquad \text{for } s \gg 1 \tag{7-71}$$

$$D(s) \approx b_m s^m \qquad \text{for } s \gg 1 \tag{7-72}$$

Thus

$$G(s) \approx \frac{a_n s^n}{b_m s^m} = \frac{a_n/b_m}{s^{m-n}} = \frac{a_n/b_m}{s^r} \qquad \text{for } s \gg 1 \tag{7-73}$$

where $r = m - n$.

In effect, Eq. (7-73) tells us that for very large s, $G(s)$ may be approximated by a constant times $1/s^r$. This means that as s approaches infinity, $G(s)$ approaches zero when $m > n$. If $r = 1$, $G(s)$ approaches zero as $1/s$, whereas if $r = 2$, $G(s)$ approaches zero as $1/s^2$, etc. Thus larger values of r indicate a more rapid decrease of $G(s)$ for large s. In a matter of speaking, then, it seems feasible to say that $G(s)$ has a zero of order r at $s = \infty$, since $G(s)$ does indeed approach zero as s approaches infinity, the basic requirement of a zero, and the quantity r reflects how rapidly $G(s)$ approaches zero.

Referring again to the *s*-plane plot of Fig. 7-11b there are four finite poles and three finite zeros. Then we may say that the transfer function has one zero at $s = \infty$, yielding an equal number of poles and zeros. Note that this latter zero is not shown. Zeros at infinity may always be determined by counting the number of finite poles and zeros.

We have thus far assumed that the degree of the denominator polynomial of a true transfer function is at least as large as the degree of the numerator polynomial. On the other hand, driving-point impedance and admittance functions do not necessarily satisfy this restriction. Quite often, the numerator polynomial of a driving-point impedance or admittance will be one degree higher than the denominator polynomial. Although it is rare to find a situation in which the numerator polynomial is more than one degree higher, we will develop the following definition in a more general manner to take into consideration any possible rare case in general systems analysis.

Consider a special function $G(s)$ in which $n > m$. *A function $G(s)$, in which $n > m$, is to said have k poles at infinity, where*

$$k = n - m$$
$$= \text{(degree of numerator polynomial)} - \text{(degree of denominator polynomial)} \tag{7-74}$$

Again, we may attach a reasoning behind this concept by looking at the asymptotic behavior of $G(s)$. From Eqs. (7-71) and (7-72) we have

$$G(s) \approx \frac{a_n s^n}{b_m s^m} = \frac{a_n s^{n-m}}{b_m} = \frac{a_n s^k}{b_m} \qquad \text{for } s \gg 1 \tag{7-75}$$

Thus $G(s)$ increases without limit as s increases without limit, the rate of increase being more significant for larger values of k. Hence it seems feasible to speak of a "kth-order pole at infinity." Impedance functions will be considered in more detail in a later section.

Assume now that all the finite zeros and finite poles of a transfer function are known. Let us designate the n finite zeros as $z_1, z_2, \ldots, z_n$ and the finite poles as $p_1, p_2, \ldots, p_m$. Since $N(s)$ and $D(s)$ may be specified within a constant multiplier by a knowledge of their roots, we may express $G(s)$ as

$$G(s) = \frac{A(s - z_1)(s - z_2) \cdots (s - z_n)}{(s - p_1)(s - p_2) \cdots (s - p_m)} \tag{7-76}$$

where A is a constant multiplier. In terms of the notation of Eqs. (7-67) and (7-68) this constant is $A = a_n/b_m$ and is essentially a single quantity even though it may appear to be composed of two separate constants.

Thus, *a transfer function may be determined within a constant multiplier by a knowledge of its poles and zeros.* The constant multiplier could then be determined if we specify one further item of information such as the magnitude of $G(s)$ at some value of s, and so on.

Referring back to the s-plane plot of Fig. 7-11b this transfer function can be written as

$$G(s) = \frac{As(s + 3)^2}{(s + 1 - j2)(s + 1 + j2)(s - j4)(s + j4)}$$
$$= \frac{As(s + 3)^2}{(s^2 + 2s + 5)(s^2 + 16)} \tag{7-77}$$

If desired, we could expand both the numerator and the denominator to obtain the overall polynomials.

7-6 STABILITY

The concept of stability is a very important subject in the design and analysis of active electronic circuits and closed-loop feedback control systems. In this section we will look at the basic concept in order to acquaint the reader with sufficient background for further study. There are several different approaches to stability, but for our purposes, we will make the following definitions:

1. *Stable system.* A system is said to be stable if the impulse response approaches zero for sufficiently large time.
2. *Unstable system.* A system is said to be unstable if the impulse response grows without bound (i.e., approaches infinity) for sufficiently large time.
3. *Marginally stable system.* A system is said to be marginally stable if the impulse response approaches a constant nonzero value or a constant-amplitude oscillation for sufficiently large time.

Examples of three possibilities are shown in Fig. 7-12. The response of (a) is stable, the response of (b) is unstable, and the response of (c) is marginally stable.

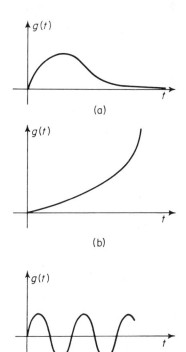

Figure 7-12   Examples of (a) stability, (b) instability, and (c) marginal stability.

We have defined the stability concept in terms of the impulse response since it is usually the simplest type of response to analyze, and it provides the basic form of the transient response. Thus if the impulse response is stable, the transient response due to any finite excitation will vanish after a sufficiently long time, leaving only a possible steady-state response. On the other hand, if the impulse response is unstable, the response due to any real excitation will usually be unstable.

How can we tell if a given transfer function represents a stable, unstable, or marginally stable situation? To begin with, we should emphasize that *a passive linear circuit without electronic feedback can never be unstable.* Our sole purpose in considering the unstable case is simply to introduce the reader to this possibility for future reference. Almost all passive *RLC* circuits fall into the stable case, although if certain idealized assumptions are made, the marginally stable case will occasionally be encountered.

To determine a criterion for stability, let us look at the relationship between pole positions and the corresponding terms of the time response. We will consider each possible type of response individually. In the discussion that follows, reference to the *left-hand half-plane* will include all points to the left of, but not including, the $j\omega$-axis. Similarly the *right-hand half-plane* will include all points to the right of, but not including, the $j\omega$-axis. The *negative real axis* will include all points on the real axis to the left of, but not including, the origin, and the *positive real axis* will include all points on the real axis to the right of, but not including, the origin. The $j\omega$-axis (including the origin) will be considered as a separate domain.

Pole on the Negative Real Axis

A pole plot of a first-order pole on the negative real axis is shown in Fig. 7-13a and the corresponding time response is shown in (b). The time response is of the form

$$f(t) = A\epsilon^{-\alpha t} \tag{7-78}$$

which approaches zero and is clearly *stable*.

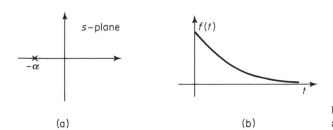

(a)   (b)

Figure 7-13   Pole on negative real axis and the corresponding time response.

If the pole were a multiple-order pole, the time response would be of the form

$$f(t) = At^k\epsilon^{-\alpha t} \tag{7-79}$$

We leave it as an exercise for the reader to show that Eq. (7-79) represents a stable response.

Complex Poles in the Left-Hand Half-Plane

A pole plot of a pair of first-order complex poles in the left-hand half-plane is shown in Fig. 7-14a and the corresponding time response is shown in (b). The time response is of the form

$$f(t) = A\epsilon^{-\alpha t} \sin(\omega t + \theta) \tag{7-80}$$

which approaches zero and is clearly *stable*.

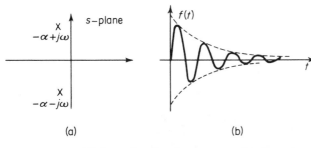

Figure 7-14   Complex poles in left-hand half-plane and the corresponding time response.

(a)   (b)

If the pair of poles were multiple-order poles, the time response would be of the form

$$f(t) = At^k\epsilon^{-\alpha t} \sin(\omega t + \theta) \tag{7-81}$$

We will leave as an exercise for the reader to show that Eq. (7-81) represents a stable response.

So far, we have seen that both simple-order and multiple-order poles in the left-hand half-plane represent stable responses. Next we look at the effect of poles in the right-hand half-plane.

Pole on the Positive Real Axis

A pole plot of a first-order pole on the positive real axis is shown in Fig. 7-15a and the corresponding time response is shown in (b). The time response is of the form

$$f(t) = A\epsilon^{\alpha t} \tag{7-82}$$

Since this function grows without bound, the response is clearly *unstable*. It can be readily shown that multiple-order poles on the positive real axis also represent unstable responses.

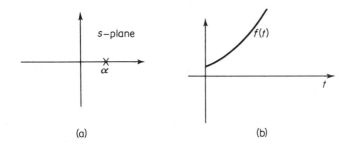

Figure 7-15   Pole on positive real axis and the corresponding time response.

(a)   (b)

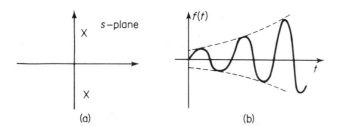

Figure 7-16   Complex poles in right-hand half-plane and the corresponding time response.

Complex Poles in the Right-Hand Half-Plane

A pole plot of a pair of first-order complex poles in the right-hand half-plane is shown in Fig. 7-16a and the corresponding time response is shown in (b). The time response is of the form

$$f(t) = A\epsilon^{\alpha t} \sin(\omega t + \theta) \tag{7-83}$$

This function grows without bound, and thus the response is *unstable*. It can be readily shown that multiple-order poles in the right-hand half-plane also represent unstable responses. Thus, poles in the right-hand half-plane always represent instability. The reader is reminded that right-hand half-plane poles can never occur in passive *RLC* networks.

Pole at the Origin

So far, we have completely ignored the $j\omega$-axis. In considering the $j\omega$-axis, let us first observe the effect of a pole at the origin. A pole plot of a first-order pole at the origin is shown in Fig. 7-17a and the corresponding time response is shown in (b). The time response is simply

$$f(t) = A \tag{7-84}$$

Since this term is, for all time, a constant, it is classified as a *marginally stable* case.

Suppose that the pole at the origin were a multiple-order pole. The response would then be of the form

$$f(t) = At^k \tag{7-85}$$

A typical response is shown in Fig. 7-18 for the case of a second-order pole [$k = 1$ in Eq. (7-85)]. This response is clearly *unstable*. This unstable response could result

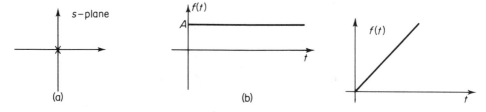

Figure 7-17   Pole at origin and the corresponding time response.

Figure 7-18   Response due to second-order poles at origin.

in two ways: (a) The system itself could contain a multiple-order pole resulting in *natural instability*; (b) The system could contain a first-order pole at the origin which in itself represents the marginally stable case. However, if the system were excited by a step (dc) function, the transform of the response would contain a second-order pole at the origin, resulting in *forced instability*.

Poles on the $j\omega$-Axis

A pole plot of a pair of first-order poles on the $j\omega$-axis is shown in Fig. 7-19a and the corresponding time response is shown in (b). The time function is of the form

$$f(t) = A \sin (\omega t + \theta) \tag{7-86}$$

Since this function oscillates for all time, it represents a *marginally stable* case.

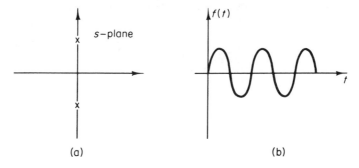

(a)   (b)

Figure 7-19   Poles on imaginary axis and the corresponding time response.

If the pair of poles were of multiple order, the response would be of the form

$$f(t) = At^k \sin (\omega t + \theta) \tag{7-87}$$

A typical response is shown in Fig. 7-20 for the case of a second-order pole. This response is *unstable* and, as in the case of the pole at the origin, could result in two ways: (a) The system itself could contain a pair of multiple-order poles on the $j\omega$-axis, resulting in natural instability; (b) The system could contain a pair of first-order poles on the $j\omega$-axis, and if the excitation were a sinusoid of the same frequency, the transform of the response would contain a pair of second-order poles on the $j\omega$-axis, resulting in forced instability.

The forced unstable response produced by exciting a system containing poles on the $j\omega$-axis with an excitation having poles at the same location is a form of

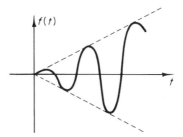

Figure 7-20   Response due to second-order poles on imaginary axis.

forced resonance. The system is excited by a source having the same frequency as a natural resonant frequency of the system, resulting in a continual growth of the response.

Now that we have observed the effect of possible pole locations on the nature of the time function, the next question concerns the effect of zero locations. This question is readily resolved if we note that, in a true transfer function (excluding driving-point impedance functions), the locations of zeros do not ordinarily affect stability, but merely the magnitudes of the various components of the time response. Thus, in general, *stable transfer functions may contain zeros in the right-hand half-plane.* A stable transfer function which does not contain any right-hand half-plane zeros is called a *minimum phase function.* If right-hand half-plane zeros are present, it is called a *non-minimum phase function.* The reasoning behind this terminology will be clearer in the next chapter.

Does this freedom on zero locations also exist for a driving-point impedance function? This question can be resolved if we recall that the impedance function is unique in that, if the roles of the excitation and response are reversed, the new "transfer function" is the reciprocal of the original. Thus if $I(s)$ is regarded as the excitation and $V(s)$ as the response, the "transfer function" is

$$\frac{V(s)}{I(s)} = Z(s) \tag{7-88}$$

However, if $V(s)$ is regarded as the excitation and $I(s)$ as the response, the transfer function is

$$\frac{I(s)}{V(s)} = Y(s) = \frac{1}{Z(s)} \tag{7-89}$$

From Eqs. (7-88) and (7-89) it is seen that the zeros of $Z(s)$ become the poles of $Y(s)$, and vice versa. Thus if a right-hand half-plane zero were possible for $Z(s)$, it would represent a natural unstable situation for $Y(s)$, which is impossible in a passive network. Thus *neither poles nor zeros of a driving-point impedance function can ever exist in the right-hand half-plane.*

A pole on the $j\omega$-axis is not unstable as long as it is a first-order pole. By the same argument as in the preceding paragraph, a simple-order zero on the $j\omega$-axis should be possible. Thus *poles and zeros of a driving-point impedance function may be located on the $j\omega$-axis, but such critical frequencies are always of simple order.*

From a design standpoint, these restrictions on the poles and zeros of impedance functions are *necessary conditions* but are not *sufficient conditions.* To be physically realizable, further conditions must be met. Such conditions will be considered in the next section.

To summarize:

1. Poles of a transfer function in the left-hand half of the *s*-plane represent stable time-response terms, regardless of the order of the poles.
2. Poles of a transfer function in the right-hand half of the *s*-plane represent unstable time-response terms, regardless of the order of the poles.

3. Poles of the transfer function of simple order on the $j\omega$-axis (including the origin) represent marginally stable time-response terms. If, however, the system is excited by a sinusoidal signal of the same frequency (dc for pole at origin), the response will be unstable.

4. Poles of the transfer function of multiple order on the $j\omega$-axis represent unstable time-response terms.

5. A system is only as stable as its least stable term. Thus all poles of a perfectly stable system must lie in the left-hand half of the s-plane.

6. Excluding driving-point impedance functions, zeros of transfer functions may lie in the right-hand half of the s-plane without affecting stability.

7. Both poles and zeros of driving-point impedance and admittance functions are never located in the right-hand half of the s-plane.

8. Poles and zeros of driving-point impedance functions located on the $j\omega$-axis of the s-plane are always of simple order.

Example 7-9

The circuit of Fig. 7-21 is an active circuit with feedback. The amplifier portion of the circuit performs the following operation:

$$V_3(s) = 4[(V_1(s) + V_2(s)]   \tag{7-90}$$

Show that the circuit is unstable, and determine the pole and zero locations. The amplifier is assumed to have infinite input impedance and zero output impedance.

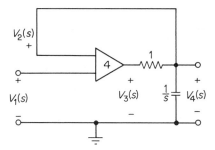

Figure 7-21   Circuit of Example 7-9.

Solution   Since the entire output voltage is fed back to the input,

$$V_2(s) = V_4(s)   \tag{7-91}$$

The RC portion of the circuit has the transfer function

$$\frac{V_4(s)}{V_3(s)} = \frac{1}{s+1}   \tag{7-92}$$

To obtain a single relationship between $V_1(s)$ and $V_4(s)$, we may solve for $V_3(s)$ in Eq. (7-92) and substitute this expression and the result of Eq. (7-91) into Eq.

(7-90). We obtain

$$(s + 1)V_4(s) = 4V_1(s) + 4V_4(s) \tag{7-93}$$

Solving for the transfer function, we have

$$G(s) = \frac{V_4(s)}{V_1(s)} = \frac{4}{s - 3} \tag{7-94}$$

Thus the single pole is located at $s = 3$, representing an *unstable* situation. The single zero is located at $s = \infty$. The impulse response is

$$g(t) = 4\epsilon^{3t} \tag{7-95}$$

which grows without bound.

This circuit is not intended to represent any useful circuit but is merely a demonstration of instability. Actually, as the response grows in magnitude, the circuit and amplifier components will saturate, resulting in decreased gain. This effect will eventually limit the magnitude of the response to a finite value. Thus whenever a system has an unstable response that appears to grow without bound, it must be remembered that the linear model holds only within regions since no real physical variable can grow indefinitely.

Instability in control systems and feeback amplifier circuits usually occurs accidentally and is not normally desirable. On the other hand, the design of oscillator circuits requires a deliberate approach toward instability. Actually the ideal sinusoidal oscillator is designed to have a pair of simple-order poles on the $j\omega$-axis. Thus the degree of stability desired depends on the application at hand.

7-7 TRANSFER FUNCTION "ALGEBRA"

At the systems level, complete linear circuit blocks are connected together to achieve a composite system function. If the complete system is linear and has a single input and a single output, a composite transfer function may be used to describe the net relationship between the input and the output. This composite transfer function may be expressed as a combination of the individual transfer functions. Certain rules will be developed in this section for combining forms that arise frequently in practice.

At the outset, an important stipulation must be made. *Each transfer function given is either assumed to be unaffected by the interconnection used, or the transfer function is defined under the loaded conditions given.* To elaborate further on this concept, assume a voltage transfer function $V_2(s)/V_1(s) = G(s)$. If the output terminals are connected to the input terminals of one other unit, there may be a loading effect that could result in a different value for $V_2(s)$ than the value determined under open-circuit conditions. If the loading effect does alter $V_2(s)$, a modified transfer function could be defined under such conditions. The point is that one cannot simply "throw together" blocks and assume that transfer functions remain unchanged. The effects of loading will be illustrated with a simple circuit in Example 7-10. Several arrangements will now be considered.

Cascade Connection

A cascade connection of several blocks having individual transfer functions is shown at the top of Fig. 7-22. To avoid "clutter" on the figure, only the net input $X(s)$ and the net output $Y(s)$ are shown. However, for the development that follows, assume that the output of $G_1(s)$ is $Y_1(s)$, the output of $G_2(s)$ is $Y_2(s)$, and so on. The input of $G_n(s)$ is $Y_{n-1}(s)$, and the output of $G_n(s)$ is the overall output $Y(s)$. The following array of equations can then be written

$$Y_1(s) = G_1(s)X(s) \tag{7-96a}$$

$$Y_2(s) = G_2(s)Y_1(s) \tag{7-96b}$$

$$\vdots$$

$$Y_{n-1}(s) = G_{n-1}(s)Y_{n-2}(s) \tag{7-96c}$$

$$Y(s) = G_n(s)Y_{n-1}(s) \tag{7-96d}$$

At the risk of redundancy, it should be stressed again that the preceding individual transfer functions are either unaffected by the connections, or the transfer functions are defined under loaded conditions.

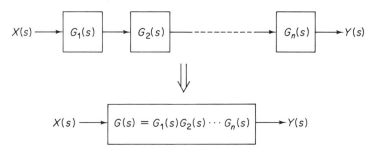

Figure 7-22   Cascade connection of transfer functions and an equivalent single transfer function.

The preceding equations must be combined in a manner that will eliminate all variables from $Y_1(s)$ through $Y_{n-1}(s)$. A systematic way of achieving this is as follows: Substitute $Y_{n-1}(s)$ expressed in terms of $Y_{n-2}(s)$ from Eq. (7-96c) in Eq. (7-96d). Next substitute $Y_{n-2}(s)$ from the assumed next equation (not shown) expressed in terms of $Y_{n-3}(s)$. This pattern continues will $Y_1(s)$ expressed in terms of $X(s)$ is reached. At that point, the resulting equation, appropriately arranged, reads

$$Y(s) = G_1(s)G_2(s) \cdots G_n(s)X(s) \tag{7-97}$$

The net transfer function is thus

$$G(s) = \frac{Y(s)}{X(s)} = G_1(s)G_2(s) \cdots G_n(s) \tag{7-98}$$

Stated in words, *the composite transfer function of a cascade connection of transfer functions is the product of all the individual transfer functions.*

Parallel Connection

A parallel connection of several individual transfer functions is shown at the top of Fig. 7-23. Only the net input $X(s)$ and the net output $Y(s)$ are shown. However, define the outputs of the n blocks as $Y_1(s)$ through $Y_n(s)$. Since $X(s)$ is common to all of these, we can write

$$Y_1(s) = G_1(s)X(s) \tag{7-99a}$$

$$Y_2(s) = G_2(s)X(s) \tag{7-99b}$$

$$\vdots$$

$$Y_n(s) = G_n(s)X(s) \tag{7-99c}$$

These n variables are assumed to be combined together at the output according to the equation

$$Y(s) = Y_1(s) + Y_2(s) + \cdots + Y_n(s) \tag{7-100}$$

When the individual forms are substituted in Eq. (7-100), $X(s)$ is a common factor, and we have

$$Y(s) = [G_1(s) + G_2(s) + \cdots + G_n(s)]X(s) \tag{7-101}$$

The net transfer function is thus

$$G(s) = \frac{Y(s)}{X(s)} = G_1(s) + G_2(s) + \cdots + G_n(s) \tag{7-102}$$

Stated in words, *the composite transfer function of a parallel connection of transfer functions is the sum of all the individual transfer functions.*

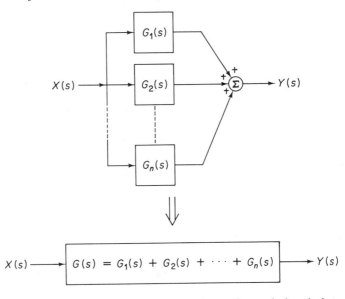

Figure 7-23   Parallel connection of transfer functions and an equivalent single transfer function.

Feedback Loop

A feedback loop, consisting of a forward transfer function $G_1(s)$ and a feedback transfer function $G_2(s)$ is shown at the top of Fig. 7-24. Let $F(s)$ represent the output of the feedback block, and this quantity is subtracted from the input $X(s)$ to yield a difference variable $D(s)$ defined as

$$D(s) = X(s) - F(s) \qquad (7\text{-}103)$$

The difference variable is multiplied by the forward transfer function to yield the output $Y(s)$, that is,

$$Y(s) = G_1(s)D(s) \qquad (7\text{-}104)$$

The output is multiplied by the feedback transfer function to produce the feedback variable.

$$F(s) = G_2(s)Y(s) \qquad (7\text{-}105)$$

The variables $F(s)$ and $D(s)$ are eliminated by substituting Eq. (7-105) in Eq. (7-103) and then substituting that result in Eq. (7-104). After some manipulation, we obtain

$$Y(s) = \left[\frac{G_1(s)}{1 + G_1(s)G_2(s)} \right] X(s) \qquad (7\text{-}106)$$

The transfer function is

$$G(s) = \frac{Y(s)}{X(s)} = \frac{G_1(s)}{1 + G_1(s)G_2(s)} \qquad (7\text{-}107)$$

The result of Eq. (7-107) is one of the most important relationships of linear system theory. It serves as the basis for much of the design work of stable linear circuits and closed-loop feedback control systems. Our approach here is simply to show this as a simplication of feedback loops that might arise in working with transfer functions.

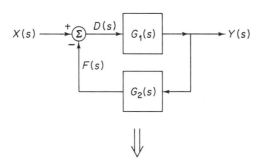

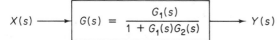

Figure 7-24 Transfer function with feedback loop and an equivalent single transfer function.

Example 7-10

A simple resistive network will be used in this problem to illustrate how loading effects can affect or alter the simplified relations of this section. Consider the simple voltage divider of Fig. 7-25a. (a) Determine the transfer function of this circuit. (b) Assume that two of these sections are connected together as shown in Fig. 7-25b. Work out the composite transfer function and compare it with the value that would be obtained by employing the cascade formula of this section.

      Solution   (a) Let $G_A(s) = V(s)/V_1(s)$ for the simple network of Fig. 7-25a. This result is immediately determined as

$$G_A(s) = \frac{1000}{1000 + 1000} = \frac{1}{2} \tag{7-108}$$

      (b) The transfer function of the circuit in Fig. 7-25b can be determined on a circuit analysis basis by several approaches. This author's favorite method for this circuit is a successive Thévenin transformation approach as illustrated in Fig. 7-25c. Looking back from the output of the first stage, the open-circuit voltage is $V_1(s)/2$, and the equivalent resistance is 500 Ω. A further application of the voltage-divider rule reads

$$G(s) = \frac{V_2(s)}{V_1(s)} = \frac{1000}{1000 + 1500} \times \frac{1}{2} = \frac{1}{5} \tag{7-109}$$

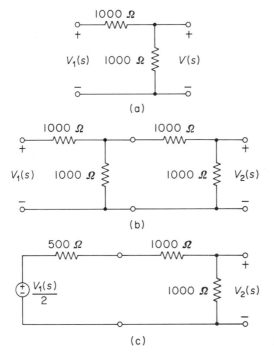

(a)

(b)

(c)

Figure 7-25 Circuits of Example 7-10.

Suppose, however, that one blindly used the result of the cascade simplification formula of Eq. (7-97) as depicted in Fig. 7-22. Since each section has an individual transfer function of $\frac{1}{2}$, the formula would suggest a net transfer function of $\frac{1}{2} \times \frac{1}{2} = \frac{1}{4}$, which is obviously different from the result obtained. What is wrong?

The answer to this difficulty lies in the interaction and loading effects of the two sections. Assume that an input voltage is applied to the single stage of Fig. 7-25a. If connecting the second stage to the output of the first stage alters the output voltage of the first stage, there will be a loading effect. For transfer functions to be multiplicative, there must be no alteration of the output variable of the first stage when the second stage is connected. In this circuit, the voltage across the output stage will no longer be $V_1(s)/2$ when the second stage is connected.

When the output voltage of a stage is the input voltage to the next stage, a necessary condition for minimal loading effect is that the input resistance looking into the second stage must be very large compared with the equivalent output resistance of the first stage (typically 100 times as great or more). In this circuit, the input resistance to the second stage is 2000 Ω and the output resistance of the first stage is 500 Ω, so the desired inequality is not satisfied. The preceding discussion explains why the ideal desired output impedance of a voltage amplifier is zero and why the ideal desired input impedance is infinite. For a current amplifier, the desired situation is opposite.

Example 7-11

Determine a single transfer function equivalent to the system of Fig. 7-26a.

Solution   As stated numerous times earlier, it is assumed that the transfer functions shown are applicable under the loaded conditions given. To simplify the notation the s-variable will be eliminated from the transfer functions in parts (b), (c), and (d) of Fig. 7-26 and in the discussion that follows.

The first step is to simplify the cascade connection of G_1 and G_2, whose net transfer function is $G_1 G_2$. The resulting system is shown in Fig. 7-26b. Next, the feedback loop consisting of $G_1 G_2$ as the forward transfer function and G_3 as the feedback transfer function is simplified, as shown in Fig. 7-26c. Finally, the resulting parallel structure is simplified as shown in Fig. 7-26d. The overall transfer function is

$$G(s) = \frac{Y(s)}{X(s)} = \frac{G_1 G_2}{1 + G_1 G_2 G_3} + G_4 \qquad (7\text{-}110a)$$

$$= \frac{G_1 G_2 + G_4 + G_1 G_2 G_3 G_4}{1 + G_1 G_2 G_3} \qquad (7\text{-}110b)$$

where the latter form utilizes a common denominator.

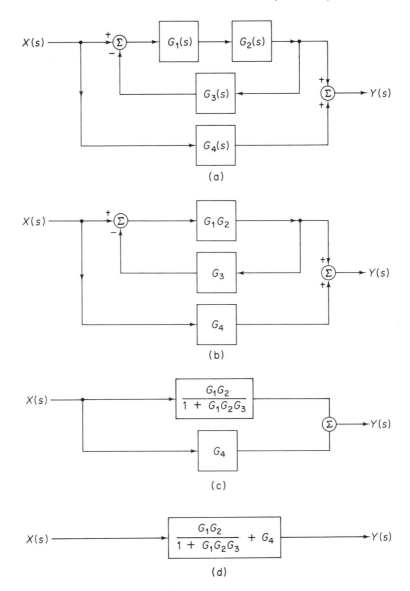

Figure 7-26　System of Example 7-11 and simplification steps.

7-8 CONVOLUTION

We have seen that the output response of a system may be determined from a knowledge of the impulse response and the input excitation by transforming all variables to the s-domain and utilizing the transfer function concept. We consider now the concept of *convolution*, which permits the response to be determined from the impulse response and the input by working directly in the time domain.

The concept of convolution will be developed through an analysis which provides some insight into the process and which provides a mechanism for later evolution into discrete-time system analysis to be considered in Chapter 10.

Consider the signal $x(t)$ shown in Fig. 7-27, which is assumed to excite a system having an impulse response $g(t)$ at $t = 0$. Assume that the signal is decomposed into a series of rectangles each of width $\Delta\tau$. The height of each rectangular "slab" will be assumed to be the value of $x(t)$ at the beginning of the interval. Eventually, the slabs will be made to be quite narrow, in which case the difference in heights between the beginning and the ending of a given rectangle will be insignificant.

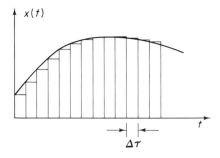

Figure 7-27 Representation of an excitation function in terms of rectangles as used in convolution development.

The process being considered is leading up to the representation of $x(t)$ as a series of narrow pulses, each of which may be approximated as an impulse function. The first pulse has an area $x(0)\,\Delta\tau$, and this value is assumed to be the weight of an impulse starting at $t = 0$. The second pulse has an area $x(\Delta\tau)\,\Delta\tau$, and this value is assumed to be the weight of an impulse starting at $t = \Delta\tau$, etc. Thus $x(t)$ can be approximated as

$$x(t) \approx x(0)\,\Delta\tau\,\delta(t) + x(\Delta\tau)\,\Delta\tau\,\delta(t - \Delta\tau) + x(2\,\Delta\tau)\,\Delta\tau\,\delta(t - 2\Delta\tau) + \cdots$$

$$= \sum_{n=0}^{\infty} x(n\,\Delta\tau)\,\Delta\tau\,\delta(t - n\,\Delta\tau) \tag{7-111}$$

A unit impulse starting at $t = 0$ would generate the impulse response $g(t)$. A unit impulse starting at $t = \Delta\tau$ would generate the impulse response $g(t - \Delta\tau)$, and so on. The excitation terms in Eq. (7-111) weighted appropriately and the corresponding response terms are summarized below.

Excitation	*Response*
$x(0)\,\Delta\tau\,\delta(t)$	$x(0)\,\Delta\tau\,g(t)$
$x(\Delta\tau)\,\Delta\tau\,\delta(t - 2\delta\tau)$	$x(\Delta\tau)\,\Delta\tau\,g(t - \Delta\tau)$
$x(2\,\Delta\tau)\,\Delta\tau\,\delta(t - 2\delta\tau)$	$x(2\,\Delta\tau)\,\Delta\tau\,g(t - 2\Delta\tau)$
$x(n\,\Delta\tau)\,\Delta\tau\,\delta(t - n\,\Delta\tau)$	$x(n\,\Delta\tau)\,\Delta\tau\,g(t - n\,\Delta\tau)$

The response $y(t)$ at an arbitrary time $t = N\,\Delta\tau$ can be determined by summing all the preceding contributions. The response is closely approximated as

$$y(t) \approx \sum_{n=0}^{N} x(n\,\Delta\tau)g(t - n\,\Delta\tau)\,\Delta\tau \tag{7-112}$$

We will now allow the widths of the individual slabs to approach zero. The quantity $\Delta\tau$ then becomes the differential $d\tau$, $n\,\Delta\tau$ is replaced by a continuous variable τ, and the summation becomes an integral. As a formal limiting process, we have

$$y(t) = \lim_{\Delta t \to 0} \sum_{n=0}^{N} x(n\,\Delta\tau)g(t - n\,\Delta\tau)\,\Delta\tau \tag{7-113a}$$

$$= \int_0^t x(\tau)g(t - \tau)\,d\tau \tag{7-113b}$$

This result is the *convolution integral.* Convolution is a commutative process, so an alternative form is

$$y(t) = \int_0^\infty g(\tau)x(t - \tau)\,d\tau \tag{7-114}$$

Note that the integration is performed in turns of a variable τ, but the final result is a function of time t. The variable τ is said to be a "dummy variable."

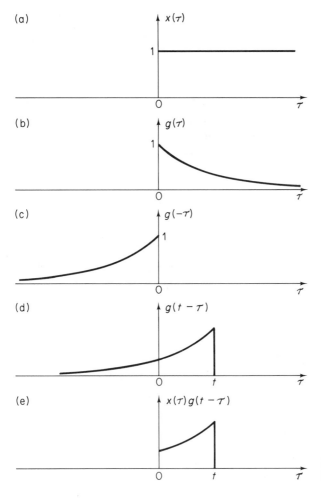

Figure 7-28 Convolution integrand functions in Example 7-12.

Example 7-12

The impulse response of a certain system is

$$g(t) = \epsilon^{-t} \qquad \text{for } t > 0 \tag{7-115}$$

Using convolution, determine the response $y(t)$ due to an input unit step excitation.

$$x(t) = u(t) \tag{7-116}$$

Solution   The form of Eq. (7-113b) will be employed, i.e.,

$$y(t) = \int_0^t x(\tau)g(t - \tau)\, d\tau \tag{7-117}$$

A series of diagrams is very helpful in visualizing the convolution process and is recommended for most cases. Refer to Fig. 7-28 in the steps that follow. First, the original time variable t in $g(t)$ and $x(t)$ is replaced by the dummy variable τ. The resulting functions have the same shapes as functions of τ as they did as functions of t and are shown in (a) and (b) of Fig. 7-28. The functional forms are $x(\tau) = u(\tau)$ and $g(\tau) = \epsilon^{-\tau}$ for $\tau > 0$.

The function $g(-\tau)$ is then formed as shown in Fig. 7-28c. Note that this operation simply rotates the function about the vertical axis and results in a new function that is nonzero only for negative τ. Next, the argument $-\tau$ is replaced by $t - \tau$ as shown in Fig. 7-28d. For positive t, this simply shifts the function to the right by t as shown.

The last step in forming the integrated is to form the product $x(\tau)g(t - \tau)$ as shown in Fig. 7-28e. The limits of the resulting integral (0 to t) are clearly visible in this sketch.

For the particular functions of this example, we have

$$\begin{aligned}
y(t) &= \int_0^t u(\tau)\epsilon^{-(t-\tau)}\, d\tau = \int_0^t \epsilon^{-t}\epsilon^{\tau}\, d\tau \\
&= \epsilon^{-t} \int_0^t \epsilon^{\tau}\, d\tau = \epsilon^{-t}\big[\epsilon^{\tau}\big]_0^t \\
&= \epsilon^{-t}(\epsilon^t - 1) = 1 - \epsilon^{-t}
\end{aligned} \tag{7-118}$$

The reader is invited to verify the correctness of this result using transform methods.

7-9 DRIVING-POINT IMPEDANCE FUNCTIONS

In considering the transfer function concept throughout this chapter, we have frequently referred to driving-point impedance or admittance functions, and we have indicated that they can often be viewed from the transfer function approach whenever the two variables of interest are the voltage and current associated with a single pair of terminals. However, we have seen that the possible pole–zero locations of a driving-point impedance or admittance are somewhat more restricted than those of the general transfer function. Impedance or admittance functions have a number of interesting properties that warrant special consideration, and this section will be devoted to such properties.

We must emphasize that the material of this section is not intended to provide a thorough treatment of this subject. The study of driving-point impedance functions constitutes a significant portion of most network synthesis texts, and much of this material is beyond the scope of this book. Most of the properties of impedance functions that we consider will be stated without proof since a complete justification of many of these properties depends heavily on the mathematics of complex-variable theory. Our purpose in stating these properties is to provide the reader with some basic insight as to the nature of driving-point impedances for reference in future work.

General Requirements

In general, all driving-point impedance or admittance functions satisfy the following basic requirements:

1. Poles and zeros lie in either the left-hand half-plane or on the $j\omega$-axis.
2. Poles and zeros on the $j\omega$-axis are of simple order. (These first two properties were developed in Section 7-6.)
3. The degrees of the numerator and denominator polynomials differ, at most, by a single degree. In effect, this statement means that as s approaches infinity, the function approximates either a constant, a constant times s, or a constant divided by s. Thus, for large s, the network is equivalent to either a resistance, an inductance, or a capacitance.
4. If $Z(s)$ or $Y(s)$ is evaluated for $s = j\omega$, the real part of the resulting expression is either positive or zero (never negative).

The preceding four requirements apply to the general *RLC* network, and they provide a means for determining whether or not a given impedance function is physically realizable. This concept is basic in network synthesis and design since, before a given synthesis can be attempted, it is necessary to determine whether or not the given function corresponds to a realizable network. The first three conditions are easily tested if the poles and zeros are known, but the fourth condition may require special considerations.

Obviously if a network is given, the impedance or admittance function will, of necessity, satisfy the requirements since it will have been derived from an actual network. Since our basic theme is network analysis and not synthesis, we will, for the most part, concern ourselves only with verifying these properties for given networks.

The preceding properties reduce to more specific requirements for cases of two-element-type circuits (i.e., *LC*, *RC*, or *RL*). In the discussion that follows it will be assumed or implied that the first three of the four general conditions must be satisfied.

The fourth condition will be automatically satisfied if the given requirements to be stated are satisfied. The value $s = \infty$ will be interpreted either as a "point" on the $j\omega$-axis or on the σ-axis, as the case may be.

LC Circuits

Driving-point impedances or admittances of circuits containing only inductance and capacitance satisfy the following restrictions on pole–zero locations:

1. All poles and zeros lie on the $j\omega$-axis, are of simple order, and alternate.
2. Either a pole or a zero must be present at $s = 0$, and either a pole or a zero must be present at $s = \infty$.

A typical LC pole–zero plot is shown in Fig. 7-29. It has become common practice to represent the point $s = \infty$ on the pole–zero plots of two-element-type circuits, as can be seen from the figure. However, the scale has been considerably compressed by necessity!

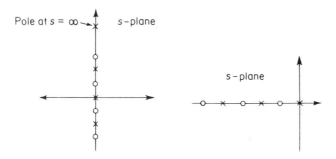

Figure 7-29 Typical pole–zero plot for LC impedance.

Figure 7-30 Typical pole–zero plot for RC impedance.

RC Circuits

Driving-point impedances or admittance of circuits containing only resistance and capacitance satisfy the following restrictions on pole–zero locations:

1. All poles and zeros lie on the σ-axis, are of simple order, and alternate.
2. The critical frequency of $Z(s)$ nearest the origin is always a pole, and the critical frequency of $Z(s)$ farthest from the origin is always a zero. However, it is not necessary that $Z(s)$ possess a pole at the origin and a zero at $s = \infty$, although it is possible to have either or both conditions. Notice that these conditions would be completely opposite for $Y(s)$.

A typical RC pole–zero plot is shown in Fig. 7-30.

RL Circuits

Driving-point impedances or admittances of circuits containing only resistance and inductance satisfy the following restrictions on pole–zero locations:

1. All poles and zeros lie on the σ-axis, are of simple order, and alternate.
2. The critical frequency of $Z(s)$ nearest the origin is always a zero, and the critical frequency of $Z(s)$ farthest from the origin is always a pole. However, it is not necessary that $Z(s)$ possess a zero at the origin and a pole at $s = \infty$, although it is possible to have either or both conditions.

Again, these conditions would be opposite for $Y(s)$. In fact, the $Z(s)$ conditions for *RC* networks are the same as the $Y(s)$ conditions for *RL* networks, and vice versa. A typical *RL* pole–zero plot is shown in Fig. 7-31.

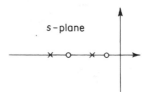

s–plane

Figure 7-31   Typical pole–zero plot for *RL* impedance.

Example 7-13

Calculate the input impedance of the *LC* network shown in Fig. 7-32a and demonstrate that it satisfies the conditions stated in this section.

    Solution   The reader is invited to show that successive series and parallel combinations yield for the input impedance

$$Z(s) = \frac{12s(s^2 + 4)}{s^4 + 10s^2 + 9} = \frac{12s(s^2 + 4)}{(s^2 + 1)(s^2 + 9)} \tag{7-119}$$

The pole–zero plot is shown in Fig. 7-32b and it is seen to satisfy the requirements for *LC* networks.

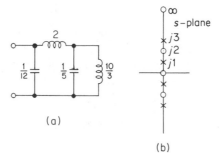

(a)

(b)

Figure 7-32   Circuit and pole–zero plot for $Z(s)$ in Example 7-13.

GENERAL PROBLEMS

7-1. Determine the transfer function and the impulse response of the *RC* circuit of Fig. P7-1. The input is v_1 and the output is v_2.

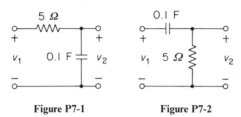

Figure P7-1   Figure P7-2

7-2. Determine the transfer function and the impulse response of the *RC* circuit of Fig. P7-2. The input is v_1 and the output is v_2.

7-3. Determine the transfer function and the impulse response of the *RL* circuit of Fig. P7-3, and show that they are the same as for the *RC* circuit of Problem 7-1. The input is v_1 and the output is v_2.

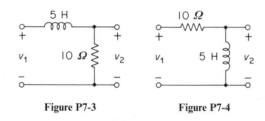

Figure P7-3   Figure P7-4

7-4. Determine the transfer function and the impulse response of the *RL* circuit of Fig. P7-4 and show that they are the same as for the *RC* circuit of Problem 7-2. The input is v_1 and the output is v_2.

7-5. Determine the transfer function and the impulse response of the circuit of Fig. P7-5. The input is v_1 and the output is v_2.

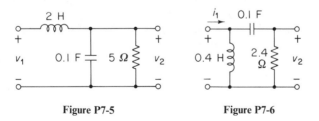

Figure P7-5   Figure P7-6

7-6. Determine the transfer function and the impulse response of the circuit of Fig. P7-6. The input is v_1 and the output is v_2.

7-7. The transfer function of a normalized second-order active low-pass Butterworth filter was determined in Example 7-3. That particular filter used a unity-gain amplifier. A different form in which the two capacitors have equal values is shown in Fig. P7-7. The

VCVS now has a gain of 1.5858. Determine the transfer function $G(s) = V_2(s)/V_1(s)$ and show that it has the same *form* as in Example 7-3. (The denominator polynomial is the same but the numerator constant is different.)

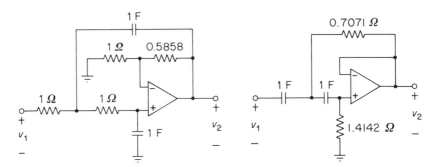

Figure P7-7   Figure P7-8

7-8. The circuit of Fig. P7-8 is one particular active form of a second-order Butterworth high-pass filter using an operational amplifier, and it is normalized to a cutoff frequency of 1 rad/s with 1-Ω resistors. Determine the transfer function $G(s) = V_2(s)/V_1(s)$. The operational amplifier is connected as a voltage follower and it is modeled in the same manner as in Example 7-3.

7-9. The impulse response of a certain system is given by

$$g(t) = 8\epsilon^{-4t}$$

Using transform methods, determine the response $y(t)$ when $x(t) = 10 \cos 4t$.

7-10. The impulse response of a certain system is given by

$$g(t) = \epsilon^{-t} \sin 2t$$

Using transform methods, determine the response $y(t)$ when $x(t) = 10u(t)$.

7-11. Determine the transfer function of the circuit of Problem 7-1 by first obtaining a differential equation and then transforming.

7-12. Determine the transfer function of the circuit of Problem 7-2 by first obtaining a differential equation and then transforming.

7-13. Determine the transfer function of the circuit of Problem 7-5 by first obtaining a differential equation and then transforming.

7-14. Determine the transfer function of the circuit of Problem 7-6 by first obtaining a differential equation and then transforming.

7-15. The input to a certain linear system is $x(t)$, and the output is $y(t)$. The relationship between input and output is given by

$$2\frac{d^2y}{dt^2} + 8\frac{dy}{dt} + 26y = 25\frac{dx}{dt} + 50x$$

Determine the transfer function.

7-16. The input–output relationship of a certain ideal delay line is given by

$$y(t) = x(t - a)u(t - a)$$

Determine the transfer function and the impulse response.

7-17. The transfer function of a certain circuit is given by

$$G(s) = \frac{V_2(s)}{V_1(s)} = \frac{2s^2 + 5}{s^2 + 3s + 4}$$

Write the differential equation describing the input–output relationship.

7-18. The transfer function of a certain system is given by

$$G(s) = \frac{Y(s)}{X(s)} = \frac{50s^2 + 30s + 25}{17s^3 + 28s^2 + 19s + 40}$$

Write the differential equation describing the input–output relationship.

7-19. Determine the step response of the circuit of Problem 7-1 from the impulse response by using Eq. (7-51).

7-20. Determine the step response of the circuit of Problem 7-2 from the impulse response by using Eq. (7-51).

7-21. The step response of a certain system is given by

$$h(t) = 5 + 10\epsilon^{-t} \sin (2t - 30°)$$

Determine the impulse response using Eq. (7-54).

7-22. The step response of a certain system is given by

$$h(t) = 8 + 4\epsilon^{-t} - 2\epsilon^{-2t}$$

Determine the impulse response using Eq. (7-54).

7-23. (a) For the circuit of Fig. P7-23, assume that the excitation is $i(t)$ and the response is $v(t)$. Show that the transfer function is $Z(s)$ and determine it.

(b) Now assume that the excitation is $v(t)$ and the response is $i(t)$. Show that the transfer function is $Y(s)$ and is the reciprocal of the result of (a).

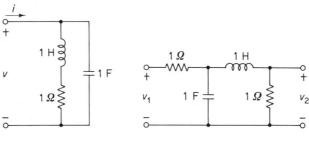

Figure P7-23   Figure P7-24

7-24. (a) Determine the transfer function of the circuit of Fig. P7-24 with $v_1(t)$ considered as the excitation and $v_2(t)$ considered as the response.

(b) Assume now that $v_2(t)$ is the excitation and $v_1(t)$ is the response. Determine the transfer function, and demonstrate that it is not simply related to the result of (a).

7-25. Determine the transfer function of the circuit of Fig. P7-25 with $i_1(t)$ considered as the excitation and $v_2(t)$ considered as the response. Show that if the network is turned end to end, the same transfer function is obtained. Explain by means of the reciprocity theorem. This type of transfer function is called a *transfer impedance* since it has the dimensions of ohms. It occurs frequently in four-terminal network theory and is designated as $Z_{21}(s)$ or $Z_{12}(s)$.

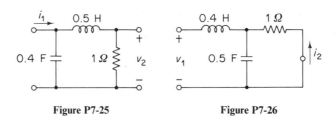

Figure P7-25   Figure P7-26

7-26. Determine the transfer function of the circuit of Fig. P7-26 with $v_1(t)$ considered as the excitation and $i_2(t)$ considered as the response. Show that if the network is turned end to end, the same transfer function is obtained. (The original input terminals are shorted, and the original output terminals are opened to allow the assumed voltage source to be inserted.) Explain by means of the reciprocity theorem. This type of transfer function is called a *transfer admittance* since it has the dimensions of siemens. It occurs frequently in four-terminal network theory and is designated as $Y_{21}(s)$ or $Y_{12}(s)$.

In Problems 7-27 through 7-31, determine the locations of all poles and zeros (including zeros at $s = \infty$). In each case, make an s-plane plot of the finite poles and zeros.

7-27. $G(s) = \dfrac{3s^2 + 9s + 6}{2s^3 + 14s^2 + 24s}$ **7-28.** $G(s) = \dfrac{10}{s^3 + 2000s^2 + 5 \times 10^6 s}$

7-29. $G(s) = \dfrac{s(s^2 - 2s + 5)}{(s^2 + 4)(s^2 + 4s + 13)}$ **7-30.** $G(s) = \dfrac{5s^2 + 20s + 15}{s(s + 3)(s^2 + 4s + 4)}$

7-31. $G(s) = \dfrac{20s(s + 2)(s - 2)}{(s^4 + 3s^2 + 2)(s^4 + 4s^2 + 3)}$

7-32. Determine the input impedance $Z(s)$ of the circuit of Fig. P7-32. Determine the poles and zeros and plot them in the s-plane. Show that $Z(s)$ contains a pole at $s = \infty$.

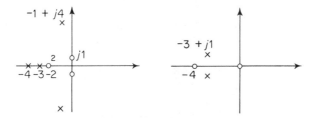

Figure P7-32

In Problems 7-33 through 7-36, determine the transfer functions from the pole–zero plots given. In each case, determine the arbitrary constant from the information provided.

7-33. See Fig. P7-33. $G(0) = 1$.   **7-34.** See Fig. P7-34. $G(\infty) = 10$.

7-35. See Fig. P7-35. $G(\infty) \approx 10/s$.   **7-36.** See Fig. P7-36. $G(\infty) \approx 1/s^2$.

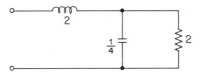

Figure P7-33   Figure P7-34

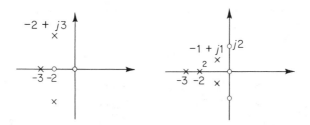

| Figure P7-35 | Figure P7-36 |

In Problems 7-37 through 7-40, determine whether the given transfer function is stable, unstable, or marginally stable. (None represent driving-point impedance functions.)

7-37. $G(s) = \dfrac{20(s^2 + 16)(s + 4)}{(s^2 + 3s + 2)(s^2 + 6s + 25)}$

7-38. $G(s) = \dfrac{100s(s - 2)}{(s^2 + 16)(s^2 + 6s + 8)}$

7-39. $G(s) = \dfrac{2s(s + 4)}{s^2 + s - 2}$

7-40. $G(s) = \dfrac{2s + 1}{s(s^4 + 2s^2 + 1)}$

7-41. Determine a single transfer function equivalent to the system of Fig. P7-41.

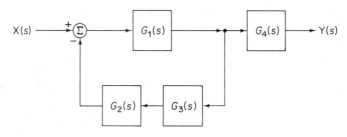

Figure P7-41

7-42. Determine a single transfer function equivalent to the system of Fig. P7-42.

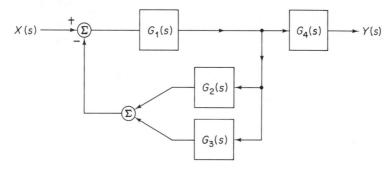

Figure P7-42

7-43. The impulse response of a certain system is

$$g(t) = 5\epsilon^{-5t}$$

Using convolution, determine the response $y(t)$ due to an input unit step function

$$x(t) = u(t)$$

7-44. For the system of Problem 7-43, use convolution to determine the response $y(t)$ due to an input ramp function $x(t) = tu(t)$

7-45. Determine the response in Problem 7-9 using convolution.

7-46. Determine the response in Problem 7-10 using convolution.

7-47. Calculate the driving-point impedance $Z(s)$ for the circuit of Fig. P7-47, and show that the four requirements for general *RLC* impedances are met.

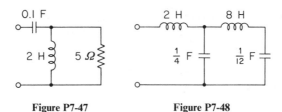

Figure P7-47   **Figure P7-48**

7-48. Calculate the driving-point impedance $Z(s)$ for the *LC* circuit of Fig. P7-48, and show that the pole–zero requirements are met.

7-49. Calculate the driving-point impedance $Z(s)$ for the *RC* circuit of Fig. P7-49, and show that the pole–zero requirements are met.

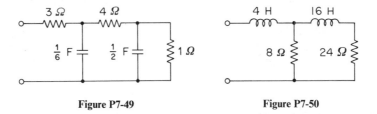

Figure P7-49   **Figure P7-50**

7-50. Calculate the driving-point impedance $Z(s)$ for the *RL* circuit of Fig. P7-50, and show that the pole–zero requirements are met.

DERIVATION PROBLEMS

7-51. Prove that the response of Eq. (7-79) is a stable response by showing that

$$\lim_{t \to \infty} f(t) = 0$$

(*Hint:* Consider L'Hôpital's rule.)

7-52. Prove that the response of Eq. (7-81) is a stable response by showing that

$$\lim_{t \to \infty} f(t) = 0$$

(See the hint in Problem 7-51.)

7-53. Consider a transfer function of the form

$$G(s) = \frac{N(s)}{s^2 + bs + c}$$

where $N(s)$ is the numerator polynomial, which is assumed to be of degree two or less. (Note that the factor of s^2 has been adjusted to be unity.) Show that necessary and sufficient conditions that $G(s)$ be stable are $b > 0$ and $c > 0$.

7-54. Consider the normalized active filter circuit shown in Fig. P7-54.

(a) Show that the transfer function is

$$G(s) = \frac{K}{s^2 + s(3 - K) + 1}$$

(b) Show that the circuit is stable for $K < 3$. (*Hint:* Refer to Problem 7-53.)
(c) Show that the circuit becomes marginally stable for $K = 3$.
(d) Determine the radian frequency of the natural oscillation under the condition of (c).

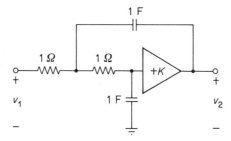

Figure P7-54

7-55. A somewhat general circuit form from which a variety of active filter types, such as considered in this chapter, can be derived is shown in Fig. P7-55. The amplifier is assumed to be an ideal VCVS with voltage gain K. Each of the blocks is a passive admittance. Show that the transfer function is

$$G(s) = \frac{V_2(s)}{V_1(s)} = \frac{K Y_1 Y_2}{Y_1 Y_2 + Y_1 Y_4 + Y_2 Y_4 + Y_3 Y_4 + Y_2 Y_3 (1 - K)}$$

(For brevity, the s arguments have been deleted from the admittances in this expression.)

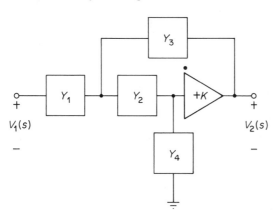

Figure P7-55

7-56. Consider the net transfer function for a feedback loop as given by Eq. (7-107). Show that the system is marginally stable if

$$G_1(j\omega_0)G_2(j\omega_0) = -1$$

where the $j\omega$-axis poles associated with the marginally stable condition are $\pm j\omega_0$.

COMPUTER PROBLEMS

7-57. Consider the transfer function

$$G(s) = \frac{Y(s)}{X(s)} = \frac{A}{(s + \alpha_1)(s + \alpha_2)}$$

Assume that the system is excited by a step function $x(t) = Xu(t)$. After obtaining a suitable expression for $y(t)$, write a computer program to evaluate $y(t)$ over an interval from t_1 to t_2 in steps of Δt.

Input data: $\alpha_1, \alpha_2, A, X, t_1, t_2, \Delta t$

Output data: $t, y(t)$

7-58. Write a computer program as discussed in Problem 7-57 for the transfer function

$$G(s) = \frac{As}{(s + \alpha_1)(s + \alpha_2)}$$

7-59. Write a computer program as discussed in Problem 7-57 for the transfer function

$$G(s) = \frac{As^2}{(s + \alpha_1)(s + \alpha_2)}$$

7-60. Write a computer program as discussed in Problem 7-57 for the transfer function

$$G(s) = \frac{A}{(s + \alpha)^2}$$

(Note that $\alpha_1 = \alpha_2$ in this case, so only one value of α is needed in the input data.)

8

THE SINUSOIDAL STEADY STATE

OVERVIEW

The analysis of electric circuits under steady-state sinusoidal conditions is one of the most important special cases of interest in circuit theory. Although the sinusoidal steady-state response of a network is obtained along with the transient response whenever Laplace transform methods are used, it is frequently desirable to determine the steady-state response directly. In this chapter the basic concepts of steady-state ac circuit theory will be presented, and it will be shown that it is very closely related to the more general transform approach. It will be shown that, in a sense, steady-state ac circuit theory may be considered as a special case of transform analysis.

The concepts of amplitude and phase response of a linear network as a function of frequency are developed. Special techniques for plotting amplitude and phase functions using breakpoint approximations are presented.

OBJECTIVES

After completing this chapter, the reader should be able to

1. Discuss the concept of the steady-state sinusoidal response of a circuit and compare it with the general transform solution.
2. Represent sinusoidals as phasors.
3. Determine phasor impedance forms for all passive circuit components.
4. Define *impedance*, *resistance*, *reactance*, *admittance*, *conductance*, and *susceptance*.
5. Use phasor techniques to solve for steady-state voltages and currents in circuits excited by sinusoidal sources.
6. Determine the steady-state transfer function and show how it is related to the Laplace transfer function.

7. Determine the *amplitude* and *phase* response functions for a circuit.

8. Discuss decibel gain and loss functions, and show how they are related.

9. Show the forms of the amplitude and phase response functions, along with the breakpoint approximation to the amplitude response, for poles and zeros on the real axis.

10. Determine the complete breakpoint amplitude response approximation for a transfer function composed of real poles and zeros.

8-1 GENERAL DISCUSSION

For the purpose of this chapter, let us assume that all circuits or systems under consideration belong to the *stable* category. Certainly, there is little point in considering the unstable situation, since its response will grow without bound, but the reader may wonder why we are excluding the marginally stable case. The reason behind this restriction is the fact that, in a marginally stable system, the natural response due to the system does not approach zero, and it is meaningless to speak of a steady-state response due to the excitation. Practically speaking, the marginally stable case is rarely encountered in an actual passive network since there is almost always some damping in the circuit that shifts the poles slightly into the left-hand half of the *s*-plane.

If the circuits or systems under consideration belong to the stable variety, all poles of transfer functions and both poles and zeros of impedance functions are located in the left-hand half of the *s*-plane, and the impulse response will approach zero after a sufficiently long time. Suppose then that the excitation is a single sinusoidal function. After the transient disappears, the resulting response at any point in the circuit will be a sinusoidal response of the same frequency as the excitation. We wish to emphasize this fact since it is a basic property of a linear system: *The steady-state response of a stable linear time-invariant circuit or system excited by a sinusoid is also a sinusoid of the same frequency.*

Thus if the sinusoidal excitation is specified, we automatically know the nature of any given steady-state response in the circuit. However, we must determine two parameters of the desired response: (a) the *amplitude* of the response, and (b) the relative *phase* of the response with respect to the excitation. The concept of steady-state ac circuit theory was developed around this property, and the use of steady-state techniques allows one to determine the magnitudes and phases of voltages and currents in a circuit without going through a complete transient analysis.

Historically, steady-state ac circuit theory was well developed long before transform methods became popular in circuit analysis. There are a number of excellent texts devoted to ac circuits, and it is felt that most readers of this text will have been exposed to this subject. The treatment of this chapter has been designed to correlate the fundamental concepts of steady-state ac circuit theory with the more general transform methods, but a complete treatment of the many "fine points" is best reserved for a text devoted to the subject.

Why are steady-state sinusoidal concepts so important in electrical engineering? To begin with, a vast number of sinusoidal waveforms actually occur in electric circuit applications. Second, in many circuits actually designed for random nonsinusoidal waveforms, sinusoidal *techniques* are used to measure and characterize the circuits. For example, the reader may be familiar with the frequency-response characteristic of a typical stereo amplifier in which the amplifier is said to have a uniform response from perhaps 20 Hz to 20 kHz. Although the amplifier is designed to handle non-sinusoidal music waveforms, the use of sinusoidal methods provides a convenient means of characterizing the amplifier. In fact, we show in this chapter that the transient and sinusoidal steady-state responses of a network are uniquely related, and if one is known, the other may be predicted.

8-2 THE SINUSOIDAL STEADY STATE

Let us assume that all transients in a given stable circuit have disappeared and that all voltages and currents in the circuit are sinusoidal. A typical current somewhere in the circuit will be of the form

$$i(t) = I \sin(\omega t + \theta)' \tag{8-1}$$

Similarly, a typical voltage will be of the form

$$v(t) = V \sin(\omega t + \phi) \tag{8-2}$$

In Laplace transform circuit analysis, all voltages and currents were replaced by their *transforms*. In a similar manner, in steady-state ac circuit analysis, we replace sinusoidal voltages and currents by *phasors*.

The phasor concept was considered briefly in Chapter 2 by means of the rotating-arm analogy. Our purpose at that time was to provide a convenient method for looking at the relative phase relationships between various sinusoidal functions and for adding sinusoids of the same frequency. We saw then that the phasor representations were obtained by "freezing" the rotating arms at $t = 0$ and considering the resulting fixed-position quantities as complex phasor representations of the time functions.

The reader may wish to review the elements of complex algebra provided in Appendix A if necessary. Using the notation of complex algebra, the phasor representations of $i(t)$ and $v(t)$ of Eqs. (8-1) and (8-2) are

$$\bar{I} = I\epsilon^{j\theta} = I\underline{/\theta} \tag{8-3}$$

and

$$\bar{V} = V\epsilon^{j\phi} = V\underline{/\phi} \tag{8-4}$$

Typical phasors representing these quantities are shown in Fig. 8-1.

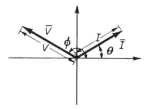

Figure 8-1 Typical phasor diagram.

Thus, a basic property of steady-state circuit theory is that voltages and currents as functions of time are replaced by complex phasors which are no longer functions of time. A voltage $v(t)$ is replaced by $\bar{V}$, and a current $i(t)$ is replaced by $\bar{I}$. In this sense we detect some close similarity between transform and steady-state analysis. This similarity will become even more evident later.

To understand this concept from a more rigorous viewpoint, let us look at the mathematical form of the *unit rotating phasor*. Consider the complex function

$$f(t) = \epsilon^{j\omega t} = 1\underline{/\omega t} \tag{8-5}$$

A complex representation of this function is shown in Fig. 8-2. The amplitude of the phasor is unity, but the phase is continually increasing linearly with time, meaning that the tip of the arm will trace out a circle as it rotates. At any given time, the projection of the phasor on the horizontal axis represents cos ωt, and the projection on the vertical axis represents sin ωt. This idea forms the basis for Euler's formula, which reads

$$\epsilon^{j\omega t} = \cos \omega t + j \sin \omega t \tag{8-6}$$

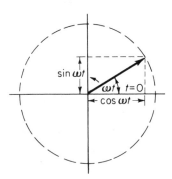

Figure 8-2 Unit rotating phasor.

If the given phasor has an initial phase angle of θ at $t = 0$, the corresponding relationship reads

$$\epsilon^{j(\omega t + \theta)} = \cos (\omega t + \theta) + j \sin (\omega t + \theta) \tag{8-7}$$

If we "freeze" this phasor at $t = 0$, it will be displaced by an angle of $\theta°$ from the horizontal axis.

If the reader reviews the geometric process of generating the sinusoidal function in Chapter 2, it will be seen that the rotating phasor concept presented here is merely a more formal mathematical statement of the original development. From Eq. (8-7) it is seen that the unit rotating phasor contains both cos $(\omega t + \theta)$ and sin $(\omega t + \theta)$ and can be used to represent either. Since our original development was based on the vertical projection, we will choose to use sin $(\omega t + \theta)$, and thus, to represent it as the imaginary part of the unit rotating phasor. Hence

$$\sin (\omega t + \theta) = \text{Im} \left[\epsilon^{j(\omega t + \theta)} \right] \tag{8-8}$$

where Im means *imaginary part of*.[†] Using $i(t)$ of Eq. (8-1) as a reference, and the definition of Eq. (8-8), we may write

$$i(t) = \text{Im} \left[I\epsilon^{j(\omega t + \theta)} \right] \tag{8-9}$$

The reader may be puzzled by the fact that we have apparently added an extra term to $i(t)$ in using the definition of Eq. (8-9). This is certainly true, but the real and imaginary parts provide a way of separating the extra term after the final circuit analysis is made. One can always separate the desired solution from the extra solution by recognition of this fact. We will see shortly that the use of the rotating arm concept provides a simple operational form for differentiation and integration in the steady-state.

The definition of Eq. (8-9) is somewhat awkward to work with because of the imaginary part designation. It has become standard practice to simply write

$$i(t) = I\epsilon^{j(\omega t + \theta)} \tag{8-10}$$

with the imaginary part (or in some cases real part) designation understood. Similarly, the voltage of Eq. (8-2) is written as

$$v(t) = V\epsilon^{j(\omega t + \phi)} \tag{8-11}$$

Looking at $i(t)$ of Eq. (8-10), we may write

$$i(t) = I\epsilon^{j\theta}\epsilon^{j\omega t} \tag{8-12}$$

Let

$$\bar{I} = I\epsilon^{j\theta} = I\underline{/\theta} \tag{8-13}$$

which agrees in form with Eq. (8-3). Now Eq. (8-12) becomes

$$i(t) = \bar{I}\epsilon^{j\omega t} \tag{8-14}$$

Similarly, $v(t)$ of Eq. (8-11) becomes

$$v(t) = \bar{V}\epsilon^{j\omega t} \tag{8-15}$$

where

$$\bar{V} = V\epsilon^{j\phi} = V\underline{/\phi} \tag{8-16}$$

agreeing with Eq. (8-4).

We see that $i(t)$ and $v(t)$ are expressed as their phasor representations times the unit rotating phasor factor. Now, how do we "freeze" a phasor at $t = 0$ to provide the proper phasor form? We simply remove the unit rotating-phasor factor from $i(t)$ or $v(t)$, leaving $\bar{I}$ or $\bar{V}$ as a fixed phasor. After a solution is obtained from phasor manipulation, we merely introduce the unit rotating phasor again, and the time-domain solution is readily obtained.

[†] Some authors prefer cos $(\omega t + \theta)$ as a reference and, consequently, use the real part of the rotating phasor. However, the final results are perfectly identical, and for reasons of clarity in correlating transform and steady-state concepts, this author prefers to use the imaginary part idea.

We will refer to the representation of sinusoids in terms of phasors as a representation in the *phasor domain* or the *steady-state domain*. Now let us observe the voltage–current relationships for the basic circuit parameters in the phasor domain.

Inductance.   Assume that the current through an inductor and the voltage across the inductor are of the forms

$$i(t) = \bar{I}\epsilon^{j\omega t} \tag{8-17}$$

$$v(t) = \bar{V}\epsilon^{j\omega t} \tag{8-18}$$

The basic voltage–current relationship is

$$v(t) = L\frac{di(t)}{dt} \tag{8-19}$$

Substitution of Eqs. (8-17) and (8-18) into Eq. (8-19) yields

$$\bar{V}\epsilon^{j\omega t} = j\omega L\bar{I}\epsilon^{j\omega t} \tag{8-20}$$

Cancellation of the rotating phasor terms results in

$$\bar{V} = j\omega L\bar{I} \tag{8-21}$$

A typical phasor diagram is shown in Fig. 8-3. An important deduction is that *the steady-state sinusoidal voltage across an inductor leads the current through it by 90°.* Another basic result is that *differentiation in the time domain corresponds to multiplication by jω in the steady-state domain.*

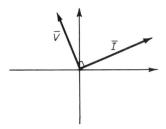

Figure 8-3   Phasor relationships for inductor.

Before considering the other circuit parameters, we need to define the concept of *steady-state impedance.* As in transform analysis, the use of phasor quantities will result in algebraic relationships for the voltages and currents associated with circuit elements, as is evident from Eq. (8-21). The basic symbol for impedance is Z. In the same manner that transform impedances are functions of s, steady-state impedances are functions of the angular frequency ω. However, we will consider them as functions of the imaginary argument $j\omega$, since steady-state and transform concepts are better correlated with this choice.

The *steady-state or phasor impedance, $Z(j\omega)$,* associated with a given passive network is defined by

$$Z(j\omega) = \frac{\bar{V}}{\bar{I}} \tag{8-22}$$

where $\bar{V}$ and $\bar{I}$ are the phasors associated with the terminals. Solving for $\bar{V}$, we may express *Ohm's law in the steady-state domain* as

$$\bar{V} = Z(j\omega)\bar{I} \tag{8-23}$$

This concept is illustrated in Fig. 8-4. In a problem dealing with a single-frequency sinusoid, $Z(j\omega)$ is evaluated at the frequency under consideration.

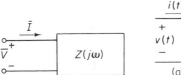

Figure 8-4 Steady-state impedance concept.

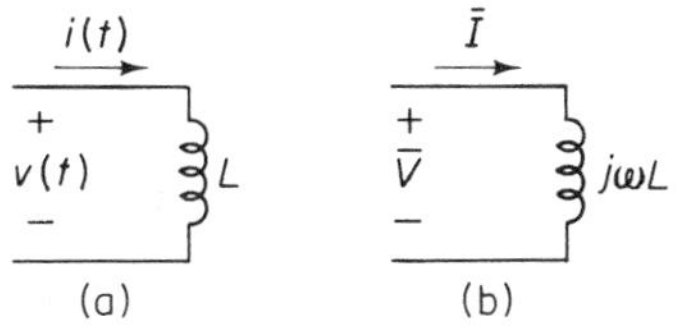

Figure 8-5 Time-domain and phasor-domain forms for inductor.

Returning to the inductor and noting Eq. (8-21), we see that the steady-state impedance of the inductor is

$$Z(j\omega) = j\omega L \tag{8-24}$$

Thus for steady-state analysis, the inductor is replaced by an impedance $j\omega L$ as shown in Fig. 8-5.

Capacitance.   The voltage-current relationship for a capacitor reads

$$v(t) = \frac{1}{C} \int_0^t i(t)\, dt \tag{8-25}$$

Using Eqs. (8-17) and (8-18) and assuming that the effect of the lower limit of integration will have disappeared in the steady state, we obtain

$$\bar{V}\epsilon^{j\omega t} = \frac{1}{j\omega C} \cdot \bar{I}\epsilon^{j\omega t} \tag{8-26}$$

Cancellation of the rotating phasor terms yields

$$\bar{V} = \frac{1}{j\omega C} \cdot \bar{I} = \frac{-j}{\omega C} \cdot \bar{I} \tag{8-27}$$

A typical phasor diagram is shown in Fig. 8-6. An important deduction is that *the steady-state voltage across a capacitor lags the current through it by 90°.*

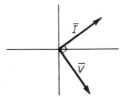

Figure 8-6 Phasor relationships for capacitor.

For the capacitor, the steady-state impedance is

$$Z(j\omega) = \frac{1}{j\omega C} = \frac{-j}{\omega C} \tag{8-28}$$

For steady-state analysis, the capacitor is replaced by its impedance as shown in Fig. 8-7.

Resistance.   The voltage–current relationship for a resistor reads

$$v(t) = Ri(t) \tag{8-29}$$

Using Eqs. (8-17) and (8-18), we obtain

$$\bar{V}\epsilon^{j\omega t} = R\bar{I}\epsilon^{j\omega t} \tag{8-30}$$

resulting in

$$\bar{V} = R\bar{I} \tag{8-31}$$

A typical phasor diagram is shown in Fig. 8-8. It is seen that *the voltage and current associated with a resistor are always in phase.* The steady-state impedance of a resistor is simply

$$Z(j\omega) = R \tag{8-32}$$

which is a constant, independent of frequency.

Driving-point impedances involving several elements of different types will, in general, contain both a real and an imaginary part. In general, we may express $Z(j\omega)$ as

$$Z(j\omega) = R(\omega) + jX(\omega) \tag{8-33}$$

The quantity $R(\omega)$ represents the resistive part of the impedance, and $X(\omega)$ is defined as the *reactance*. (Do not confuse the X symbol here with the X used to represent an arbitrary excitation in considering the transfer function concept.) Both $R(\omega)$ and $X(\omega)$ are themselves real functions of frequency [the j in front of $X(\omega)$ takes care of the imaginary interpretation], and there is no convenient reason for carrying along the j factor in the arguments. Both R and X are measured in *ohms*.

In many problems, a single radian frequency is specified, in which case $Z(j\omega)$ is simply a single complex number. However, if the frequency dependency of the impedance of a circuit is a matter of interest, the quantity ω is considered as a variable,

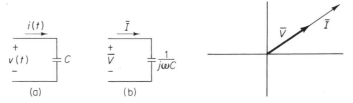

Figure 8-7 Time-domain and phasor-domain forms for a capacitor.

Figure 8-8 Phasor relationships for resistor.

and both $R(\omega)$ and $X(\omega)$ are functions of the frequency. A large number of imped-
ance-measuring devices such as ac impedance bridges are designed to measure $R(\omega)$
and $X(\omega)$ as the reference frequency of the device is varied.

Referring back to Eqs. (8-24), (8-28), and (8-32), we see that the impedance of
a resistor contains only a real part, whereas the impedances of an ideal inductor and
an ideal capacitor contain only imaginary parts. As a result of the latter fact, inductors and capacitors are referred to as *reactive elements*.

The reciprocal of steady-state impedance is called steady-state admittance, and
it is designated as $Y(j\omega)$. Thus

$$Y(j\omega) = \frac{1}{Z(j\omega)} \tag{8-34}$$

The admittance, in general, will also consist of a real part and an imaginary part,
and it can be expressed as

$$Y(j\omega) = G(\omega) + jB(\omega) \tag{8-35}$$

$G(\omega)$ represents the conductance and $B(\omega)$ is defined as the *susceptance*. Both G and
B are measured in *siemens*.

Although $Y(j\omega)$ is the reciprocal of $Z(j\omega)$, it should be emphasized that, in
general, $G(\omega)$ is *not* the reciprocal of $R(\omega)$ and $B(\omega)$ is *not* the reciprocal of $X(\omega)$.
Some widely used definitions are:

1. *Inductive reactance* (X_L). The inductive reactance of an inductor is defined as

$$X_L = \omega L \tag{8-36}$$

2. *Capacitive reactance* (X_C). The capacitive reactance of a capacitor is defined as

$$X_C = \frac{-1}{\omega C} \tag{8-37}$$

The choice of these definitions results in the property that the impedance for
either an inductive or a capacitive reactance can be expressed as

$$Z = jX \tag{8-38}$$

since the negative sign for a capacitor is absorbed in the definition of X_C. In this
sense, a positive reactance represents an inductive reactance, while a negative reactance represents a capacitive reactance.

Now let us illustrate some very important relationships between the transform
domain and the steady-state domain. A comparison between the Laplace transform
domain and the steady-state domain quantities is shown in Fig. 8-9. It is readily
deduced that *to convert from the transform domain to the steady-state domain, the
transform voltages and currents are replaced by phasor voltages and currents, and the
quantity s is replaced by the quantity $j\omega$ in all impedances.*

We have seen that the quantity s may be considered as a complex frequency
of the form $s = \sigma + j\omega$. Thus for steady-state ac analysis, we restrict s to lie along
the $j\omega$-axis (i.e., $s = j\omega$). In this sense, steady-state ac circuit analysis may be considered to be a special case of the more general transform analysis.

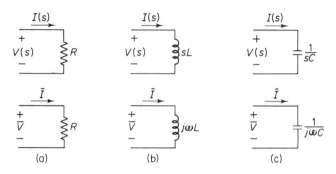

Figure 8-9 Comparison of transform and steady-state quantities.

To determine a steady-state response within a network, one transforms the circuit to the steady-state domain by replacing all components by their steady-state impedances and sources by steady-state phasors. The circuit is then solved by conventional algebraic methods in conjunction with network theorems to obtain any desired response. When the desired phasor response is obtained, the rotating phasor is introduced, and the appropriate real or imaginary part is formed to yield the desired time function.

The development of this section has been designed to provide the reader with a basic understanding of the theory of the phasor concept. It is not necessary to go through this development each time a problem is worked. Both the transformation from the time domain to the phasor domain and the inverse transformation from the phasor domain to the time domain can be done by a simple inspection.

The close similarity between the Laplace transform and phasor methods should now be clear. If desired, one could refer to a phasor as a *steady-state transform*, but we have chosen to use the more established phasor definitions.

We might point out one more basic fact that may or may not be clear to the reader. In a general sense, dc may be considered as a special case of ac with $\omega = 0$. Thus the phasor concept may be used to determine the steady-state dc response of a circuit if we simply let $s = 0$ in all impedances. Of course, it seems absurd to call the voltages and currents phasors in this case. Instead they are merely dc values.

Note that letting $s = 0$ results in zero impedance for an inductor (short-circuit) and infinite impedance for a capacitor (open-circuit). These limiting forms are in perfect agreement with the steady-state dc concepts of Chapter 4.

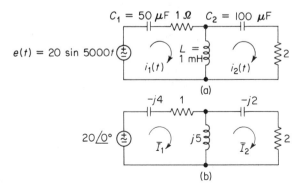

Figure 8-10 Circuit of Example 8-1.

Example 8-1

The circuit of Fig. 8-10a is operating under steady-state conditions. Determine $i_2(t)$ by steady-state phasor analysis.

Solution   We must first transform the circuit to the steady-state domain. Since the radian frequency is 5000 rad/s, the reactances are

$$X_{C_1} = \frac{-1}{5000 \times 50 \times 10^{-6}} = -4\,\Omega \tag{8-39}$$

$$X_{C_2} = \frac{-1}{5000 \times 100 \times 10^{-6}} = -2\,\Omega \tag{8-40}$$

$$X_L = 5000 \times 1.0 \times 10^{-3} = 5\,\Omega \tag{8-41}$$

The transformed circuit is shown in Fig. 8-10b. Note that the transform of the excitation $e(t)$ is simply a phasor $\bar{E} = 20\underline{/0^\circ} = 20$.

The simultaneous mesh current equations are

$$(1 + j)\bar{I}_1 \qquad - j5\bar{I}_2 = 20 \tag{8-42}$$

$$-j5\bar{I}_1 + (2 + j3)\bar{I}_2 = 0 \tag{8-43}$$

Simultaneous solution by either determinants or substitution results in

$$\bar{I}_2 = \frac{j100}{24 + j5} = \frac{100\underline{/90^\circ}}{24.52\underline{/11.77^\circ}} \tag{8-44}$$

$$= 4.078\underline{/78.23^\circ}$$

A phasor diagram is shown in Fig. 8-11 with the phasor E chosen along the reference positive real-axis.

To obtain the steady-state time function, we will proceed in two ways. The formal mathematical approach is to introduce the rotating phasor term and write

$$i_2(t) = \bar{I}_2 \epsilon^{j\omega t}$$

$$= 4.078\epsilon^{j78.23^\circ}\epsilon^{j5000t} \tag{8-45}$$

$$= 4.078\epsilon^{j(5000t + 78.23^\circ)}$$

However, we remember that this equation really means that we use the imaginary part of the result since the excitation was a basic sine function. Thus we have

$$i_2(t) = \text{Im}\left[4.078\epsilon^{j(5000t + 78.23^\circ)}\right]$$
$$= 4.078\sin(5000t + 78.23^\circ) \tag{8-46}$$

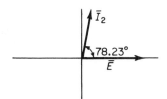

Figure 8-11   Phasor diagram of Example 8-1.

The second and shorter method is to recognize that the magnitude of $\bar{I}_2$ is simply the amplitude of the response $i_2(t)$, and the phase angle of $\bar{I}_2$ is simply the relative phase between the excitation and $i_2(t)$. Looking at $i_2(t)$ which we just calculated, we readily see that conversion from $\bar{I}_2$ to $i_2(t)$ can be done by a simple inspection.

With this simpler approach, it is not necessary that we keep up with the details of the real and imaginary parts of the exponential whenever the excitation has some different phase angle, as long as the reference phase of the excitation is known. For example, suppose that the excitation in this problem had been

$$e(t) = 20 \cos (5000t - 20°) \tag{8-47}$$

Then the response would be

$$
\begin{aligned}
i_2(t) &= 4.078 \cos (5000t - 20° + 78.23°) \\
&= 4.078 \cos (5000t + 58.23°)
\end{aligned} \tag{8-48}
$$

Example 8-2

The circuit of Fig. 8-12a is operating under steady-state conditions. Determine the steady-state current response (a) from the Laplace transform approach, and (b) from the phasor approach.

  Solution (a) The Laplace transform domain circuit is shown in Fig. 8-12b. The current is

$$I(s) = \frac{40/(s^2 + 4)}{2s + 4} = \frac{20}{(s + 2)(s^2 + 4)} \tag{8-49}$$

The complete solution would contain an exponential term, which is transient in nature, and the steady-state sinusoidal term. We could easily obtain the entire solution if desired, but since we are interested only in the steady-state part, the "trick formula" of Section 5-7 will readily determine this quantity. Although normally, $i(t)$ would be used to represent the total response, we will define $i(t)$ here to mean only the steady-state response. The reader is invited to show that the steady-state response is

$$i(t) = \frac{5}{\sqrt{2}} \sin (2t - 45°) \tag{8-50}$$

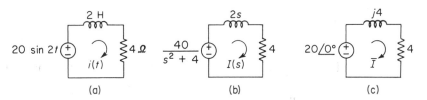

(a)   (b)   (c)

Figure 8-12 Circuits of Example 8-2.

(b) To determine the response by means of the phasor concept, the reader may be tempted to simply insert $s = j\omega$ into Eq. (8-49). However, *it is not correct to substitute $s = j\omega$ into an expression for a transform current or a transform voltage.* The reason is that a transform current or voltage will contain terms representing the transform of the excitation. Since a phasor excitation is not derived from a transform excitation by means of this substitution, such a procedure would not be correct. We have seen that impedances may be converted from the transform domain to the steady-state domain by substituting $s = j\omega$, and we will see in the next section that transfer functions may be converted by this substitution. However, voltages and currents are converted by *replacing* transforms by phasors.

It is best in this problem to simply go back to the original circuit and transform it to the steady-state domain, as shown in Fig. 8-12c. Note that the impedance is evaluated at $\omega = 2$. The reader is invited to verify that the solution is the same as obtained previously.

8-3 THE STEADY-STATE TRANSFER FUNCTION

In Section 8-2 we observed that steady-state or phasor impedances could be derived from transform impedances if we simply substitute $s = j\omega$. In this section we will look at the transfer function concept and show that a steady-state transfer function may also be derived by this substitution.

First, let us assume as a reference a lumped, linear, time-invariant, stable circuit or system with transfer function $G(s)$, transform excitation $X(s)$, and transform response $Y(s)$. (Do not confuse the X with reactance.) The basic relationship is

$$\frac{Y(s)}{X(s)} = G(s) \tag{8-51}$$

If the time-domain excitation $x(t)$ is assumed to be a sinusoid of radian frequency ω, the steady-state response will also be a sinusoid of the same frequency. *For the purposes of this chapter, we define $y(t)$ to be the steady-state response.*

To obtain the steady-state relationships for the circuit or system, we would substitute $s = j\omega$ in all circuit impedances and replace transforms by phasors. Since the resulting transfer function does not contain the transforms of any sources, but only operations involving impedances, *the steady-state transfer function is obtained from the transform transfer function by substituting $s = j\omega$.* Thus, while $G(s)$ represents the transform transfer function, $G(j\omega)$ represents the steady-state transfer function.

In Section 8-2, we designated phasors representing single-frequency sinusoids by bars. In many cases it is desirable to emphasize the frequency dependency of a phasor excitation or response. The notations $X(j\omega)$ and $Y(j\omega)$ will be used often to indicate general input and output phasors. A bar will no longer be needed, since the argument $j\omega$ will serve to identify the quantities as phasors. Thus, in the sinusoidal steady state, the system relationships become

$$\frac{Y(j\omega)}{X(j\omega)} = G(j\omega) \tag{8-52}$$

or

$$Y(j\omega) = G(j\omega)X(j\omega) \tag{8-53}$$

Equations (8-52) and (8-53) provide a very important set of relationships. Equation (8-52) states that the ratio of the output phasor to the input phasor is simply the steady-state transfer function. Equation (8-53) states that we obtain the output phasor by multiplying the input phasor by the steady-state transfer function.

In determining the steady-state transfer function, we may make the substitution $s = j\omega$ in the actual circuit before the transfer function is obtained, or $G(s)$ may be determined first and the substitution made as a final step. If ω is a specified frequency, then $G(j\omega)$ reduces to a single complex number. If, however, ω is kept as a variable, then $G(j\omega)$ will, in general, be a complex function of frequency.

Since $G(j\omega)$ is a complex number, it may be written in the form

$$G(j\omega) = A(\omega)\epsilon^{j\beta(\omega)}$$
$$= A(\omega)/\beta(\omega) \tag{8-54}$$

where

$$A(\omega) = |G(j\omega)| \tag{8-55}$$

and $\beta(\omega)$ is the phase angle of $G(j\omega)$. The quantity $A(\omega)$ is called the *amplitude response* of the circuit or system, and $\beta(\omega)$ is called the *phase response*. Note that the use of the word *response* has a somewhat different meaning here than in previous usage. The functions $A(\omega)$ and $\beta(\omega)$ are both real functions of ω, and there is no point in using the j factor in their arguments.

Suppose $x(t)$ is a single-frequency sinusoid of the form

$$x(t) = X \sin(\omega t + \theta) \tag{8-56}$$

The phasor representation of $x(t)$ is

$$X(j\omega) = X/\theta \tag{8-57}$$

By means of Eqs. (8-53) and (8-54) the output phasor is

$$Y(j\omega) = [A(\omega)/\beta(\omega)] \cdot [X/\theta\ ]$$
$$= A(\omega)X/\beta(\omega) + \theta \tag{8-58}$$

For convenience, the output phasor may be written as

$$Y(j\omega) = Y/\phi \tag{8-59}$$

Comparing Eqs. (8-58) and (8-59), we deduce

$$Y = A(\omega)X \tag{8-60}$$

and

$$\phi = \beta(\omega) + \theta \tag{8-61}$$

Equations (8-60) and (8-61) provide two very important properties of the steady-state transfer function. They are:

1. *The magnitude of the output phasor is equal to the magnitude of the input phasor times the amplitude response of the system.*

2. *The phase of the output phasor is equal to the phase of the input phasor plus the phase response of the system.*

Another way to look at this concept is that the amplitude response is the ratio of the magnitudes of the output steady-state signal to the input steady-state signal, and the phase response is the amount by which the signal is shifted in phase from input to output. These ideas form the basis for *frequency response measurements*. A diagram illustrating the approach for measuring the frequency response of a circuit with input and output voltages is shown in Fig. 8-13. The input is usually held at some constant amplitude, and the output amplitude and relative phase between input and output are measured as the frequency of the generator is varied. At any given frequency, the ratio of the output voltage to the input voltage is the amplitude response, and the relative phase shift between input and output is the phase response.

If $A(\omega) > 1$ at a given frequency, the output phasor has a larger magnitude than the input phasor. This does not necessarily imply that an active amplifier is contained in the network, as many passive circuits provide a voltage or a current gain (but never a power gain) to a signal. If $A(\omega) < 1$, the output phasor is smaller in magnitude than the input phasor, and the signal has undergone *attenuation*. If $A(\omega) = 0$ for a particular frequency, there is no output, and the circuit is said to have a *zero of transmission* or a *null* at the given frequency. It can be shown that this condition results from the presence of a pair of conjugate zeros on the $j\omega$-axis of the s-plane at the same frequency as the excitation.

In measuring the relative voltage or current amplitude response of a circuit, it is often convenient to employ the concepts of *relative decibel gain* and *relative decibel loss*. Assume that some particular amplitude response A_0, evaluated at some particular frequency, is used as a reference. The *relative decibel gain* at any other frequency is given by

$$\text{relative dB gain} = 20 \log_{10} \frac{A(\omega)}{A_0} \qquad (8\text{-}62)$$

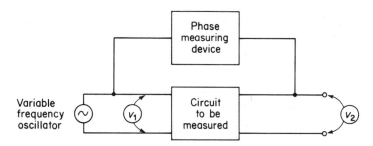

Figure 8-13   Frequency response measurements.

The relative decibel loss is given by

$$\text{relative dB loss} = 20 \log_{10} \frac{A_0}{A(\omega)} \tag{8-63}$$

A fundamental identity of logarithms is

$$\log a = -\log \frac{1}{a} \tag{8-64}$$

Thus the dB gain and dB loss expressions are related by

$$\text{relative dB gain} = -\text{relative dB loss} \tag{8-65}$$

The logarithm of a number greater than 1 is positive, whereas the logarithm of a number less than 1 is negative. For the reason that positive numbers are more desirable to work with, the dB-gain concept might be preferred whenever $A(\omega) > A_0$ at most frequencies, and the dB-loss concept might be preferred whenever $A(\omega) < A_0$ at most frequencies. We might note that a negative dB gain implies an actual loss or attenuation, and a negative dB loss implies an actual gain.

As a final point of interest, a driving-point steady-state impedance function $Z(j\omega)$ obeys many of the same properties as a steady-state transfer function and can often be considered as a special type of transfer function. However, the terminology associated with a steady-state impedance function is somewhat different than that for a transfer function. For example, the real and imaginary part implications of impedance functions (resistance and reactance) are rarely of importance in a transfer function. On the other hand, the terminology of amplitude response used with transfer functions is not normally used in describing impedance functions.

When it is desired to express an impedance function in polar form, one usually writes an expression of the form

$$Z(j\omega) = |Z(j\omega)| \underline{/\theta(\omega)} \tag{8-66}$$

in which $|Z(j\omega)|$ is simply called the *magnitude* of the impedance. Of course, $\theta(\omega)$ is the *phase angle* of the impedance.

Example 8-3

(a) Referring back to Example 7-1, determine the steady-state transfer function of the circuit for any arbitrary ω.

(b) Determine the steady-state response due to the excitation given in the problem by means of the phasor concept.

(c) Determine the steady-state response if the angular frequency of the excitation is changed to 10 rad/sec.

(d) Using the amplitude response at dc as a reference, determine the relative dB loss at $\omega = 2$ and at $\omega = 10$.

Solution   (a) From the notation of Example 7-1, the general transfer function was shown to be

$$G(s) = \frac{V_2(s)}{V_1(s)} = \frac{2}{s+2} \tag{8-67}$$

Letting $s = j\omega$, we have the steady-state transfer function:

$$G(j\omega) = \frac{2}{2 + j\omega} \qquad (8\text{-}68)$$

As a matter of interest, the amplitude and phase responses for the network are

$$A(\omega) = \frac{2}{\sqrt{4 + \omega^2}} \qquad (8\text{-}69)$$

$$\beta(\omega) = -\tan^{-1}\frac{\omega}{2} \qquad (8\text{-}70)$$

(b) Since the excitation is

$$v_1(t) = 5 \sin 2t \qquad (8\text{-}71)$$

the input phasor is

$$V_1(j2) = 5\underline{/0^\circ} = 5 \qquad (8\text{-}72)$$

Either Eqs. (8-68) or (8-69) and (8-70) may be used in the determination of $V_2(j2)$. Using Eq. (8-68) in this part, we have

$$G(j2) = \frac{2}{2 + j2} = \frac{1}{\sqrt{2}}\underline{/-45^\circ} = 0.7071\underline{/-45^\circ} \qquad (8\text{-}73)$$

and since

$$V_2(j\omega) = G(j\omega)V_1(j\omega) \qquad (8\text{-}74)$$

and

$$V_2(j2) = G(j2)V_1(j2) = \left[\frac{1}{\sqrt{2}}\underline{/-45^\circ}\right] \cdot 5$$

$$= \frac{5}{\sqrt{2}}\underline{/-45^\circ} \qquad (8\text{-}75)$$

The steady-state response is

$$v_2(t) = \frac{5}{\sqrt{2}} \sin(2t - 45^\circ) = 3.536 \sin(2t - 45^\circ) \qquad (8\text{-}76)$$

which agrees with the steady-state result of Example 7-1.

(c) For $\omega = 10$ rad/sec, we will use Eqs. (8-69) and (8-70). Thus

$$A(10) = \frac{2}{\sqrt{4 + 100}} = 0.196 \qquad (8\text{-}77)$$

and

$$\beta(10) = -\tan^{-1} 5 = -78.69^\circ \qquad (8\text{-}78)$$

The output phasor is

$$V_2(j10) = G(j10)V_1(j10) = [0.196\underline{/-78.69^\circ}] \cdot 5$$
$$= 0.981\underline{/-78.69^\circ} \qquad (8\text{-}79)$$

The steady-state response in this case is

$$v_2(t) = 0.981 \sin(10t - 78.69°) \qquad (8\text{-}80)$$

The steady-state output at the higher frequency is seen to be smaller, and the phase shift is larger. The decreased output is readily explained by the fact that this circuit is a very simple *low-pass filter*, since the shunt capacitive reactance decreases as the frequency increases.

(d) As specified, we will use the dc steady-state amplitude response as a reference. It is

$$G(0) = \frac{2}{2 + j0} = 1 = A(0) \qquad (8\text{-}81)$$

This result could also have been predicted by inspection of the circuit since the capacitor is an open-circuit in the dc steady state.

At $\omega = 2$, the relative dB loss is

$$\text{relative dB loss } (\omega = 2) = 20 \log_{10} \frac{A(0)}{A(2)}$$

$$= 20 \log_{10} \frac{1}{0.7071} = 3.01 \text{ dB} \qquad (8\text{-}82)$$

Since the dB loss is positive, this implies that the output is less than the reference output, and if desired we may write

$$\text{relative dB gain } (\omega = 2) = -3.01 \text{ dB} \qquad (8\text{-}83)$$

At $\omega = 10$, we have

$$\text{relative dB loss } (\omega = 10) = 20 \log_{10} \frac{A(0)}{A(10)}$$

$$= 20 \log_{10} \frac{1}{0.196} = 14.5 \text{ dB} \qquad (8\text{-}84)$$

8-4 FREQUENCY RESPONSE PLOTS

In this chapter we have considered the steady-state representations of voltages, currents, impedances, and transfer functions. In general, all of these quantities are functions of the angular frequency ω, whenever frequency is considered as a variable. Furthermore, each steady-state quantity is a complex value and, as such, can be represented as a magnitude and a phase angle. Both the magnitude and the phase will, in general, be functions of ω.

In this section we investigate some basic procedures, including the *breakpoint* approximation technique, that will allow us to predict the magnitude and phase curves for a frequency-dependent steady-state quantity. Although this approach is not always exact, it is close enough for many engineering purposes. Furthermore, even if an exact analysis is required, the use of these methods will often help one avoid unnecessary calculations in the final analysis. The plots to be obtained are referred to as *Bode plots*.

To provide a convenient base for developing the techniques, we will consider a transfer function $G(j\omega)$ for reference in subsequent expressions. However, we wish to emphasize that the approach is equally suited for a driving-point steady-state impedance $Z(j\omega)$, a phasor voltage $V(j\omega)$, or a phasor current $I(j\omega)$. All one need do is change the notation to match the appropriate quantity.

Since we will use a transfer function for reference, the two quantities of interest are $A(\omega)$ and $\beta(\omega)$. Furthermore, $A(\omega)$ may be plotted without modification, or the relative decibel gain or loss associated with $A(\omega)$ may be plotted. We will refer to a curve of $A(\omega)$ as an *absolute* amplitude response curve and a curve of some decibel measure of $A(\omega)$ as a dB amplitude response curve.

To effectively use these methods, it is necessary to employ a *log-log* scale for plotting an absolute amplitude response curve and a *semilog* scale for plotting a dB amplitude response curve. A typical log-log plot is shown in Fig. 8-14, and a possible

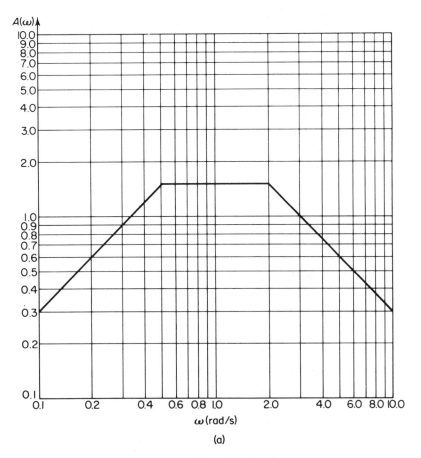

(a)

Figure 8-14 Typical log-log plot.

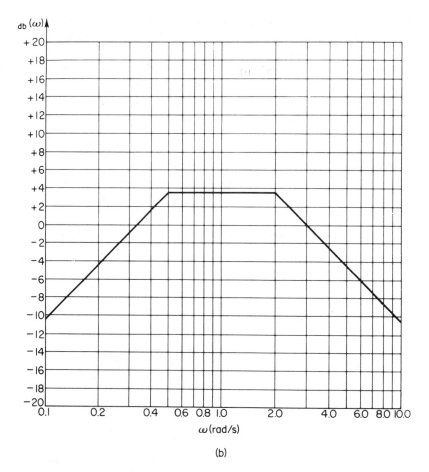

Figure 8-15   Typical semilog plot.

corresponding semilog plot is shown in Fig. 8-15. A basic point of interest here is that *the general shape of the absolute amplitude response curve on a log-log scale is identical to the general shape of the dB amplitude response curve on a semilog scale.*

The reason behind this fact is that a logarithmic amplitude scale corresponds to a linear decibel scale. Thus the choice between an absolute and a dB plot is often a matter of individual choice, although the experimentalist often tends to favor the dB plot since many measurements are made in dB. However, since the shapes are the same, if either plot is given, it may be converted to the other by placing an appropriate scale on the vertical axis.

Why is it desirable to employ a logarithmic scale for plotting the frequency response? The basic reason is that multiplication and division are accomplished by the addition and subtraction of logarithmic quantities, and consequently, the curve-plotting process will be greatly simplified. Also, many functions that are curved appear as straight lines on a logarithmic plot.

We now wish to look at each of the possible forms for numerator and denominator factors individually, and then we will look at the form of an overall response.

To effectively illustrate the amplitude response concept, we will employ both absolute and dB scales in all plots. In all cases, the absolute scale will be on the left and the dB scale will be on the right.

A full analytical justification of the properties of these plots is somewhat cumbersome to develop. Instead, since all of the properties that we state can be justified by simply plotting the curves, we will concern ourselves mainly with the presentation of the procedures for predicting and plotting the curves.

The decibel quantity of interest will be defined as

$$A_{dB}(\omega) = 20 \log_{10} A(\omega) \tag{8-85}$$

The quantity A_{dB} can be considered as a *relative dB gain* with the reference gain chosen at unity.

Zero at Origin (Simple-Order)

Consider a function of the form

$$G(s) = s \tag{8-86}$$

The steady-state function is

$$G(j\omega) = j\omega \tag{8-87}$$

The absolute amplitude response is

$$A(\omega) = \omega \tag{8-88}$$

The dB amplitude response is

$$A_{dB}(\omega) = 20 \log_{10} \omega \tag{8-89}$$

A plot showing A and A_{dB} is given in Fig. 8-16a. It can be seen that *the amplitude response of a simple-order zero at the origin is a straight line with slope $+1$ on an absolute plot, and increase of 6 dB/octave or 20 dB/decade on a dB plot.* The frequency is said to be increased by an *octave* when it is *doubled*, and it is said to be increased by a *decade* when it is multiplied by *10*.

The phase response of the first-order zero is simply

$$\beta(\omega) = 90° \tag{8-90}$$

meaning that the phase shift is constant at all frequencies. The phase shift is shown in Fig. 8-16b.

Zero at Origin (Multiple-Order)

Consider the function

$$G(s) = s^n \tag{8-91}$$

The absolute amplitude response is

$$A(\omega) = \omega^n \tag{8-92}$$

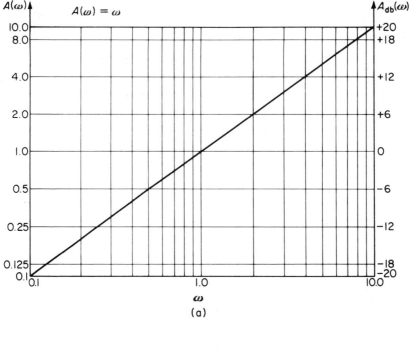

Figure 8-16   Amplitude and phase curves for simple-order zero at origin.

The dB amplitude response is

$$A_{dB}(\omega) = 20 \log_{10} \omega^n = 20n \log_{10} \omega \qquad (8\text{-}93)$$

Comparing Eqs. (8-89) and (8-93), we see that the response curve on a logarithmic scale in this case is also a straight line, but with a slope n times the slope of the simple-order zero response.

A plot showing A and A_{dB} for $n = 2$ is given in Fig. 8-17a. It can be shown that *the amplitude response of an nth-order zero at the origin is a straight line with slope $+n$ on an absolute plot and an increase of 6n dB/octave or 20n dB/decade on a dB plot.*

The phase response of the multiple-order zero is simply

$$\beta(\omega) = 90n^\circ \qquad (8\text{-}94)$$

A plot of β for $n = 2$ is shown in Fig. 8-17b.

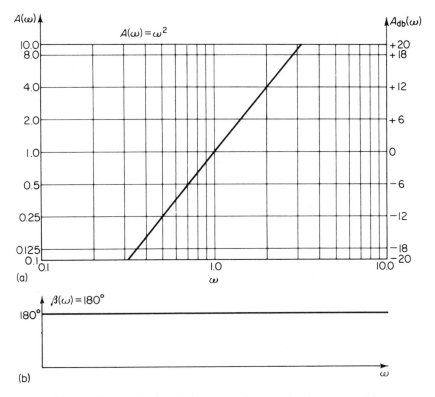

Figure 8-17 Amplitude and phase curves for second-order zero at origin.

Pole at Origin (Simple-Order)

Consider the function

$$G(s) = \frac{1}{s} \tag{8-95}$$

The absolute amplitude response is

$$A(\omega) = \frac{1}{\omega} \tag{8-96}$$

The dB amplitude response is

$$A_{dB}(\omega) = 20 \log_{10} \frac{1}{\omega} = -20 \log_{10} \omega \tag{8-97}$$

A plot showing A and A_{dB} is given in Fig. 8-18a. It can be seen that *the amplitude response of a simple-order pole at the origin is a straight line with slope -1 on an absolute plot, and a decrease of 6 dB/octave or 20 dB/decade on a dB plot.*

The phase response of the simple-order pole is

$$\beta(\omega) = -90° \tag{8-98}$$

which is shown in Fig. 8-18b.

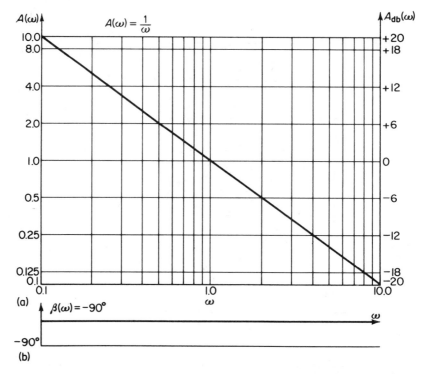

Figure 8-18   Amplitude and phase curves for simple-order pole at origin.

Pole at Origin (Multiple-Order)

Consider the function

$$G(s) = \frac{1}{s^n} \tag{8-99}$$

The absolute amplitude response is

$$A(\omega) = \frac{1}{\omega^n} \tag{8-100}$$

The dB amplitude response is

$$A_{dB}(\omega) = 20 \log_{10} \frac{1}{\omega^n} = -20n \log_{10} \omega \tag{8-101}$$

A plot showing A and A_{dB} for $n = 2$ is given in Fig. 8-19a. We see that *the amplitude response of an nth-order pole at the origin is a straight line with slope* $-n$ *on an absolute scale and a decrease of 6n dB/octave or 20n dB/decade on a dB scale.*

The phase response of the multiple-order pole is

$$\beta(\omega) = -90n° \tag{8-102}$$

This function is shown in Fig. 8-19b for $n = 2$.

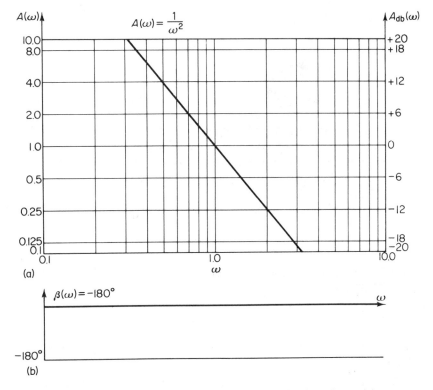

Figure 8-19 Amplitude and phase curves for second-order pole at origin.

Before going further, we need to clarify some notation that will be used. In developing the basic exponential function in Chapter 2, we defined two parameters: *the damping constant* (α) and *the time constant* (τ). In dealing with transform manipulations, we have used factors of the form ($s + \alpha$) corresponding to a critical frequency at $s = -\alpha$, which is on the negative real axis if $\alpha > 0$. If the critical frequency had been a *pole*, there would be a corresponding exponential term in the time response, with a damping factor exactly equal in magnitude to the pole location.

Since α and τ are related by

$$\alpha = \frac{1}{\tau} \tag{8-103}$$

we can always rearrange an expression such that all factors are in the form ($1 + s/\alpha$), or equivalently ($1 + s\tau$). These forms are the most convenient for dealing with frequency response evaluations.

Zero on Real Axis (First-Order)

Consider a function of the form

$$G(s) = 1 + \frac{s}{\alpha} \qquad \text{with } \alpha > 0 \tag{8-104}$$

The absolute amplitude response is

$$A(\omega) = \sqrt{1 + (\omega/\alpha)^2} \tag{8-105}$$

The dB amplitude response is

$$A_{dB}(\omega) = 20 \log_{10} \sqrt{1 + \left(\frac{\omega}{\alpha}\right)^2} = 10 \log_{10} \left[1 + \left(\frac{\omega}{\alpha}\right)^2\right] \tag{8-106}$$

An exact plot showing A and A_{dB} is given by curve E of Fig. 8-20a. Since α may assume any value, we have chosen the normalized variable ω/α on the horizontal axis, which makes the curve somewhat universal in nature. Notice that for $\omega = \alpha$, the curve is 3 dB higher than the unity reference level. This corresponds to an absolute relative gain of $\sqrt{2}$.

Now let us look at some upper and lower approximations for A_{dB}. First consider that $\omega \ll \alpha$. In this case, Eq. (8-106) can be approximated by

$$A_{dB}(\omega) \approx 10 \log_{10}(1) = 0 \text{ dB} \qquad \text{for } \omega \ll \alpha \tag{8-107}$$

Thus for very low frequencies, the curve is very close to the 0 dB or unity absolute response line.

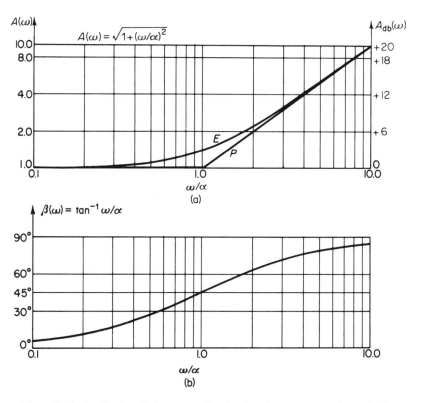

Figure 8-20 Amplitude and phase curves for simple-order zero on negative real axis.

Next consider that $\omega \gg \alpha$. Equation (8-106) can now be approximated by

$$A_{dB}(\omega) \approx 20 \log_{10} \frac{\omega}{\alpha} \qquad \text{for } \omega \gg \alpha \tag{8-108}$$

The high-frequency approximation is a straight line on a logarithmic scale, with a slope +1 on an absolute plot and an increase of 6 dB/octave or 20 dB/decade on a dB plot.

The so-called *breakpoint approximation* to the response is shown by curve P of Fig. 8-20a. The breakpoint approximation is seen to be quite accurate for $\omega \ll \alpha$ and for $\omega \gg \alpha$. The worst approximation occurs at $\omega = \alpha$ where the error is 3 dB.

If a given α had been specified and if it were desired to simply use ω on the horizontal scale, the shape of the curve would be exactly the same. However, the 3 dB angular frequency corresponding to the "break" in the approximate curve would occur at $\omega = \alpha$. The angular frequency $\omega = \alpha$ is called a *break frequency*. The trick to using this approach is to first draw the breakpoint approximation and then, if desired, add a smooth curve taking into consideration the effect of the actual curve near the break frequency.

The phase response is

$$\beta(\omega) = \tan^{-1} \frac{\omega}{\alpha} \tag{8-109}$$

which is shown in Fig. 8-20b. Note that the phase slowly increases from $0°$ at dc to $90°$ at infinite frequency. The phase shift at $\omega = \alpha$ is $45°$.

In the event that a zero is located on the positive real axis, an expression of the following form is considered:

$$G(s) = -\left(1 - \frac{s}{\alpha}\right) \qquad \text{with } \alpha > 0 \tag{8-110}$$

A quick calculation reveals that A and A_{dB} are identical to Eqs. (8-105) and (8-106), respectively. In general, this type of equality can be shown to be true for complex zeros in the right-hand half-plane also. Thus *the amplitude response of right-hand half-plane zeros is identical to the response of the corresponding left-hand half-plane zeros.*

On the other hand, the phase response for right-hand half-plane zeros is not the same. For the case described by Eq. (8-110), the phase is

$$\beta(\omega) = \tan^{-1} \frac{\omega}{-\alpha} \tag{8-111}$$

This phase shift is shown in Fig. 8-21.

By definition, a function with no zeros in the right-hand half-plane is a *minimum phase* function. The reasoning behind this definition is that, while a right-hand half-plane zero contributes the same amplitude response as a left-hand half-plane zero, the contribution to the phase shift is larger in magnitude.

Zero on Real Axis (Multiple-Order)

Consider the function

$$G(s) = \left[1 + \frac{s}{\alpha}\right]^n \tag{8-112}$$

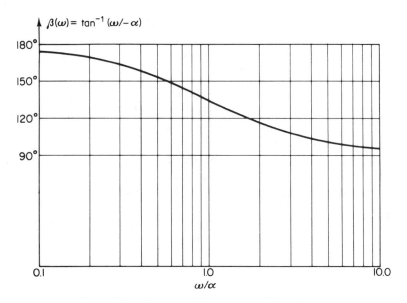

Figure 8-21 Phase curve for zero on positive real axis.

The amplitude response is

$$A(\omega) = \left[1 + \left(\frac{\omega}{\alpha} \right)^2 \right]^{n/2} \tag{8-113}$$

The dB response is

$$A_{dB}(\omega) = 10n \log_{10} \left[1 + \left(\frac{\omega}{\alpha} \right)^2 \right] \tag{8-114}$$

An exact plot showing A and A_{dB} for $n = 2$ is given by curve E of Fig. 8-22a. The low-frequency approximation can be readily shown to be the same as Eq. (8-107), whereas the high-frequency approximation is

$$A_{dB}(\omega) \approx 20n \log_{10} \frac{\omega}{\alpha} \qquad \text{for } \omega \gg \alpha \tag{8-115}$$

The high-frequency approximation is a straight line on a logarithmic scale, with slope $+n$ *on an absolute plot and an increase of* $6n$ *dB/octave or* $20n$ *dB/decade on a dB plot.*

The breakpoint approximation is given by curve P of Fig. 8-22a. The break frequency is the same in this case, namely $\omega = \alpha$. However, the error at $\omega = \alpha$ is $3n$ dB, which for $n = 2$ is 6 dB.

The phase response in this case is

$$\beta(\omega) = n \tan^{-1} \frac{\omega}{\alpha} \tag{8-116}$$

A typical curve is shown in Fig. 8-22b for $n = 2$. The phase shift increases from $0°$ at $\omega = 0$ to $90n°$ at $\omega = \infty$. The phase shift at $\omega = \alpha$ is $45n°$.

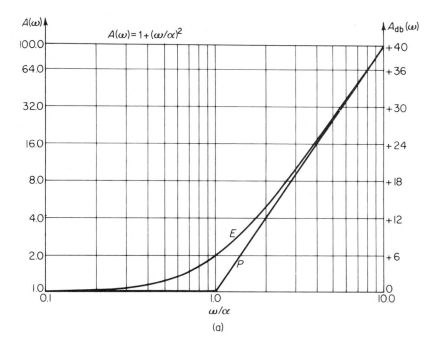

(a)

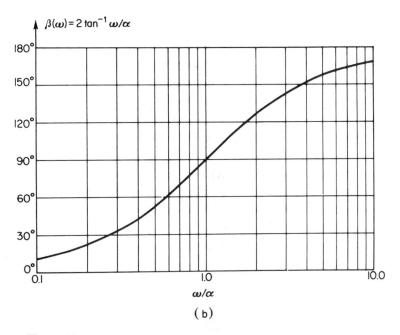

(b)

Figure 8-22   Amplitude and phase curves for second-order zero on negative real axis.

Pole on Negative Real Axis (Simple-Order)

Consider the function

$$G(s) = \frac{1}{1 + s/\alpha} \tag{8-117}$$

The absolute amplitude response is

$$A(\omega) = \frac{1}{\sqrt{1 + (\omega/\alpha)^2}} \tag{8-118}$$

The dB amplitude response is

$$A_{dB}(\omega) = -10 \log\left[1 + \left(\frac{\omega}{\alpha}\right)^2\right] \tag{8-119}$$

An exact plot showing A and A_{dB} is given by curve E of Fig. 8-23a. The breakpoint approximation is given by curve P. Note that in this case, the breakpoint approximation is 3 dB higher than the exact curve at $\omega = \alpha$, representing the largest error. It can be readily seen that *the high-frequency approximation is a straight line*

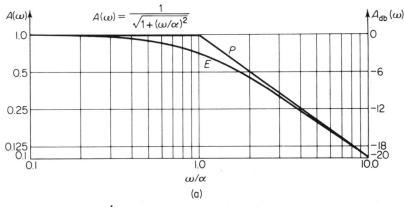

(a)

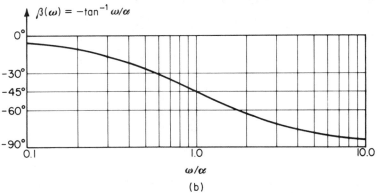

(b)

Figure 8-23 Amplitude and phase curves for pole on negative real axis.

on a logarithmic scale, with slope -1 *on an absolute plot and a decrease of 6 dB/octave or 20 dB/decade on a dB plot.*

The phase response is

$$\beta(\omega) = -\tan^{-1}\frac{\omega}{\alpha} \tag{8-120}$$

which is shown in Fig. 8-23b. The phase shift decreases from $0°$ to $-90°$ with a value of $-45°$ at $\omega = \alpha$.

Pole on Negative Real Axis (Multiple-Order)

Consider the function

$$G(s) = \frac{1}{(1 + s/\alpha)^n} \tag{8-121}$$

The absolute amplitude response is

$$A(\omega) = \frac{1}{[1 + (\omega/\alpha)^2]^{n/2}} \tag{8-122}$$

The dB amplitude response is

$$A_{dB} = -10n \log_{10}\left[1 + \left(\frac{\omega}{\alpha}\right)^2\right] \tag{8-123}$$

An exact plot showing A and A_{dB} for $n = 2$ is given by curve E of Fig. 8-24a. The breakpoint approximation is given by curve P. The breakpoint approximation is $3n$ dB higher than the exact curve at $\omega = \alpha$. It can be seen that *the high-frequency asymptote is a straight line on a logarithmic scale, with slope* $-n$ *on an absolute plot and a decrease of 6n dB/octave or 20n dB/decade on a dB plot.*

The phase response is

$$\beta(\omega) = -n \tan^{-1}\frac{\omega}{\alpha} \tag{8-124}$$

which is shown in Fig. 8-24b for $n = 2$. The phase shift decreases from $0°$ to $-90n°$ with a value of $-45n°$ at $\omega = \alpha$.

Before considering complex roots, it is necessary to define and explain some standard notation that is used in conjunction with quadratic factors with complex-conjugate roots. For simplicity in previous work, we have indicated a quadratic factor in the simple form $(s^2 + as + b)$. Furthermore, we designated the roots by

$$\begin{cases} s_1 \\ s_2 \end{cases} = -\alpha \pm j\omega_d \tag{8-125}$$

where α is the *damping constant* and ω_d is the *damped natural angular frequency*. If ω_0 is defined as the *undamped natural angular frequency*, a further relationship reads

$$\omega_d^2 = \sqrt{\omega_0^2 - \alpha^2} \tag{8-126}$$

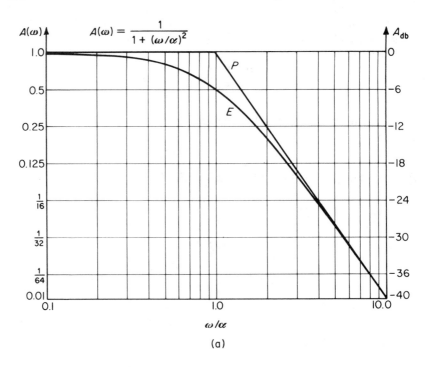

(a)

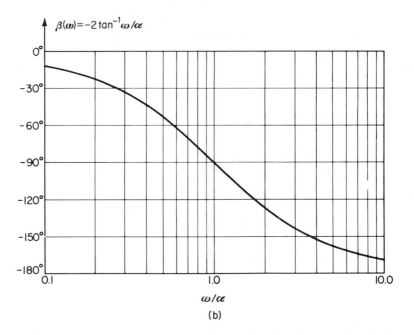

(b)

Figure 8-24   Amplitude and phase curves for second-order pole on negative real axis.

or

$$\omega_0^2 = \omega_d^2 + \alpha^2 \tag{8-127}$$

If desired, a quadratic factor can be written in the form

$$F(s) = s^2 + 2\alpha s + \omega_0^2 \tag{8-128}$$

A well-established form of notation is to employ the definition of the *damping ratio* ζ. It is defined as

$$\zeta = \frac{\alpha}{\omega_0} \tag{8-129}$$

Substitution of Eq. (8-129) into Eq. (8-128) yields

$$F(s) = s^2 + 2\zeta\omega_0 s + \omega_0^2 \tag{8-130}$$

The roots are then located at

$$\begin{cases} s_1 \\ s_2 \end{cases} = -\zeta\omega_0 \pm j\omega_0\sqrt{1 - \zeta^2} \tag{8-131}$$

As long as $\zeta < 1$, the roots are clearly complex.

For the purpose of plotting frequency response curves, it is desirable to rearrange Eq. (8-130) into the modified form

$$F_m(s) = 1 + \left(\frac{2\zeta}{\omega_0}\right)s + \left(\frac{s}{\omega_0}\right)^2 \tag{8-132}$$

Thus, in dealing with quadratic factors with complex roots, one should rearrange such factors into the form of Eq. (8-132).

Complex Zeros (Simple-Order)

Consider the function

$$G(s) = 1 + \left(\frac{2\zeta}{\omega_0}\right)s + \left(\frac{s}{\omega_0}\right)^2 \tag{8-133}$$

The absolute amplitude response is

$$A(\omega) = \sqrt{\left[1 - \left(\frac{\omega}{\omega_0}\right)^2\right]^2 + \left(\frac{2\zeta\omega}{\omega_0}\right)^2} \tag{8-134}$$

The dB amplitude response is

$$A_{dB}(\omega) = 10 \log_{10}\left\{\left[1 - \left(\frac{\omega}{\omega_0}\right)^2\right]^2 + \left(\frac{2\zeta\omega}{\omega_0}\right)^2\right\} \tag{8-135}$$

Plots showing A and A_{dB} for several values of ζ are given in Fig. 8-25a. For $\zeta < 0.707$, there is an *undershoot* near $\omega = \omega_0$. It can be seen that *the high-frequency approximation is a straight line on a logarithmic scale, with slope +2 on an absolute plot and an increase of 12 dB/octave or 40 dB/decade on a dB plot.*

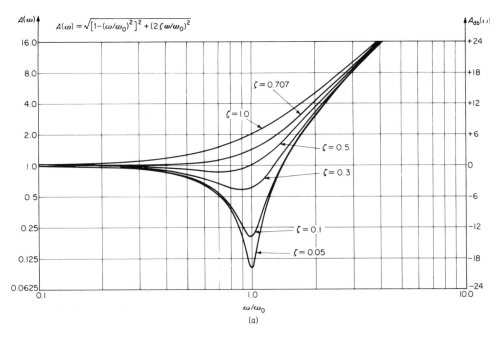

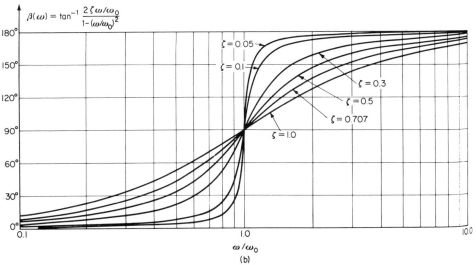

Figure 8-25   Amplitude and phase curves for complex zeros in left-hand half-plane.

The phase response is

$$\beta(\omega) = \tan^{-1} \frac{2\zeta\omega/\omega_0}{1 - (\omega/\omega_0)^2} \qquad (8\text{-}136)$$

Some curves are shown in Fig. 8-25b.

Complex Poles (Simple-Order)

Consider the function

$$G(s) = \frac{1}{1 + (2\zeta/\omega_0)s + (s/\omega_0)^2} \tag{8-137}$$

The absolute amplitude response is

$$A(\omega) = \frac{1}{\sqrt{[1 - (\omega/\omega_0)^2] + (2\zeta\omega/\omega_0)^2}} \tag{8-138}$$

The dB amplitude response is

$$A_{dB}(\omega) = -10 \log_{10}\left\{\left[1 - \left(\frac{\omega}{\omega_0}\right)^2\right]^2 + \left(\frac{2\zeta\omega}{\omega_0}\right)^2\right\} \tag{8-139}$$

Plots showing A and A_{dB} for different values of ζ are shown in Fig. 8-26a. For $\zeta < 0.707$, there is an overshoot near $\omega = \omega_0$. In this case, *the high-frequency asymptote is a straight line on a logarithmic scale, with slope -2 on an absolute plot and a decrease of 12 dB/octave or 40 dB/decade on a dB plot.*

The phase response is

$$\beta(\omega) = -\tan^{-1}\frac{2\zeta\omega/\omega_0}{1 - (\omega/\omega_0)^2} \tag{8-140}$$

Some curves are shown in Fig. 8-26b.

We will not present a development of multiple-order complex roots. From preceding work, it should be clear that the effect of multiple-order roots is to multiply the slope of the amplitude curve and the scale of the phase curve by the order of the roots.

Now that we have looked at the possible basic forms for factors, let us turn our attention to a complete response. In general, the numerator and denominator polynomials of a given function can be factored into "elementary polynomials" of the forms considered earlier. On the logarithmic scales, multiplication is accomplished by addition of dB quantities and division is accomplished by subtraction of dB quantities. The complete frequency-response curve is represented by the algebraic sum of the individual factors of the type considered earlier. This procedure is best illustrated by an example.

Example 8-4

Plot the approximate amplitude response curve for the transfer function

$$G(s) = \frac{80s}{(s + 2)(s + 20)} \tag{8-141}$$

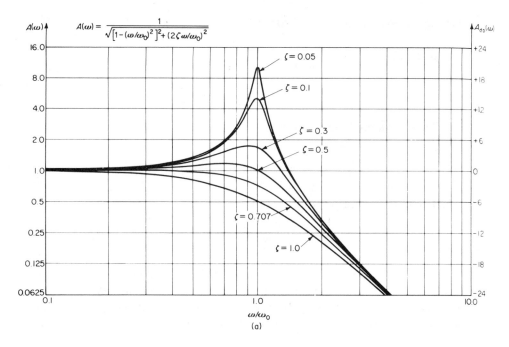

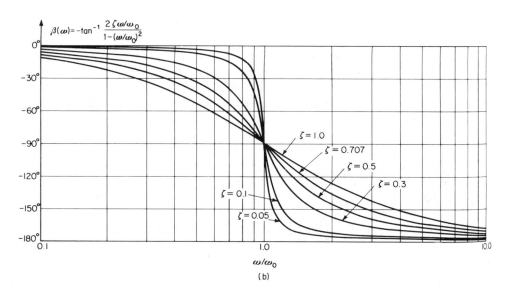

Figure 8-26   Amplitude and phase curves for complex poles in left-hand half-plane.

Solution   We must first place the denominator factors in the proper form.   We have

$$G(s) = \frac{2s}{(1 + s/2)(1 + s/20)} \tag{8-142}$$

A complete development reads as follows:

$$A(\omega) = \frac{2\omega}{\sqrt{1 + (\omega/2)^2}\sqrt{1 + (\omega/20)^2}} \tag{8-143}$$

$$A_{dB}(\omega) = 20 \log_{10} + 20 \log_{10} \omega$$

$$-10 \log_{10}\left[1 + \left(\frac{\omega}{2}\right)^2\right] - 10 \log_{10}\left[1 + \left(\frac{\omega}{20}\right)^2\right] \tag{8-144}$$

The first term above is merely a constant (6 dB), and the second, third, and fourth terms are forms considered previously. A plot of the breakpoint approximations of each of the four component parts is shown in Fig. 8-27a, and the resultant is given by curve P in (b). A smooth approximation to the actual curve is then made by "educated guessing," resulting in curve E.

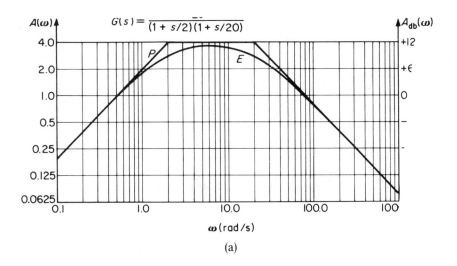

(a)

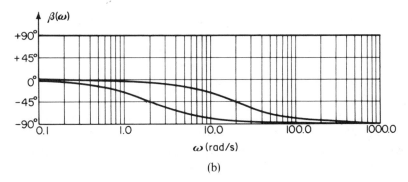

(b)

Figure 8-27   Amplitude response curves for Example 8-4.

As long as the break frequencies are reasonably far apart, the response can be determined fairly accurately. However, if some break frequencies are fairly close together, it may be necessary to actually evaluate the function at some points.

With some practice, it is not always necessary to plot the various components of the overall response, as the complete breakpoint response may be drawn in one stage. One merely "moves" along the frequency axis and notes the effect of various numerator and denominator factors. The net slope at any point is the algebraic sum of all previous slopes. This process is almost identical to the technique for adding ramp functions considered in Chapter 2, and the reader might find it advantageous to review Section 2-12 if necessary.

GENERAL PROBLEMS

8-1. Using phasor methods, solve for the steady-state variables $i(t)$, $v_R(t)$, $v_L(t)$, and $v_C(t)$ in the circuit of Fig. P8-1.

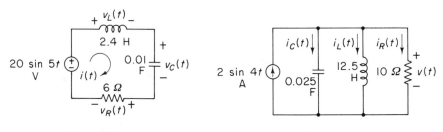

Figure P8-1   **Figure P8-2**

8-2. Using phasor methods, solve for the steady-state variables $v(t)$, $i_R(t)$, $i_C(t)$, and $i_L(t)$ in the circuit of Fig. P8-2.

8-3. Using phasor methods, solve for the steady-state variables $i_R(t)$, $i_C(t)$, $i_L(t)$, and $i_0(t)$ in the circuit of Fig. P8-3.

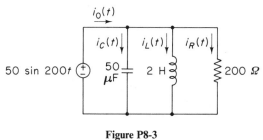

Figure P8-3

8-4. Using phasor methods, solve for the steady-state currents $i_1(t)$ and $i_2(t)$ in the circuit of Fig. P8-4.

8-5. At the terminals of a certain passive circuit, the net impedance (in ohms) at a specific frequency is

$$Z = 60 + j25$$

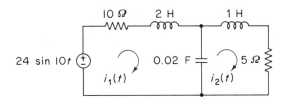

Figure P8-4

(a) Identify the resistance and the reactance.
(b) Construct an equivalent circuit for the impedance at the given frequency.
(c) Determine the admittance Y, and identify the conductance and the susceptance.
(d) Determine an equivalent parallel circuit at the given frequency.

8-6. At the terminals of a certain passive circuit, the net admittance (in siemens) at a specific frequency is

$$Y = 35 + j120$$

(a) Identify the conductance and the susceptance.
(b) Construct an equivalent circuit for the admittance at the given frequency.
(c) Determine the impedance Z, and identify the resistance and the reactance.
(d) Determine an equivalent series circuit at the given frequency.

8-7. Determine the impedance $Z(j\omega)$ of the circuit of Fig. P8-7 with ω considered as a variable. Determine $R(\omega)$ and $X(\omega)$.

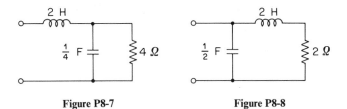

Figure P8-7   **Figure P8-8**

8-8. Determine the impedance $Z(j\omega)$ of the circuit of Fig. P8-8 with ω considered as a variable. Determine $R(\omega)$ and $X(\omega)$.

8-9. Determine the phasor Thévenin equivalent circuit at the terminals a-a' for the circuit of Fig. P8-9.

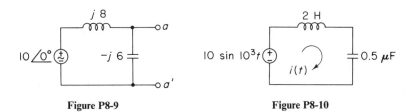

Figure P8-9   **Figure P8-10**

8-10. Attempt to determine the steady-state current $i(t)$ in the circuit of Fig. P8-10. Explain the difficulties encountered.

8-11. Consider the *RC* circuit of Fig. P8-11.
  (a) Assuming that the circuit is initially relaxed, determine the *s*-domain model, and use it to determine the complete time-domain current $i(t)$ and voltage $v(t)$.
  (b) Determine the phasor domain model, and use it to determine the steady-state current $i_{ss}(t)$ and voltage $v_{ss}(t)$.

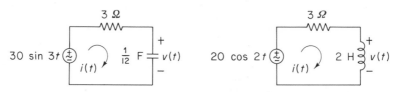

Figure P8-11   **Figure P8-12**

8-12. Consider the *RL* circuit of Fig. P8-12.
  (a) Assuming that the circuit is initially relaxed, determine the *s*-domain model, and use it to determine the complete time domain current $i(t)$ and voltage $v(t)$.
  (b) Determine the phasor-domain model, and use it to determine the steady-state current $i_{ss}(t)$ and voltage $v_{ss}(t)$.

8-13. For the circuit of Fig. P8-13, determine mathematical expressions for the steady-state transfer function, the absolute amplitude response, the decibel amplitude response, and the phase responses. The input is v_1 and the output is v_2.

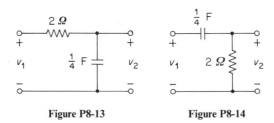

Figure P8-13   **Figure P8-14**

8-14. For the circuit of Fig. P8-14, determine mathematical expressions for the steady-state transfer function, the absolute amplitude response, the decibel amplitude response, and the phase response. The input is v_1 and the output is v_2.

In Problems 8-15 through 8-20 plot the breakpoint approximations to the amplitude response of the functions given.

8-15. $G(s) = \dfrac{10}{1 + s/5}$

8-16. $G(s) = \dfrac{10s}{1 + s/5}$

8-17. $G(s) = \dfrac{0.1(1 + s/2)}{1 + s/20}$

8-18. $G(s) = \dfrac{0.5(1 + s/20)}{1 + s/2}$

8-19. $G(s) = \dfrac{50}{(1 + s/10)(1 + s/100)}$

8-20. $G(s) = \dfrac{500s}{s^2 + 120s + 2000}$

DERIVATION PROBLEMS

8-21. In deriving the steady-state impedance for inductance in the text, the derivative form expressing voltage in terms of current as given by Eq. (8-19) was used as the starting point. Provide an alternative derivation starting with the integral relationship expressing current in terms of voltage. Ignore the lower limit of the integral because it would be a constant and would disappear in the steady state.

8-22. In deriving the steady-state impedance for capacitance in the text, the integral form expressing voltage in terms of current as given by Eq. (8-25) was used as the starting point. Provide an alternative derivation starting with the derivative relationship expressing current in terms of voltage.

8-23. Steady-state resonance is a process in which the net capacitive and inductive reactances cancel at a given frequency, thus yielding a purely resistive impedance. Consider the series *RLC* circuit of Fig. P8-23.
  (a) Determine the impedance $Z(j\omega)$
  (b) Impose the condition that the net reactance be zero at a given frequency f_0 and show that

$$f_0 = \frac{1}{2\pi\sqrt{LC}}$$

By a simple argument show that the magnitude $|Z(j\omega)|$ of the impedance is *minimum* at f_0.

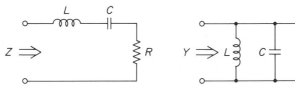

Figure P8-23   **Figure P8-24**

8-24. Problem 8-23 dealt with a *series* resonant circuit, while in this problem we consider a parallel resonant circuit. Consider the circuit of Fig. P8-24.
  (a) Determine the admittance Y.
  (b) Impose the condition that the net susceptance be zero at a given frequency f_0 and show that

$$f_0 = \frac{1}{2\pi\sqrt{LC}}$$

  (c) By a simple argument, show that the magnitude $|Z(j\omega)|$ of the impedance is *maximum* at f_0.

8-25. The circuit of Fig. P8-25 can be used as a high-frequency *preemphasis* network. It has the property that the amplitude response has a larger magnitude for higher-frequency components of a given signal. When proper compensation is used, the signal-to-noise ratio in certain processes such as frequency modulation can be improved.

(a) For the circuit as given, derive an expression for the transfer function $G(s) = V_2(s)/V_1(s)$ and arrange in the form most suitable for a Bode plot.

(b) Show that the amplitude response has a numerator break frequency f_{bn} and a denominator break frequency f_{bd}. Obtain expressions for these quantities, and show that $f_{bn} < f_{bd}$.

(c) Sketch the form of the breakpoint decibel amplitude response approximation, and label the levels of the flat portions of the response.

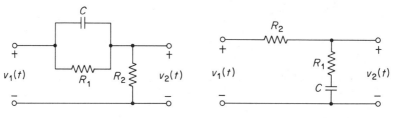

Figure P8-25   **Figure P8-26**

8-26. The circuit of Fig. P8-26 can be used as a high-frequency *deemphasis* network. It has the property that the amplitude response has a larger magnitude for lower-frequency components of a given signal. It can be used to compensate for the preemphasis characteristics of a circuit such as that of Problem 8-25.

(a) For the circuit as given, derive an expression for the transfer function $G(s) = V_2(s)/V_1(s)$ and arrange in the form most suitable for a Bode plot.

(b) Show that the amplitude response has a numerator break frequency f_{bn} and a denominator break frequency f_{bd}. Obtain expressions for these quantities and show that $f_{bd} < f_{bn}$.

(c) Sketch the form of the breakpoint decibel amplitude response approximation, and label the levels of the flat portions of the response.

8-27. Consider a transfer function having n_i zeros at $s = \infty$. It was shown in Chapter 7 by the work leading up to Eq. (7-73) that $G(s)$ could then be approximated as

$$G(s) \simeq \frac{K}{s^r} \qquad \text{for } |s| \gg 1$$

and $K = a_n/b_m$ in the notation of Eq. (7-73). Show in the range where this result is valid that the slope of the decibel amplitude response is very nearly $-6r$ dB/octave.

8-28. Consider the second-order low-pass amplitude response $A(\omega)$ given in general form by Eq. (8-138).

(a) Applying differential calculus, show that a maximum in the amplitude response occurs at a frequency ω_{max} given by

$$\omega_{max} = \omega_0 \sqrt{1 - 2\zeta^2}$$

[*Hint:* It is easier to define a new function $y = 1/A^2(\omega)$ and determine the frequency at which this response has a minimum level.]

(b) Show that there is a maximum in the vicinity of ω_0 when $\zeta < 0.707$. (This corresponds to a peaked response.)

(c) Show that the value of the peak A_{max} is

$$A_{max} = \frac{1}{2\zeta\sqrt{1 - \zeta^2}}$$

(d) Check the result of (c) by computing A_{max} for one of the curves of Fig. 8-26.

COMPUTER PROBLEMS

8-29. Consider the simple low-pass transfer function

$$G(s) = \frac{A_0}{1 + s/2\pi f_b}$$

where the input break frequency f_b in hertz has been specified. Write a computer program to evaluate the amplitude response and the phase response over a frequency range from f_1 to f_2 in steps of Δf.

Input data: $A_0, f_b, f_1, f_2, \Delta f$

Output data: $f, A(\omega), A_{dB}(\omega), \beta(\omega)$

8-30. Consider the simple high-pass transfer function

$$G(s) = \frac{A_0 s}{1 + s/2\pi f_b}$$

Write a computer program to evaluate the amplitude response and the phase response over a frequency range from f_1 to f_2 in steps of Δf.

Input data: $A_0, f_b, f_1, f_2, \Delta f$

Output data: $f, A(\omega), A_{dB}(\omega), \beta(\omega)$

8-31. Consider the transfer function

$$G(s) = \frac{A_0}{(1 + s/2\pi f_{b1})(1 + s/2\pi f_{b2})}$$

where the two break frequencies f_{b1} and f_{b2} in hertz have been specified. Write a computer program to evaluate the amplitude response and the phase response over a frequency range from f_1 to f_2 in steps of Δf.

Input data: $A_0, f_{b1}, f_{b2}, f_1, f_2, \Delta f$

Output data: $f, A(\omega), A_{dB}(\omega), \beta(\omega)$

8-32. Consider the transfer function

$$G(s) = \frac{A_0 \left(1 + \dfrac{s}{2\pi f_{b1}}\right)}{\left(1 + \dfrac{s}{2\pi f_{b2}}\right)}$$

Write a computer program to evaluate the amplitude response and the phase response over a frequency range from f_1 to f_2 in steps of Δf.

Input data: $A_0, f_{b1}, f_{b2}, f_1, f_2, \Delta f$

Output data: $f, A(\omega), A_{dB}(\omega), \beta(\omega)$

8-33. Consider the second-order low-pass transfer function of Eq. (8-137). With $f_0 = \omega_0/2\pi$ and ζ specified, write a computer program to evaluate the amplitude response and the phase response over a frequency from f_1 to f_2 in steps of Δf.

Input data: $\zeta, \omega_0, f_1, f_2, \Delta f$

Output data: $f, A(\omega), A_{dB}(\omega), \beta(\omega)$

9

FOURIER ANALYSIS

OVERVIEW

Fourier analysis is the process of representing an arbitrary time function as the sum of sinusoidal functions. The number of sinusoidal functions may be finite, or the representation may require an infinite number of terms. The frequencies involved may be integer multiples of a *fundamental* frequency, or they may represent a continuum of frequencies.

The process of determining the Fourier components of a signal is called *spectral analysis*. The Fourier components that represent the signal are referred to collectively as the *frequency spectrum* of the signal.

Fourier analysis as developed in this chapter is divided into (a) the *Fourier series* and (b) the *Fourier transform*. The concept of *Fourier series* is applied to *periodic* signals, and the *Fourier transform* is applied to *nonperiodic* signals. The spectrum of a periodic signal appears only at integer multiples of the fundamental frequency, while the spectrum of a nonperiodic signal is a continuous function of frequency.

OBJECTIVES

After completing this chapter, the reader should be able to

1. Determine the values of the fundamental frequency and the harmonic frequencies contained in a periodic signal.
2. State the three forms of the Fourier series.
3. For a given periodic signal, determine the three forms of the Fourier series.
4. Convert between the various forms of the Fourier series.
5. Construct both one-sided and two-sided frequency spectrum plots from a given Fourier series.

310

6. Inspect a given periodic signal to determine any possible symmetry conditions.

7. Indicate the implications of any symmetry conditions as determined in (6) in terms of the resulting spectral content and simplification of the spectrum computation.

8. State the definitions of the Fourier transform and inverse transform.

9. Discuss the difference between the Fourier series and transform in terms of applicable time functions and the resulting spectral properties.

10. Inspect a given nonperiodic function to determine any possible symmetry condition.

11. Indicate the implications of any symmetry condition as determined in (10) in terms of the resulting spectral content and simplification of the spectrum computation.

12. Discuss the various Fourier transform operation pairs and the effect on the spectrum in each case.

13. Determine the roll-off rate of the spectrum of a given time function from a knowledge of the relative continuity of the function and its derivatives.

9-1 FOURIER SERIES[†]

The concept of the Fourier series is based on representing a periodic signal as the sum of harmonically related sinusoidal functions. As we are already aware, sinusoidal functions are among the most important waveforms that arise in electrical systems. With the aid of Fourier series, complex periodic waveforms can be represented in terms of sinusoidal functions, whose properties are very familiar to us. Thus, the response of a system to a complex waveform can be viewed as the response to a series of sinusoidal functions.

Without realizing it, the reader may already be familiar with some aspects of Fourier theory through possible interests in audio or stereo systems. Consider an audio amplifier in which the amplitude response is specified as flat from, say, 20 Hz to 20 kHz. This specification is based on a test of the amplitude response using a sinusoidal input as discussed in Chapter 8. A sinusoidal function certainly does not represent a signal of any interest for listening. However, the use of the sinusoid for testing is based on the fact that a complex signal can be represented as a combination of sinusoids. By studying the response of the system to individual frequencies, it is possible to infer properties of the system for complex waveforms such as music.

The Fourier series is best applied to a periodic function, and the treatment in this text will be limited to that case. Consider the arbitrary periodic signal $x(t)$ shown in Fig. 9-1. The function could represent either a voltage or a current waveform. [The symbol $f(t)$ is not used here since f is the designation for frequency.] According

---

[†]Portions of this chapter have been adapted from the author's text, *Electronic Communications Systems*, Reston Publishing Co., Reston Va., 1982.

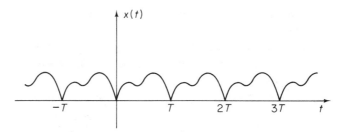

Figure 9-1   Example of an arbitrary periodic signal for which a Fourier series representation could be obtained.

to Fourier theory, this signal may be represented by the sum of a series of sine and/or cosine functions plus a dc term. The resulting series is called a *Fourier series*.

The lowest frequency (other than dc) of the sinusoidal components is a frequency f_1 given by

$$f_1 = \frac{1}{T} \tag{9-1}$$

This frequency is referred to as the *fundamental* component, and it is the same as that of the waveform itself. Thus a periodic signal with a period of 1 millisecond (ms) will have a fundamental component with a frequency $f_1 = 1$ kHz.

All other frequencies in the signal will be integer multiples of the fundamental. These various components are referred to as *harmonics*, with the *order* of a given harmonic indicated by the ratio of its frequency to the fundamental frequency. Thus for the waveform suggested in the preceding paragraph, frequencies appearing in the signal above 1 kHz would, in general, be a second harmonic at 2 kHz, a third harmonic at 3 kHz, a fourth harmonic at 4 kHz, and so on.

Any transmission system through which a given signal passes must have a bandwidth sufficiently large to pass all significant frequencies of the signal.

In a purely mathematical sense, many common waveforms theoretically contain an infinite number of harmonics. The preceding two sentences would then lead one to believe that an infinite bandwidth would be required to process such signals, which is impossible. The key to this apparent contradiction is the term "significant" in the first sentence.

It is easy to predict the theoretical frequencies present in a given periodic signal, as we have just seen. However, it is much more difficult to predict the relative magnitudes of the different components in order to determine which components are significant and which are not. From a signal transmission point of view, if we can predict the frequency range over which the magnitudes are significant in size, we can then estimate the bandwidth requirements. In this sense, we simply ignore all harmonics of the signal that are outside the range and assume that their exclusion produces no noticeable degradation in the signal quality.

We will now consider the mathematical structure of the Fourier series. There are three forms of the Fourier series. They are (a) the *sine-cosine* form, (b) the *amplitude-phase* form, and (c) the *complex exponential* form. Each form will now be considered individually.

Sine–Cosine Form

The sine–cosine form is the one most commonly presented first in circuits and mathematics texts. This form represents a periodic signal $x(t)$ as a sum of sines and cosines in the form

$$x(t) = A_0 + \sum_{n=1}^{\infty} (A_n \cos n\omega_1 t + B_n \sin n\omega_1 t) \tag{9-2}$$

where

$$\omega_1 = 2\pi f_1 = \frac{2\pi}{T} \tag{9-3}$$

is the fundamental *angular* frequency in radians per second (rad/s). The nth harmonic *cyclic* frequency (in hertz) is $n f_1$, and the corresponding angular frequency is $n\omega_1$. As a result of the equalities in Eq. (9-3), the argument for either the sine or cosine function in Eq. (9-2) can be expressed in either of the following forms:

$$n\omega_1 t = 2\pi n f_1 t = \frac{2\pi n t}{T} \tag{9-4}$$

The term A_0 represents the dc term and, as we will see shortly, it is simply the average value of the signal over one cycle. (*Note:* Some authors define the dc value as $A_0/2$ in order to make some of the later formulas apply to this case, but the A_0 form is easier to interpret.) Other than dc, there are two components appearing at a given harmonic frequency in the most general case: (a) a cosine term with an amplitude A_n and (b) a sine term with an amplitude B_n.

The major task involved in a Fourier analysis is in determining the A_n and B_n coefficients. The dc term A_0 is simply the average value and is given by

$$A_0 = \frac{1}{T} \int_0^T x(t)\, dt = \frac{\text{area under curve in one cycle}}{\text{period } T} \tag{9-5}$$

as discussed in Section 2-16. Formulas for A_n and B_n are derived in applied mathematics books and are summarized here as follows:

$$A_n = \frac{2}{T} \int_0^T x(t) \cos n\omega_1 t\, dt \tag{9-6}$$

for $n \geq 1$ but *not* for $n = 0$.

$$B_n = \frac{2}{T} \int_0^T x(t) \sin n\omega_1 t\, dt \tag{9-7}$$

for $n \geq 1$. In all three of the preceding integrals, the limits of integration may be changed for convenience provided that the interval of integration is over one complete cycle in a positive sense. A common alternative range is from $-T/2$ to $T/2$.

The formulas for A_n and B_n indicate that the quantities are determined by first multiplying the signal by a cosine or sine term of the corresponding frequency at which the coefficient is desired and then determining the area of the resulting product function over one cycle. In many cases, a general expression for A_n or B_n can be determined from one integration with n appearing as a parameter. In other cases, particularly for experimental data, a separate integration may be required at each frequency. Computational means have been developed for performing such operations efficiently on a computer, and many scientific computer systems have programs available for determining the Fourier series of experimental data signals. If a signal $x(t)$ is defined in different forms over different parts of a cycle, the evaluation of either of the integrals may require expansion into several integrals over shorter portions of the cycle.

Amplitude-Phase Form

The sine–cosine form of the Fourier series given in Eq. (9-2) is usually the easiest form from which to evaluate the coefficients, and it is the form most commonly tabulated in reference books. However, it suffers from the fact that there are, in a sense, two separate components at a given frequency, each of which has a separate amplitude. When we measure the magnitude of a given spectral component with a frequency-selective instrument, which component do we obtain? The fact is that the actual magnitude that would be measured with most instruments would be neither A_n nor B_n, but rather a special combination of the two, as will be seen shortly.

The amplitude-phase form of the Fourier series is based on the concept developed in Section 2-9 in which the sum of two or more sinusoids of a given frequency is equivalent to a single sinusoid at the same frequency. Specifically, the sine–cosine form involves a single sine function plus a single cosine function at each frequency in the series. The approach in Section 2-9 emphasized expressing the sum as a sine function since that form is usually more convenient in analyzing circuit problems. Traditionally, the cosine function has been probably used more in Fourier analysis, so we will consider both variations in this development.

The amplitude-phase form of the Fourier series can be expressed as either

$$x(t) = C_0 + \sum_{n=1}^{\infty} C_n \cos(n\omega_1 t + \phi_n) \tag{9-8}$$

or

$$x(t) = C_0 + \sum_{n=1}^{\infty} C_n \sin(n\omega_1 t + \theta_n) \tag{9-9}$$

The amplitudes of corresponding components in either of the preceding variations are the same, but the phase angles ϕ_n and θ_n differ since one representation involves cosine functions and the other involves sine functions. The first term C_0 is the dc value and is the same as given by Eq. (9-5); that is, $C_0 = A_0$. It has been redefined here as C_0 in order to maintain a consistent form of notation.

In the form of either Eq. (9-8) or (9-9) a given C_n represents the *net* amplitude of a given component at the frequency nf_1. Since sine and cosine phasor forms are always perpendicular to each other, the net amplitude is

$$C_n = \sqrt{A_n^2 + B_n^2} \tag{9-10}$$

The angles ϕ_n or θ_n are best determined from the procedures developed in Section 2-8 and Section 2-9. The quantity θ_n represents the angle measured from the positive sine axis, and ϕ_n represents the angle measured from the positive cosine axis. If the exact phasor form established in those sections is used, θ_n will represent the exact phasor angle, while ϕ_n would be determined by adding $-90°$ to the exact phasor angle based on the sine reference.

We return now to the question of which component is measured with a frequency-selective instrument. With the majority of common instruments of this type, the reading is proportional to C_n, the net amplitude. (The instrument may actually be calibrated to read the rms value of C_n, which is $C_n/\sqrt{2}$.) There are, however, certain special phase-sensitive instruments that can be used to obtain A_n and B_n separately, so the peculiarities of such instruments should be understood before measurements are made.

There are a number of cases of well-known waveforms in which either A_n or B_n (but obviously not both) is identically zero at all possible frequencies for the signal. Indeed, as we will see later, it is possible in some cases to choose the time origin in a way that will force this result. With such signals, the sine–cosine form is identical with one of the variations of the amplitude-phase form.

Complex Exponential Form

The complex exponential form of the Fourier series is the most difficult form for many persons to perceive, primarily because it represents a step away from the domain of real signals into a domain of complex mathematical representations. However, there are some developments that can be done much easier with the exponential form than with either of the earlier forms. In particular, the exponential form is a direct link with the concept of the Fourier transform, which is very important in dealing with nonperiodic signals, as we will see later.

The exponential form of the Fourier series is related to the fact that both the sine and cosine functions can be expressed in terms of exponential functions with purely imaginary arguments. The basis for this is Euler's formula, which is written in two separate forms as

$$\epsilon^{jn\omega_1 t} = \cos n\omega_1 t + j \sin n\omega_1 t \qquad (9\text{-}11)$$

$$\epsilon^{-jn\omega_1 t} = \cos n\omega_1 t - j \sin n\omega_1 t \qquad (9\text{-}12)$$

Alternate addition and subtraction of Eqs. (9-11) and (9-12) result in the following two expressions for cosine and sine functions:

$$\cos n\omega_1 t = \frac{\epsilon^{jn\omega_1 t} + \epsilon^{-jn\omega_1 t}}{2} \qquad (9\text{-}13)$$

$$\sin n\omega_1 t = \frac{\epsilon^{jn\omega_1 t} - \epsilon^{-jn\omega_1 t}}{2j} \qquad (9\text{-}14)$$

Note that both forms contain an exponential function with a $(jn\omega_1 t)$ argument and an exponential function with a $(-jn\omega_1 t)$ argument. The first term may be thought of as a "positive frequency" term corresponding to a frequency nf_1 (assuming n is positive), and the second term may be considered as a "negative frequency" term corresponding to a frequency $-nf_1$. Both terms are required to completely describe the sine or cosine function.

The exponential form can be developed by expanding the sine and cosine functions according to these exponential definitions and regrouping. This general process is somewhat detailed and will not be given here. We will concentrate here on the results and the corresponding interpretations.

The general form of the complex exponential form of the Fourier series can be expressed as

$$x(t) = \sum_{n=-\infty}^{\infty} \bar{X}_n \epsilon^{jn\omega_1 t} \tag{9-15}$$

where the bar above $\bar{X}_n$ indicates that it is, in general, a complex value. The Fourier coefficient at a given frequency nf_1 is the complex quantity $\bar{X}_n$. An expression for determining $\bar{X}_n$ is

$$\bar{X}_n = \frac{1}{T} \int_0^T x(t) \epsilon^{-jn\omega_1 t} \, dt \tag{9-16}$$

Some interpretation of the preceding results is in order. Note in Eq. (9-15) that the exponential series is summed over both negative and positive frequencies (or negative and positive values of n), as previously discussed. At a given real frequency kf_1, $(k > 0)$, the spectral representation consists of

$$\bar{X}_k \epsilon^{jk\omega_1 t} + \bar{X}_{-k} \epsilon^{-jk\omega_1 t}$$

The first term is thought of as the "positive frequency" contribution, while the second is the corresponding "negative frequency" contribution. Although either one of the two terms is a complex quantity, they add together in such a manner as to create a real function, and this is why both terms are required to make the mathematical form complete. On the other hand, all the spectral information can be deduced from either the $\bar{X}_k$ term or the $\bar{X}_{-k}$ term, since there is a direct relationship between them. Let $\tilde{X}$ represent the complex conjugate of $\bar{X}$. Then it can be shown that

$$\bar{X}_{-n} = \tilde{X}_n \tag{9-17}$$

Thus the negative frequency coefficient is the complex conjugate of the corresponding positive frequency coefficient.

While the coefficient $\bar{X}_n$ can be calculated from Eq. (9-16) directly, it turns out that $\bar{X}_n$ can also be calculated directly from A_n and B_n of the sine–cosine form. The relationship reads

$$\bar{X}_n = \frac{A_n - jB_n}{2}, \qquad \text{for } n \neq 0 \tag{9-18}$$

Even though A_n and B_n are interpreted only for positive n in the sine–cosine form, their functional forms may be extended for both positive and negative n in applying Eq. (9-18). Alternatively, Eq. (9-18) may be applied for positive n, and Eq. (9-17) may be used for determining the corresponding coefficients for negative n. The dc component X_0 is simply

$$\bar{X}_0 = \frac{1}{T} \int_0^T x(t) \epsilon^{-jn\omega_1 t} \, dt = A_0 = C_0 \qquad (9\text{-}19)$$

which is the same in all the Fourier forms.

In many situations involving the complex form of the Fourier series, it is desirable to express $\bar{X}_n$ as a magnitude and an angle. Since $\bar{X}_n$ is a complex quantity having a real and an imaginary part, this can be readily achieved. Thus $\bar{X}_n$ can be expressed in polar form as

$$\bar{X}_n = X_n \epsilon^{j\phi_n} = X_n \underline{/\phi_n} \qquad (9\text{-}20)$$

The quantity $X_n = |\bar{X}_n|$ represents the magnitude of the complex Fourier coefficient at a frequency nf_1, and ϕ_n is the corresponding angle or phase expressed in the polar form. For example, if $\bar{X}_n = 3 + j4$ at a given frequency, it can also be expressed as $\bar{X}_n = 5\underline{/53.13^\circ}$ with $X_n = 5$ and $\phi_n = 53.13^\circ$.

Example 9-1

This example will have as a primary objective the conversion between different forms of the Fourier series. A certain periodic bandlimited signal has only three frequencies in its Fourier series representation: dc, 1 kHz, and 2 kHz. The signal can be expressed in sine–cosine form as

$$x(t) = 18 + 40 \cos 2000\pi t - 30 \sin 2000\pi t$$
$$- 24 \cos 4000\pi t + 10 \sin 4000\pi t \qquad (9\text{-}21)$$

Express the signal in (a) amplitude-phase form and (b) complex exponential form.

          Solution   (a) The amplitude-phase form desired for the signal will read

$$x(t) = 18 + C_1 \cos (2000\pi t + \phi_1) + C_2 \cos (4000\pi t + \phi_2) \qquad (9\text{-}22)$$

or

$$x(t) = 18 + C_1 \sin (2000\pi t + \theta_1) + C_2 \sin (4000\pi t + \theta_2) \qquad (9\text{-}23)$$

where $C_0 = 18$ has been noted by obvious inspection.

The amplitude and angles will be determined from the phasor forms. Let C_1 represent the phasor associated with 1000 Hz. Using the sine function as the basis as established in Chapter 2, $\bar{C}_1$ is shown in Fig. 9-2a, and it can be expressed as

$$\bar{C}_1 = -30 + j40 = 50\underline{/126.87^\circ} \qquad (9\text{-}24)$$

Let $\bar{C}_2$ represent the corresponding phasor associated with 2000 Hz as shown in Fig. 9-2b. We have

$$\bar{C}_2 = 10 - j24 = 26\underline{/-67.38^\circ} \qquad (9\text{-}25)$$

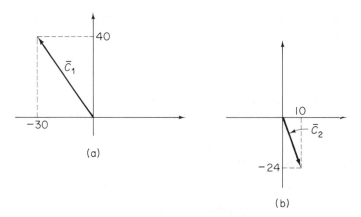

Figure 9-2   Phasor diagrams of Example 9-1.

The angles associated with Eqs. (9-24) and (9-25) are θ_1 and θ_2 respectively. The angles ϕ_1 and ϕ_2 are $\phi_1 = 126.87° - 90° = 36.87°$ and $\phi_2 = -67.38° - 90° = -157.38°$. The two variations of the amplitude-phase form are

$$x(t) = 18 + 50 \cos(2000\pi t + 36.87°) + 26 \cos(4000\pi t - 157.38°) \qquad (9\text{-}26)$$

and

$$x(t) = 18 + 50 \sin(2000\pi t + 126.87°) + 26 \sin(4000\pi t - 67.38°) \qquad (9\text{-}27)$$

(b) The complex exponential form will be determined from the formula (9-18) in conjunction with the conjugate relationship of Eq. (9-17). Thus

$$\bar{X}_1 = \frac{40 - j(-30)}{2} = 20 + j15 = 25\underline{/36.87} \qquad (9\text{-}28)$$

$$\bar{X}_{-1} = \tilde{\bar{X}}_1 = 20 - j15 = 25\underline{/-36.87°} \qquad (9\text{-}29)$$

$$\bar{X}_2 = \frac{-24 - j10}{2} = -12 - j5 = 13\underline{/-157.38°} \qquad (9\text{-}30)$$

$$\bar{X}_{-2} = \tilde{\bar{X}}_2 = -12 + j5 = 13\underline{/157.38°} \qquad (9\text{-}31)$$

The series may then be expressed as

$$\begin{aligned} x(t) = 18 &+ (20 + j15)\epsilon^{j2000\pi t} + (20 - j15)\epsilon^{-j2000\pi t} \\ &+ (-12 - j5)\epsilon^{j4000\pi t} + (-12 + j5)\epsilon^{-j4000\pi t} \end{aligned} \qquad (9\text{-}32)$$

in which the coefficients have been expressed in rectangular forms. Alternatively, the polar forms of the coefficients may be used in the expansion, in which the expression becomes

$$\begin{aligned} x(t) = 18 &+ 25\epsilon^{j(2000\pi t + 36.87°)} + 25\epsilon^{-j(2000\pi t + 36.87°)} \\ &+ 13\epsilon^{j(4000\pi t - 157.38°)} + 13\epsilon^{-j(4000\pi t - 157.38°)} \end{aligned} \qquad (9\text{-}33)$$

Example 9-2

Determine the Fourier series representation for the waveform shown in Fig. 9-3. Express in each of the following forms: (a) sine–cosine, (b) amplitude-phase, and (c) complex exponential.

Solution   (a) The sine–cosine form is usually the easiest form for determining the coefficients directly. The signal $x(t)$ over one complete cycle can be expressed as

$$x(t) = \begin{cases} A & \text{for } 0 < t < T/2 \\ 0 & \text{for } T/2 < t < T \end{cases} \tag{9-34}$$

The dc component A_0 can be determined by inspection since it is the average value. Thus,

$$A_0 = \frac{\text{area under curve in one cycle}}{T} = \frac{AT/2}{T} = \frac{A}{2} \tag{9-35}$$

The coefficients A_n can be determined from Eq. (9-6). Note that, while integration over a complete cycle is required, the function is zero over half of the cycle, so the integral reduces to

$$A_n = \frac{2}{T} \int_0^{T/2} A \cos n\omega_1 t \, dt$$

$$= \frac{2A}{n\omega_1 T} \sin n\omega_1 t \Big]_0^{T/2} = \frac{2A}{n\omega_1 T} \left(\sin \frac{n\omega_1 T}{2} - 0 \right) \tag{9-36}$$

In performing Fourier coefficient evaluations, the product $n\omega_1 T$ appears in virtually every case. When this occurs, the substitution $n\omega_1 T = 2\pi n$ is recommended to the reader as a suggested simplification. This equality is readily verified by noting that $n\omega_1 T = n2\pi f_1 T = n2\pi (1/T)T = 2\pi n$. This substitution will be made in various examples that follow without further justification.

The preceding evaluation now reduces to

$$A_n = \frac{2A}{2\pi n} \sin n\pi = 0 \qquad \text{for } n \neq 0 \tag{9-37}$$

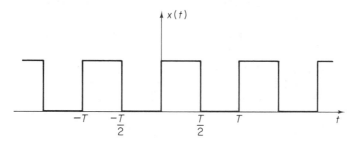

Figure 9-3   Waveform of Example 9-2.

since $\sin n\pi = 0$ for n an integer. Thus, in this particular example, the coefficients of all the cosine terms in the sine-cosine form are zero (except dc).

The B_n coefficients are now determined from the following development:

$$
\begin{aligned}
B_n &= \frac{2}{T} \int_0^{T/2} A \sin n\omega_1 t \, dt \\
&= \frac{-2A}{n\omega_1 T} \cos n\omega_1 t \Big]_0^{T/2} \\
&= \frac{-2A}{2\pi n} \left(\cos \frac{n\omega_1 T}{2} - 1 \right) = \frac{A}{n\pi} (1 - \cos n\pi)
\end{aligned}
\tag{9-38}
$$

The quantity $\cos n\pi$ satisfies

$$
\cos n\pi = \begin{cases} -1 & \text{for } n \text{ odd} \\ +1 & \text{for } n \text{ even} \end{cases}
\tag{9-39}
$$

When n is even, there is a cancellation inside the parentheses of the last term in Eq. (9-38) while for n odd, the term $-(-1) = +1$ adds to the other term. Thus the expression for B_n reduces to

$$
B_n = \begin{cases} \dfrac{2A}{n\pi} & \text{for } n \text{ odd} \\ 0 & \text{for } n \text{ even} \end{cases}
\tag{9-40}
$$

The resulting sine–cosine form of the Fourier series representation of $x(t)$ can be expressed as

$$
\begin{aligned}
x(t) = \frac{A}{2} &+ \frac{2A}{\pi} \sin \omega_1 t + \frac{2A}{3\pi} \sin 3\omega_1 t \\
&+ \frac{2A}{5\pi} \sin 5\omega_1 t + \frac{2A}{7\pi} \sin 7\omega_1 t + \cdots
\end{aligned}
\tag{9-41a}
$$

$$
= \frac{A}{2} + \sum_{\substack{n=1 \\ n \text{ odd}}}^{\infty} \frac{2A}{n\pi} \sin n\omega_1 t
\tag{9-41b}
$$

(b) Since the A_n coefficients in the sine–cosine form are zero, the sine–cosine series in this case is identical to one form of the amplitude-phase Fourier series. Specifically, the result of (a) is identical to the series form of Eq. (9-9) with $\theta_n = 0$.

(c) The complex exponential form will now be developed. While we could determine the coefficients $\bar{X}_n$ directly from the sine–cosine terms making use of Eq. (9-18), it will be more instructive for the reader's sake to start over again using the defining relationship for $\bar{X}_n$ as given by Eq. (9-16). First, we note that the dc component is again

$$
\bar{X}_0 = A_0 = \frac{A}{2}
\tag{9-42}
$$

The general coefficient $\bar{X}_n$ is given by

$$\bar{X}_n = \frac{1}{T} \int_0^{T/2} A\epsilon^{-jn\omega_1 t}\, dt = \frac{-A}{jn\omega_1 T} \epsilon^{-jn\omega_1 t} \Big]_0^{T/2}$$

$$= \frac{-A}{j2n\pi} (\epsilon^{-jn\omega_1 T/2} - 1) = \frac{A}{j2n\pi} (1 - \epsilon^{-jn\pi}) \qquad (9\text{-}43)$$

$$= \frac{A}{j2n\pi} (1 - \cos n\pi + j \sin n\pi)$$

where Euler's formula was used in the last step. This result can be readily simplified by first noting that $\sin n\pi = 0$ for n an integer. Furthermore, $1 - \cos n\pi = 2$ for n odd and $1 - \cos n\pi = 0$ for n even. Thus, $\bar{X}_n$ reduces to

$$\bar{X}_n = \frac{A}{jn\pi} = \frac{-jA}{n\pi} \qquad \text{for } n \text{ odd} \qquad (9\text{-}44)$$

The reader can readily verify that this same result is obtained quite quickly by applying Eq. (9-18) to the previous results of the sine–cosine form.

The complex Fourier series for $x(t)$ can now be expressed as

$$x(t) = \frac{A}{2} - j\frac{A}{\pi} \epsilon^{j\omega_1 t} - j\frac{A}{3\pi} \epsilon^{j3\omega_1 t} - \cdots$$

$$= +j\frac{A}{\pi} \epsilon^{-j\omega_1 t} + j\frac{A}{3\pi} \epsilon^{-j3\omega_1 t} + \cdots \qquad (9\text{-}45)$$

9-2 FREQUENCY SPECTRUM PLOTS

One of the most useful forms for displaying the Fourier series of a signal is by means of a graphical plot showing the relative strengths of the components as a function of frequency. Such a plot is loosely referred to as the *frequency spectrum* of the given signal. A frequency spectrum plot permits a quick visual determination of the frequencies present in the signal and their relative magnitudes. This is the basis for the spectrum analyzer, which is a widely used instrument in communications systems work providing a cathode-ray tube (CRT) display of the spectrum in much the same form as the graphical technique that will be discussed in this section.

In principle, any of the three forms could be shown graphically. In practice, however, the sine–cosine form is less desirable for this purpose since both A_n and B_n would have to be plotted separately (unless, of course, one of the two is zero). For the same reason, spectral displays involving the complex exponential form generally focus on the polar form (magnitude and angle) rather than the real and imaginary representation. Consequently, we focus on the amplitude-phase form and the magnitude and angle representation of the complex exponential form in our developments.

Both the amplitude-phase form and the complex exponential form have two quantities to be specified at each frequency (i.e., the amplitude and phase of the spectral components). While both quantities are necessary for mathematical reconstruction of the signal, in practical spectral displays, the amplitude is almost always the quantity that is emphasized. The reason for this is the simple fact that the relative amplitudes of spectral components are most significant in determining the bandwidth, while the phase only indicates the relative time shift of a given component relative to others. In fact, a simple time shift of the signal will readily change the phase terms without affecting the amplitude terms. The amplitude form will be referred to as an *amplitude frequency spectrum plot*.

One point about the terms "amplitude" and "magnitude" should be noted. By convention, these terms normally indicate a positive, real value. There are instances, however, when it is convenient to allow a given amplitude or magnitude spectrum to assume negative real values, and this concept will appear later in the chapter. This will be done only as a convenience in simplifying the mathematical form of the function. In the sense of complex numbers, this is permissible provided that the phase is adjusted accordingly. For example, $5/\!\!-\!150°$ can be expressed as $-5/30°$. The first form involves a "positive amplitude," and the second involves a "negative amplitude," but the two results are identical in the sense of complex numbers. Thus, the reader should not be disturbed by negative values appearing in some of the amplitude spectra in the book.

As a result of the discussion this far, it would seem then that a frequency-spectrum plot would most likely consist of a graph of either C_n from the amplitude-phase form or X_n from the complex exponential form. It will be recalled that the C_n terms are defined only for $n \geq 0$, while the X_n terms are defined for both positive and negative n (as well as $n = 0$). For this reason, the plot of C_n as a function of frequency is called a *one-sided spectrum*, while the plot of X_n as a function of frequency is called a *two-sided spectrum*. The terms "one-sided" and "two-sided" will be used extensively in this book since they are easy to remember and they quickly alert the reader to the required mathematical form.

Referring back to Eq. (9-18) momentarily and evaluating the magnitude of X_n, the result obtained is

$$X_n = |\bar{X}_n| = \sqrt{\frac{A_n^2 + B_n^2}{4}} = \frac{\sqrt{A_n^2 + B_n^2}}{2} \qquad \text{for } n \neq 0 \qquad (9\text{-}46)$$

Comparing this result with Eq. (9-10), we see that

$$X_n = \frac{C_n}{2} \qquad \text{for } n \neq 0 \qquad (9\text{-}47)$$

and, of course, $X_0 = C_0$ as previously noted.

The result of Eq. (9-47) may clear away at least a portion of the mystery surrounding the complex exponential form of the Fourier series. On an amplitude-spectrum basis, the magnitudes of the components in the two-sided spectral form (except for dc) are exactly one-half the values in the one-sided form. An artificial, but rather easy way to remember this is that, when the terms are displayed on both

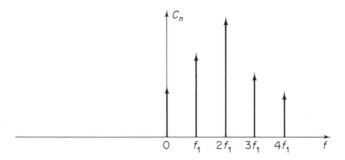

Figure 9-4 Typical one-sided amplitude frequency spectrum.

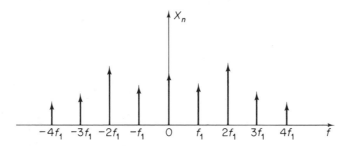

Figure 9-5 Two-sided amplitude spectrum corresponding to Fig. 9-4.

sides, the one-sided terms are "cut in half" in order to provide the components for the other side. The dc term appears only in one place, so its value does not change.

When a signal is described as a function of time, that is, as $x(t)$, the result is said to be a *time-domain* representation. Conversely, when the spectral information is provided, the result is said to be a *frequency-domain* representation. Either form completely describes the signal provided that both amplitude and phase are given in the frequency domain.

A typical example of a one-sided frequency spectrum is shown in Fig. 9-4. The corresponding two-sided spectrum for the same signal is shown in Fig. 9-5. Note that the lengths of all the components except dc in the second case are half the values in the first case, as expected. The dc component, however, is the same in both cases.

An important summary point concerning the frequency spectrum will now be made. The frequency spectrum of a periodic signal is a *discrete* or *line* spectrum; that is, it contains components only at integer multiples of the repetition frequency of the signal (including dc).

Example 9-3

Consider the bandlimited signal of Example 9-1. Plot (a) one-sided and (b) two-sided frequency spectra for this signal.

   Solution (a) The one-sided spectrum is obtained from the amplitude-phase form of the series, which was developed in Eqs. (9-26) and (9-27). The

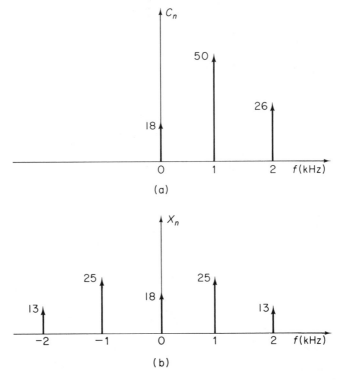

Figure 9-6   One-sided and two-sided amplitude frequency spectra of Examples 9-1 and 9-3.

amplitudes of the components on a one-sided basis and their frequencies are summarized as follows:

Frequency (Hz)	0	1000	2000
Amplitude	18	50	26

The one-sided plot is shown in Fig. 9-6a.

(b) While the two-sided plot could be readily determined directly from the one-sided plot, for convenience, the results will be tabulated again. These results could be deduced from Eq. (9-47) or from the expansion of Eq. (9-33). The results are summarized as follows:

Frequency (Hz)	0	± 1000	± 2000
Amplitude	18	25	13

The two-sided plot is shown in Fig. 9-6b.

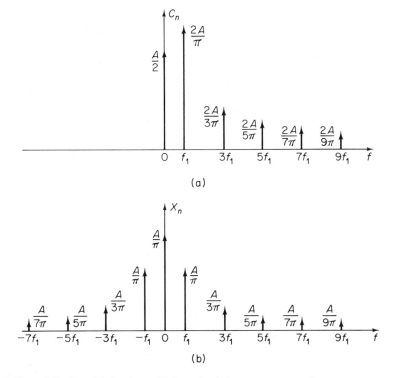

Figure 9-7 One-sided and two-sided amplitude frequency spectra of square wave (see Example 9-4).

Example 9-4

Plot (a) the one-sided amplitude frequency spectrum and (b) the two-sided amplitude frequency spectrum for the square-wave signal considered in Example 9-2.

       Solution   (a) The one-sided form is determined by noting the coefficients in Eq. (9-41a). The one-sided spectrum out to the ninth harmonic is shown in Fig. 9-7a.

       (b) The two-sided form can be determined either directly from the one-sided form or from the results of Eq. (9-45). The form of the two-sided spectrum is shown in Fig. 9-7b.

9-3 FOURIER SERIES SYMMETRY CONDITIONS

The various equations developed in the preceding several sections may, in theory, be applied to any signal to determine its spectrum. On the other hand, there are certain properties that may be used to simplify the computation of the spectrum in

TABLE 9-1 FOURIER SERIES SYMMETRY CONDITIONS

Sine–cosine form: $x(t) = A_0 + \sum_{n=1}^{\infty} (A_n \cos n\omega_1 t + B_n \sin n\omega_1 t)$, $\qquad \omega_1 = 2\pi f_1 = \dfrac{2\pi}{T}$

Amplitude–phase form: $x(t) = C_0 + \sum_{n=1}^{\infty} C_n \cos (n\omega_1 t + \phi_n) = C_0 + \sum_{n=1}^{\infty} C_n \sin (n\omega_1 t + \theta_n)$, $\qquad C_n = \sqrt{A_n^2 + B_n^2}$

Complex exponential form: $x(t) = \sum_{n=-\infty}^{\infty} \bar{X}_n \epsilon^{jn\omega_1 t}$, $\quad \bar{X}_n = \dfrac{A_n - jB_n}{2}$, $\quad$ for $n \neq 0$ $\quad \bar{X}_0 = A_0$

Condition	A_n (except $n=0$)	B_n	$\bar{X}_n$	Comments
General	$\dfrac{2}{T}\int_0^T x(t) \cos n\omega_1 t\, dt$	$\dfrac{2}{T}\int_0^T x(t) \sin n\omega_1 t\, dt$	$\dfrac{1}{T}\int_0^T x(t)\epsilon^{-jn\omega_1 t}\, dt$	
Even function $x(-t) = x(t)$	$\dfrac{4}{T}\int_0^{T/2} x(t) \cos n\omega_1 t\, dt$	0	$\dfrac{2}{T}\int_0^{T/2} x(t) \cos n\omega_1 t\, dt$	One-sided forms have only cosine terms $\bar{X}_n$ terms are real
Odd function $x(-t) = -x(t)$	0	$\dfrac{4}{T}\int_0^{T/2} x(t) \sin n\omega_1 t\, dt$	$\dfrac{-2j}{T}\int_0^{T/2} x(t) \sin n\omega_1 t\, dt$	One-sided forms have only sine terms $\bar{X}_n$ terms are imaginary
Half-wave symmetry $x\left(t+\dfrac{T}{2}\right) = -x(t)$	$\dfrac{4}{T}\int_0^{T/2} x(t) \cos n\omega_1 t\, dt$	$\dfrac{4}{T}\int_0^{T/2} x(t) \sin n\omega_1 t\, dt$	$\dfrac{2}{T}\int_0^{T/2} x(t)\epsilon^{-jn\omega_1 t}\, dt$	Odd-numbered harmonics only
Full-wave symmetry $x\left(t+\dfrac{T}{2}\right) = x(t)$	$\dfrac{4}{T}\int_0^{T/2} x(t) \cos n\omega_1 t\, dt$	$\dfrac{4}{T}\int_0^{T/2} x(t) \sin n\omega_1 t\, dt$	$\dfrac{2}{T}\int_0^{T/2} x(t)\epsilon^{-jn\omega_1 t}\, dt$	Even-numbered harmonics only

many cases. Furthermore, some of these conditions permit many important properties of certain waveforms to be obtained by simple inspection procedures. For some applications, this information might be sufficient without having to compute the spectrum at all. In most cases, the use of these conditions will at least provide some information about the spectrum from a direct inspection.

The types of criteria to be studied in this section are the *symmetry* conditions. All the various symmetry conditions to be considered in this book are summarized in Table 9-1. The proofs of these conditions are given in various texts on applied mathematics, and some are given as exercises at the end of this chapter. The emphasis in the book will be directed toward the practical interpretation and application of these various properties.

The equations at the top of the table are the various forms of the Fourier series that were developed in the preceding sections along with the relationships for converting from one form to another. The first row of the table provides the general relationships developed in the preceding sections, which can be used in all cases. Subsequent cases apply whenever the waveform possesses one or more symmetry conditions. Let us consider each case individually.

Even Function

A function $x(t)$ is said to be *even* if

$$x(-t) = x(t) \tag{9-48}$$

An example of an even function is shown in Fig. 9-8. It can be shown that in this case only cosine terms appear in the spectrum. Furthermore, the integral used in determining A_n has the property that the area under the curve in one half of a cycle is the same as that in the other half. Hence we need only integrate over half a cycle and double the result as indicated in the table. Finally, the integral for $\bar{X}_n$ reduces to an integral involving the product of the time signal and the cosine function rather than the complex exponential. The coefficients $\bar{X}_n$ are all *real* in this case.

Odd Function

A function $x(t)$ is said to be *odd* if

$$x(-t) = -x(t) \tag{9-49}$$

An example of an odd function is shown in Fig. 9-9. It can be shown that in this case, only sine terms appear in the spectrum. The integral for B_n need be evaluated

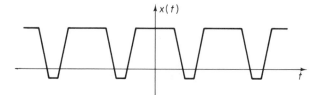

Figure 9-8   Example of an even function.

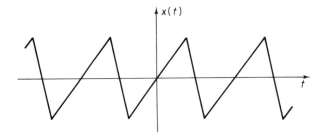

Figure 9-9 Example of an odd function.

only over half a cycle and doubled, and the integral used in determining $\bar{X}_n$ reduces to the product of the time signal and a sine function. Note that all of the $\bar{X}_n$ terms are purely imaginary in this case.

This type of symmetry is one of two types that can be "disguised" by the presence of a dc component. Consider the waveform shown in Fig. 9-10, which is identical to that of Fig. 9-9 except that a dc component has been added. Certainly, all logic tells us that the dc component should not affect any other portion of the spectrum except at zero frequency, so there should be only sine terms in the remaining part of the spectrum. However, the basic odd function condition of Eq. (9-49) is not satisfied. The dc component can be thought of as a limiting case of a cosine function of zero frequency, and this one "cosine" function obscures the symmetry.

The way around this problem is to inspect each waveform by mentally shifting it up or down to see if the symmetry condition can be achieved by this process. If so, a new function $x_1(t)$ can be formed as follows:

$$x_1(t) = x(t) - A_0 \qquad (9\text{-}50)$$

where A_0 is the dc component of the signal. The function $x_1(t)$ now satisfies the pertinent symmetry condition and can be integrated according to the form given in the table. Note that if the symmetry condition is employed, the new function $x_1(t)$ should be integrated rather than the original function.

Half-Wave Symmetry

A function is said to possess half-wave symmetry if

$$x\left(t + \frac{T}{2}\right) = -x(t) \qquad (9\text{-}51)$$

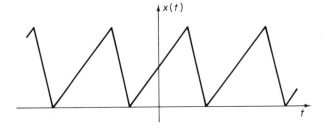

Figure 9-10 Function of Fig. 9-9 with dc component added.

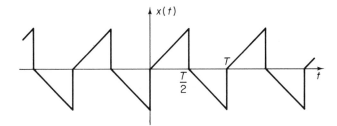

Figure 9-11 Example of function with half-wave symmetry.

A typical function having half-wave symmetry is shown in Fig. 9-11. It can be shown that a function satisfying this condition will have only *odd-numbered harmonics* (i.e., $n = 1, 3, 5, 7, \ldots$). However, unless one of the previous two conditions is also satisfied, there will be both sine and cosine terms in the expansion. As in the case of even and odd functions, integration need only be performed over half a cycle and the result is doubled.

One point of confusion regarding terms should be considered. The words *even* and *odd* were used in a different sense entirely for the previous two symmetry conditions as compared with the present condition and the next one. In the former case, even and odd referred to definitions regarding the image of a function projected around the vertical axis. In this case and in the next one, even and odd refer to the numbers of the harmonics. The two meanings are entirely different. For example, we can have an *even* function that has only *odd*-numbered harmonics.

As in the case of an odd function, the presence of a dc component can disguise half-wave symmetry. In this case the dc component is an even-numbered harmonic ($n = 0$), and its presence obscures the symmetry. A function satisfying this property is shown in Fig. 9-12. The procedure for handling this case is the same as for the previous condition. The dc component is subtracted, and the symmetry condition is applied to the new function.

Full-Wave Symmetry

A function is said to possess full-wave symmetry if

$$x\left(t + \frac{T}{2}\right) = x(t) \tag{9-52}$$

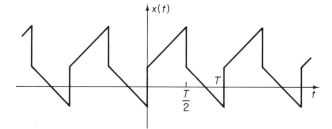

Figure 9-12 Function of Fig. 9-11 with dc component added.

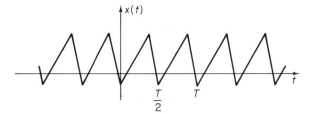

Figure 9-13 Example of function with full-wave symmetry.

A typical function having full-wave symmetry is shown in Fig. 9-13. It can be shown that a function satisfying this condition will have only *even-numbered harmonics* (i.e., $n = 0, 2, 4, 6, \ldots$).

Actually, the reader may see a flaw in the preceding discussion. From Eq. (9-52) and Fig. 9-13, the question arises as to whether the period is really T as assumed, or whether the period is, in fact, $T/2$. To get to the point, the period is really $T/2$, and we could avoid discussing the concept of this symmetry condition altogether by redefining the period as $T/2$. However, this situation frequently arises in conjunction with nonlinear operations on signals where the period is effectively halved. In such cases, it is often desirable to maintain the original base period as T. If we redefined the period as $T/2$, the fundamental frequency would be $2/T$, which is the second harmonic of the original reference fundamental frequency. However, with respect to the original fundamental, there are only even-numbered harmonics.

A given waveform may possess no more than two of the preceding symmetry conditions. The function may be either even or odd (but not both), and the function may possess either half-wave or full-wave symmetry (but not both).

Other properties pertaining to Fourier series will be discussed after the Fourier transform is introduced. The reason we postpone such considerations now is that most of them apply equally well to the Fourier transform.

Example 9-5

Consider again the square wave that was analyzed in Examples 9-2 and 9-4. Analyze this waveform to determine any symmetry conditions present, and use such symmetry conditions to simplify the computation of the spectrum.

Solution   The waveform is repeated in Fig. 9-14a for convenience. As it appears, the reader should verify that neither of the four symmetry conditions is satisfied. However, suppose the dc value $A/2$ is subtracted, and a new signal $x_1(t)$ is formed as shown in Fig. 9-14b; that is,

$$x_1(t) = x(t) - \frac{A}{2} \tag{9-53}$$

In performing this shift, we understand, of course, that the new function $x_1(t)$ will not have a dc component, so the dc component of $x(t)$ has been recognized already and should be tabulated. All other terms in $x_1(t)$ should be the same as those in $x(t)$ since only the dc level has been affected.

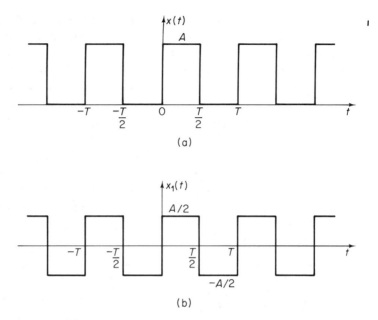

Figure 9-14   Removal of dc component from square wave to allow use of symmetry condition.

Let us now check the symmetry conditions. We quickly establish that $x_1(t)$ is an odd function since it satisfies Eq. (9-49). This indicates that only sine terms will appear in the spectrum for $x_1(t)$, so $A_n = 0$.

It is also noted that the half-wave symmetry condition of Eq. (9-51) is satisfied. This means that only odd-numbered harmonics will appear in the spectrum for $x_1(t)$.

Relating the observation for $x_1(t)$ back to $x(t)$, it can be concluded that the Fourier series for $x(t)$ consists of a dc component and sine components at odd-numbered harmonic frequencies (including the fundamental, i.e., $n = 1$). This conclusion is in obvious agreement with the result of Example 9-2, as the reader may quickly verify by referring back to that example. However, this inspection process provides valuable information in advance of a detailed calculation and can save some of the steps involved.

As a final step in this example, let us verify that application of the symmetry condition in the computation of the B_n coefficients for the modified function $x_1(t)$ produces the same results as were obtained in Example 9-2 for $x(t)$. According to Table 9-1 we should integrate only over half a cycle and double the result. However, the amplitude of $x_1(t)$ is half that of $x(t)$, so the integral is

$$B_n = \frac{4}{T} \int_0^{T/2} \frac{A}{2} \sin n\omega_1 t \, dt \tag{9-54a}$$

$$= \frac{2A}{T} \int_0^{T/2} \sin n\omega_1 t \, dt \tag{9-54b}$$

Comparison of Eq. (9-54b) with the first expression of Eq. (9-38) reveals that the two expressions are exactly the same, so we need not proceed further. The "doubling" in front of the integral cancelled the "halving" of the function amplitude after shifting, so that the B_n coefficients are the same as before.

TABLE 9-2 SOME COMMON PERIODIC SIGNALS AND THEIR FOURIER SERIES

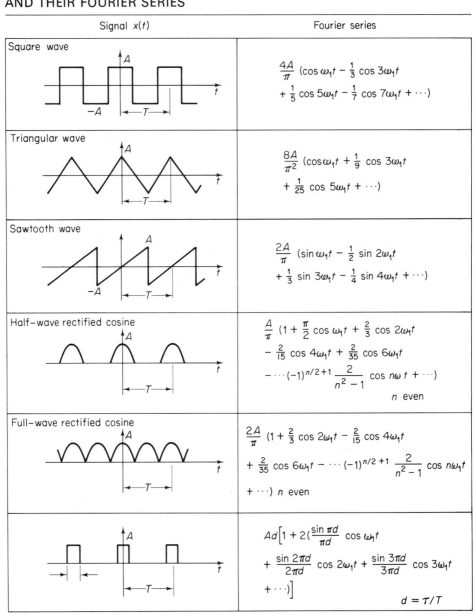

Signal $x(t)$	Fourier series
Square wave	$\dfrac{4A}{\pi}(\cos\omega_1 t - \frac{1}{3}\cos 3\omega_1 t + \frac{1}{5}\cos 5\omega_1 t - \frac{1}{7}\cos 7\omega_1 t + \cdots)$
Triangular wave	$\dfrac{8A}{\pi^2}(\cos\omega_1 t + \frac{1}{9}\cos 3\omega_1 t + \frac{1}{25}\cos 5\omega_1 t + \cdots)$
Sawtooth wave	$\dfrac{2A}{\pi}(\sin\omega_1 t - \frac{1}{2}\sin 2\omega_1 t + \frac{1}{3}\sin 3\omega_1 t - \frac{1}{4}\sin 4\omega_1 t + \cdots)$
Half-wave rectified cosine	$\dfrac{A}{\pi}(1 + \frac{\pi}{2}\cos\omega_1 t + \frac{2}{3}\cos 2\omega_1 t - \frac{2}{15}\cos 4\omega_1 t + \frac{2}{35}\cos 6\omega_1 t - \cdots (-1)^{n/2+1}\dfrac{2}{n^2-1}\cos n\omega t + \cdots)$ n even
Full-wave rectified cosine	$\dfrac{2A}{\pi}(1 + \frac{2}{3}\cos 2\omega_1 t - \frac{2}{15}\cos 4\omega_1 t + \frac{2}{35}\cos 6\omega_1 t - \cdots (-1)^{n/2+1}\dfrac{2}{n^2-1}\cos n\omega_1 t + \cdots)$ n even
	$Ad\left[1 + 2(\dfrac{\sin\pi d}{\pi d}\cos\omega_1 t + \dfrac{\sin 2\pi d}{2\pi d}\cos 2\omega_1 t + \dfrac{\sin 3\pi d}{3\pi d}\cos 3\omega_1 t + \cdots)\right]$ $d = \tau/T$

9-4 COMMON PERIODIC WAVEFORMS
AND THEIR FOURIER SERIES

To finalize the discussion of Fourier series, some common periodic waveforms and their Fourier series are summarized in Table 9-2. Carefully note the location of the origin in each case since a shift to the left or to the right of any waveform will change the exact form of the spectrum (but not the net magnitude). It turns out that all the waveforms as shown are either even or odd so that each series has only sine or cosine terms, but not both. Note also the presence or lack of presence of a dc component in the different forms. The derivations of some of these functions will be given as problems at the end of the chapter.

9-5 FOURIER TRANSFORM

The emphasis on spectral analysis thus far has centered on periodic functions, whose spectra consist of discrete components at integer multiples of the fundamental repetition frequency. We consider next the process of spectral analysis for nonperiodic signals, which is achieved with the *Fourier transform*. To satisfy certain mathematical restrictions, assume that all nonperiodic signals of interest will have finite energy.

To illustrate qualitatively the concept of the Fourier transform, refer to Fig. 9-15. Some arbitrary periodic pulse-type signal $x(t)$ and its assumed amplitude spectrum X_n are shown in (a). The time function is periodic with period T, and the spectrum is, therefore, discrete. The fundamental component is $f_1 = 1/T$, and spectral components appear at integer multiples of that frequency.

In Fig. 9-15b, the pulse width and shape remain the same, but the period is doubled by inserting a space between successive pulses. An expression for $\bar{X}_n$ would be the same as before since the integrand has not changed. What has changed, however, is the fundamental frequency f_1. When T is doubled, f_1 is halved, so the spacing between spectral lines is halved as shown.

In Fig. 9-15c, the pulse width and shape again remain the same, but the period is doubled again. The spacing between spectral lines is again halved.

The effect of this trend is that the relative shape of the envelope of the $\bar{X}_n$ coefficients remains the same, but the number of components in a given frequency interval increases as the period increases. (The level of the envelope changes but the *relative shape* remains constant. Each spectral plot is assumed to be adjusted in amplitude accordingly.) Simultaneously, the increment f_1 between successive frequency components decreases. In the limit as $T \to \infty$, as shown in Fig. 9-15d, the frequency difference approaches zero.

In this limiting form, the spectral lines all merge together, so it no longer is desirable to display the spectrum as a group of lines. Instead, the points representing the amplitudes of the lines effectively all merge together and form a continuous curve. Thus in the limit as the period approaches infinity, the spectrum becomes a continuous spectrum. While the spectrum of a *periodic* signal is *discrete*, and components appear only at integer multiples of the repetition rate, the spectrum of a *nonperiodic* signal is *continuous* and could, in theory, appear at any frequency. Of course, a bandlimited

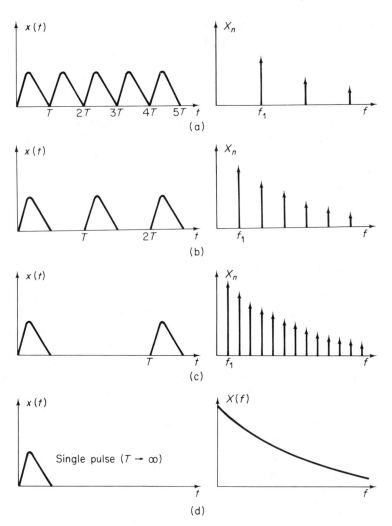

Figure 9-15 Concept of the Fourier transform as a limiting case of the Fourier series when the period becomes infinite.

signal would necessarily have a spectrum only over a finite frequency range, but the point is that there are no restrictions of the type that exist for the periodic signal.

The *Fourier transform* is the commonly used name for the mathematical function that provides the frequency spectrum of a nonperiodic signal. Assume that a nonperiodic time signal $x(t)$ is given. The Fourier transform is designated as $\bar{X}(f)$, and f is the cyclic frequency in hertz. [The overbar on $\bar{X}(f)$ emphasizes that it is a complex quantity.] Thus $\bar{X}(f)$ is the mathematical function expressed as a function of frequency that indicates the relative spectrum, which could exist at any arbitrary frequency f. $\bar{X}(f)$ for the nonperiodic signal corresponds to the two-sided spectrum $\bar{X}_n$ for the periodic case. While it would be possible to obtain Fourier transform forms that would relate more directly to the one-sided discrete spectral forms, almost all the available results are more directly related to the two-sided exponential form, and that approach will be followed here.

The process of Fourier transformation of a time function is designated symbolically as

$$\bar{X}(f) = \mathscr{F}[x(t)] \qquad (9\text{-}55)$$

The inverse operation is designated symbolically as

$$x(t) = \mathscr{F}^{-1}[\bar{X}(f)] \qquad (9\text{-}56)$$

The actual mathematical processes involved in these operations are as follows:

$$\bar{X}(f) = \int_{-\infty}^{\infty} x(t)\epsilon^{-j\omega t}\, dt \qquad (9\text{-}57)$$

$$x(t) = \int_{-\infty}^{\infty} \bar{X}(f)\epsilon^{j\omega t}\, df \qquad (9\text{-}58)$$

Note that the argument of $\bar{X}(f)$ and the differential of Eq. (9-58) are both expressed in terms of the cyclic frequency f (in hertz), but the arguments of the exponentials in Eqs. (9-57) and (9-58) are expressed in terms of the radian frequency ω, where $\omega = 2\pi f$. These are the most convenient forms for manipulating and expressing the given functions. Most spectral displays are made in terms of f, while the analytical expressions are often easier to deal with in terms of ω. This should present no serious problem as long as the 2π scale factor is understood: $\omega = 2\pi f$.

The Fourier transform $\bar{X}(f)$ is, in general, a complex function and has both a magnitude and an angle. Thus, $\bar{X}(f)$ can be expressed as

$$\bar{X}(f) = X(f)\epsilon^{j\phi(f)} = X(f)\underline{/\phi(f)} \qquad (9\text{-}59)$$

where $X(f)$ represents the *amplitude spectrum* and $\phi(f)$ is the *phase spectrum*. A typical amplitude spectrum is shown in Fig. 9-16.

Again, a point that should be stressed for the *nonperiodic* signal is that its spectrum is *continuous*, and, in general, it consists of components at *all* frequencies in the range over which the spectrum is present.

As in the case of periodic signals, certain symmetry conditions can be applied to aid in the computation of the Fourier transform. However, not all the periodic signal properties are applicable to a nonperiodic signal. Furthermore, because of established convention, the results will be interpreted in slightly different forms. For example, the coefficients A_n and B_n are widely used in dealing with Fourier series, but the corresponding forms in the Fourier transform (which would be proportional to

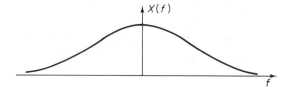

Figure 9-16 Typical amplitude spectrum of nonperiodic signal.

TABLE 9-3 FOURIER TRANSFORM SYMMETRY CONDITIONS

Condition	$\bar{X}(f)$	Comment
General	$\int_{-\infty}^{\infty} x(t)\epsilon^{-j\omega t}\,dt$	
Even function $x(-t) = x(t)$	$2\int_{0}^{\infty} x(t)\cos\omega t\,dt$	$\bar{X}(f)$ is an even real function of f
Odd function $x(-t) = -x(t)$	$-2j\int_{0}^{\infty} x(t)\sin\omega t\,dt$	$\bar{X}(f)$ is an odd imaginary function of f

the real and imaginary parts of the transform) are not used nearly as often in signal analysis and will not be discussed here.

The symmetry conditions to be considered are summarized in Table 9-3. These results indicate that for either an even or an odd function, one need integrate only over half the total interval and double the result. Furthermore, the form of the integrand is different in each case. Notice the similarity between the forms of the Fourier transform and Fourier series integrals when the function is either even or odd.

Example 9-6

Derive the Fourier transform of the rectangular pulse function shown in Fig. 9-17a.

    Solution   The pulse can be defined as

$$x(t) = \begin{cases} A & \text{for } -\tau/2 < t < \tau/2 \\ 0 & \text{elsewhere} \end{cases} \tag{9-60}$$

Since the given signal is even, a symmetry condition from Table 9-3 may be applied. Using the integral form as given in Table 9-3, we have

$$\bar{X}(f) = 2\int_{0}^{\tau/2} A\cos\omega t\,dt$$

$$= \frac{2A}{\omega}\sin\omega t\,\Big]_{0}^{\tau/2} \tag{9-61}$$

$$= \frac{2A}{\omega}\sin\frac{\omega\tau}{2}$$

By setting $\omega = 2\pi f$ and performing some additional manipulations, we can write

$$\bar{X}(f) = A\tau\,\frac{\sin\pi f\tau}{\pi f\tau} \tag{9-62}$$

The form of the amplitude spectrum $X(f)$ is shown in Fig. 9-17b. This function is very important in communications signal analysis and transmission. [The phase response is $\Phi(f) = 0$.]

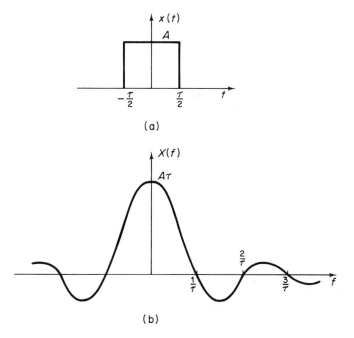

Figure 9-17   Rectangular pulse and its Fourier transform as considered in Example 9-6.

Example 9-7

Derive the Fourier transform of the exponential function given by

$$x(t) = \begin{cases} A\epsilon^{-\alpha t} & \text{for } t > 0 \\ 0 & \text{for } t < 0 \end{cases} \tag{9-63}$$

where $\alpha > 0$.

   Solution   The function is shown in Fig. 9-18a. Application of the definition of the Fourier transform as given by Eq. (9-57) yields

$$\begin{aligned} \bar{X}(f) &= \int_0^\infty A\epsilon^{-\alpha t}\epsilon^{-j\omega t}\,dt \\ &= \frac{A\epsilon^{-(\alpha+j\omega)t}}{-(\alpha+j\omega)}\Bigg]_0^\infty = 0 + \frac{A}{\alpha+j\omega} \end{aligned} \tag{9-64}$$

The result is a complex function as expected. The amplitude and phase functions are found by determining the magnitude and angle associated with the final result of Eq. (9-64). These functions are

$$X(f) = \frac{A}{\sqrt{\alpha^2 + \omega^2}} = \frac{A}{\sqrt{\alpha^2 + (2\pi f)^2}} \tag{9-65}$$

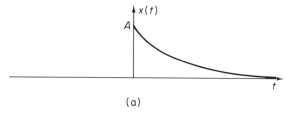

(a)

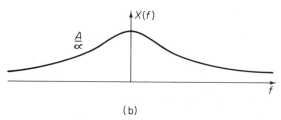

(b)

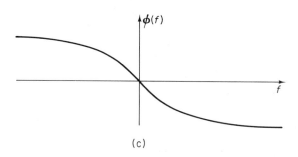

(c)

Figure 9-18 Exponential function of Example 9-7 and its amplitude and phase spectra.

and

$$\phi(f) = -\tan^{-1}\frac{\omega}{\alpha} = -\tan^{-1}\frac{2\pi f}{\alpha} \tag{9-66}$$

These functions are shown in Fig. 9-18b and c for some arbitrary value of α. As in the case of discrete spectra, the phase spectrum is not as useful as the amplitude spectrum and will usually not be shown in most developments. However, it was included here to make the development complete.

Example 9-8

One property of the impulse function not considered earlier in the book is

$$\int_{-\infty}^{\infty} g(t)\delta(t)\,dt = g(0) \tag{9-67}$$

where $g(t)$ is any continuous function. Derive the Fourier transform of the impulse function.

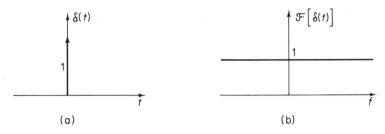

Figure 9-19   Impulse function and its spectrum as developed in Example 9-8.

Solution   The impulse function occurring at $t = 0$ is illustrated in Fig. 9-19a. Application of the definition of the Fourier transform yields

$$\mathscr{F}[\delta(t)] = \int_{-\infty}^{\infty} \delta(t)\epsilon^{-j\omega t}\, dt \qquad (9\text{-}68)$$

This integral is readily evaluated by the property of the impulse function as given by Eq. (9-67).

$$\mathscr{F}[\delta(t)] = 1 \qquad (9\text{-}69)$$

This simple result is shown in Fig. 9-19b.

This result is rather interesting and is one that should catch the reader's attention. Although the impulse function is more of an ideal mathematical model than a real physical waveform, it is often used to represent noise phenomena (e.g., a function that appears as a sharp pulse with near-zero duration). The spectrum of such a signal is extremely wide and, in the theoretical limit, would contain components at all frequencies over an infinite frequency interval.

To illustrate one more point regarding this type of phenomena, the reader has probably observed interference produced on a radio receiver due to "impulse" sources in the immediate area (e.g., electrical appliances, automobile ignition systems). These "impulses" create broad spectra that can be heard on radio receivers over a wide frequency range, as noted from the nature of the Fourier transform of the function.

9-6 COMMON NONPERIODIC WAVEFORMS AND THEIR FOURIER TRANSFORMS

Some common nonperiodic waveforms and their Fourier transforms are summarized in Table 9-4. Note the location of the origin in each case since a shift to the left or to the right of any waveform will change the exact form of the spectrum (but not the net magnitude). The derivations of some of these functions will be given as problems at the end of the chapter.

TABLE 9-4 SOME COMMON SIGNALS AND THEIR FOURIER TRANSFORM

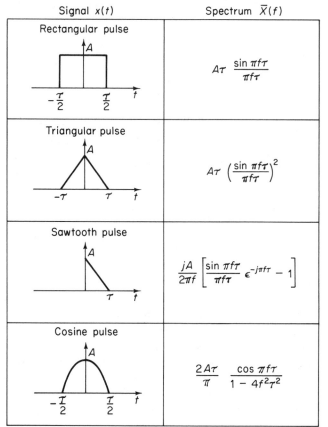

Signal $x(t)$	Spectrum $\bar{X}(f)$
Rectangular pulse	$A\tau \dfrac{\sin \pi f\tau}{\pi f\tau}$
Triangular pulse	$A\tau \left(\dfrac{\sin \pi f\tau}{\pi f\tau}\right)^2$
Sawtooth pulse	$\dfrac{jA}{2\pi f}\left[\dfrac{\sin \pi f\tau}{\pi f\tau}\, \epsilon^{-j\pi f\tau} - 1\right]$
Cosine pulse	$\dfrac{2A\tau}{\pi}\dfrac{\cos \pi f\tau}{1 - 4f^2\tau^2}$

9-7 FOURIER TRANSFORM OPERATIONS AND SPECTRAL ROLL-OFF

Time functions are altered in form as they pass through various stages of a signal-processing system or a signal transmission channel. These operations may be used to deliberately change the form of the signal in some cases, or certain of the operations may arise as a result of natural limitations of a system and could distort the signal. In either event, it is worthwhile to study the effects on the resulting signal from a spectral point of view.

The primary Fourier transform operation pairs of interest for our purposes are summarized in Table 9-5. Certain of these pairs will be derived in example problems at the end of this section; others will be left as exercises for analytically inclined readers.

The practical significance of all these operation pairs will now be discussed. The following notational form will be used here and in certain subsequent sections:

$$x(t) \leftrightarrow \bar{X}(f) \tag{9-70}$$

TABLE 9-5 FOURIER TRANSFORM OPERATION PAIRS

$x(t)$	$\bar{X}(f) = \mathscr{F}[x(t)]$	
$ax_1(t) + bx_2(t)$	$a\bar{X}_1(f) + b$	(O-1)
$\dfrac{dx(t)}{dt}$	$j2\pi f\bar{X}(f)$	(O-2)
$\displaystyle\int_{-\infty}^{t} x(t)\,dt$	$\dfrac{\bar{X}(f)}{j2\pi f}$	(O-3)
$x(t - \tau)$	$\epsilon^{-j2\pi f\tau}\bar{X}(f)$	(O-4)
$\epsilon^{j2\pi f_0 t}x(t)$	$\bar{X}(f - f_0)$	(O-5)
$x(at)$	$\dfrac{1}{a}\bar{X}\left(\dfrac{f}{a}\right)$	(O-6)

This notation indicates that $x(t)$ and $\bar{X}(f)$ are a corresponding transform pair; that is, $\bar{X}(f) = \mathscr{F}[x(t)]$. The equation numbers of the operation pairs will correspond to those of Table 9-5.

Superposition principle (O-1).   The first transform pair in the table is

$$ax_1(t) + bx_2(t) \leftrightarrow a\bar{X}_1(f) + b\bar{X}_2(f) \qquad\text{(O-1)}$$

This result specifies the basic property that the Fourier transform integral is a linear operation and thus obeys the principle of superposition as far as the level of a signal and the combination of several signals are concerned.

Differentiation (O-2).   The differentiation Fourier transform pair is

$$\frac{dx(t)}{dt} \leftrightarrow j2\pi f\bar{X}(f) \qquad\text{(O-2)}$$

This theorem, which will be derived in Example 9-9, indicates that each time a signal is differentiated the spectrum is multiplied by $j2\pi f$. Multiplication by $j2\pi f$ has the effect of decreasing the relative level of the spectrum at low frequencies and increasing the relative level at higher frequencies. Note that a pure dc component is eliminated. A sketch illustrating the general effect on the amplitude spectrum resulting from differentiating a time signal is shown in Fig. 9-20.

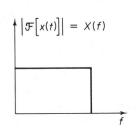

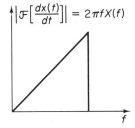

Figure 9-20 Effect on the spectrum of differentiating a time signal.

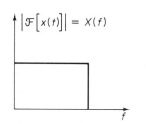

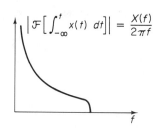

Figure 9-21 Effect on the spectrum of integrating a time signal.

Integration (O-3). The integration Fourier transform pair is

$$\int_{-\infty}^{t} x(t)\, dt \leftrightarrow \frac{\bar{X}(f)}{j2\pi f} \tag{O-3}$$

This theorem, which is the reverse of (O-2), indicates that when a signal is integrated the amplitude spectrum is divided by $2\pi f$. Division by $2\pi f$ has the effect of increasing the relative level of the spectrum at low frequencies and decreasing the relative level at higher frequencies. A sketch illustrating the general effect on the amplitude spectrum resulting from integrating a time signal is shown in Fig. 9-21. In a sense, integration is a form of low-pass filtering in that high-frequency components of the spectrum are attenuated. However, a pure integrator is not often used as a low-pass filter due to the pronounced accentuation effect at very low frequencies. Nevertheless, there are situations in which the integrator is considered as a form of a low-pass filter.

Time delay (O-4). The time delay transform operation, which is derived in Example 9-10, is

$$x(t - \tau) \leftrightarrow \epsilon^{-j2\pi f \tau} \bar{X}(f) \tag{O-4}$$

The function $x(t - \tau)$ represents the delayed version of a signal $x(t)$ as illustrated in Fig. 9-22. This operation could occur as a result of passing a signal through an ideal delay line with delay τ, for example. It can be readily shown that the amplitude spectrum is not changed by the shifting operation, but the phase spectrum is shifted by $-2\pi f \tau$ radians. Certainly, one would expect the amplitude spectrum of a given fixed signal to be independent of the time at which the signal occurs, but the phase shifts of all the components are increased to reflect the result of the time delay.

Modulation (O-5). This operation bears the same relationship to the frequency domain as the time-delay theorem does to the time domain. This theorem, whose deviation will be left as an exercise (Problem 9-27), is

$$\epsilon^{j2\pi f_0 t} x(t) \leftrightarrow \bar{X}(f - f_0) \tag{O-5}$$

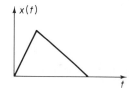

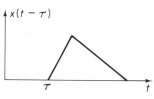

Figure 9-22 Time signal and its delayed form.

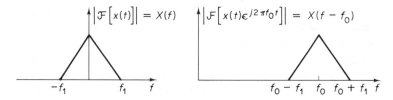

Figure 9-23   Effect on the spectrum of the modulation operation.

This result has profound implications in the study of amplitude modulation in communications theory. If a time signal is multiplied by a complex exponential, the spectrum is translated to the right by the frequency of the exponential, as shown in Fig. 9-23. In practical cases, complex exponentials occur in pairs with a term of the form of (O-5) along with its conjugate.

Time scaling (O-6).   The time-scaling operation transfer pair is

$$x(at) \leftrightarrow \frac{1}{a}\bar{X}\left(\frac{f}{a}\right) \tag{O-6}$$

The derivation of this transform pair will be left as an exercise (Problem 9-28). If $a > 1$, $x(at)$ represents a "faster" version of the original signal, while if $a < 1$, $x(at)$ represents a "slower" version. In the former case, the spectrum is broadened, while in the latter case, it is narrowed. These concepts are illustrated in Fig. 9-24.

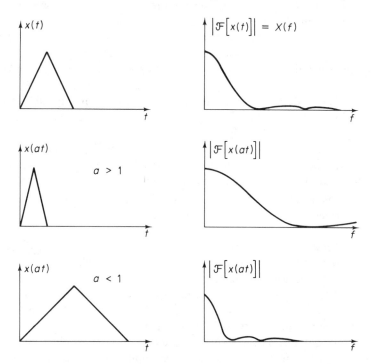

Figure 9-24   Effect on the spectrum of the time scaling operation.

We now turn to the problem of determining the *spectral roll-off rate* or, in more formal mathematical terms, the *convergence* of the Fourier spectrum. We have seen that the Fourier transforms and series for many common waveforms appear to be infinitely wide. Obviously, in the "real world" this result cannot be true. In reality, spectral components above a certain frequency range are simply ignored since their contributions are negligible, and the practical bandwidth is established at the minimum level for reasonable reproduction of the signal.

An important factor that can be used qualitatively in estimating the relative bandwidths of different signals is the spectral roll-off rate. The roll-off rate is an upper-bound (worst-case) measure of the rate at which the spectral components diminish with increasing frequency. The basic way to specify the rolloff rate is a $1/f^k$ variation for a Fourier transform or a $1/n^k$ variation for a Fourier series, where k is an integer. As k increases, the spectrum diminishes more rapidly. Thus a signal with a $1/f^3$ roll-off rate would normally have a narrower bandwidth than a signal with a $1/f^2$ rate.

A common practical way to specify the roll-off rate is in decibels/octave in somewhat the same manner as for Bode plots. A roll-off rate of $1/f^k$ can be readily shown to correspond to a slope of $-6k$ dB/octave, where an octave corresponds to a doubling of the frequency. Thus a $1/f^3$ roll-off rate corresponds to -18 dB/octave, and a $1/f^2$ rate corresponds to -12 dB/octave.

The roll-off rate refers to the worst case or upper bound of the magnitude spectrum of the signal. It is an estimate of the worst-case effect and should not be interpreted as an exact formula for predicting spectral components. For example, the spectrum of a signal may contain components that have a -6 dB/octave roll-off rate, plus components that have a -12 dB/octave roll-off rate. Eventually, the latter components will be so small that they may be ignored, and the -6 dB/octave components will dominate. The tests we will present would predict a -6 dB/octave roll-off rate, and the result has to be interpreted as a worst-case bound.

Before discussing the particular tests, the following qualitative points are very worthwhile for dealing with signals in a general sense and should be carefully noted:

1. Time functions that are relatively "smooth" (i.e., no discontinuities or jumps in the signal or its lower-order derivatives) tend to have higher roll-off rates and corresponding narrower bandwidths.

2. Time functions with discontinuities in the signal tend to have lower roll-off rates and corresponding wider bandwidths.

An example of a very smooth signal is the sinusoid whose bandwidth is so narrow that it has only one component. Conversely, a square wave has finite discontinuities in each cycle, and its spectrum is very wide.

The roll-off rates for a wide variety of common signals may be estimated from the information provided in Table 9-6. The left column provides a test to apply to the signal, and the right columns provide the appropriate roll-off rates. Observe that one column on the right is applicable to nonperiodic signals analyzed with the Fourier transform, while the other is applicable to periodic signals analyzed with the Fourier series. Some of the information in the table will now be discussed.

TABLE 9-6 SPECTRAL ROLL-OFF RATES OF FOURIER TRANSFORMS AND FOURIER SERIES

Smoothness of function	Roll-off rate	
	Fourier transform	Fourier series
$x(t)$ has impulses	No spectral roll-off	No spectral roll-off
$x(t)$ has finite discontinuities	$\dfrac{1}{f}$ or -6 dB/octave	$\dfrac{1}{n}$ or -6 dB/octave
$x(t)$ is continuous, $x'(t)$ has finite discontinuities	$\dfrac{1}{f^2}$ or -12 dB/octave	$\dfrac{1}{n^2}$ or -12 dB/octave
$x(t)$ is continuous, $x'(t)$ is continuous, $x''(t)$ has finite discontinuities	$\dfrac{1}{f^3}$ or -18 dB/octave	$\dfrac{1}{n^3}$ or -18 dB/octave

The first condition is hypothetical and applies when the signal is assumed to contain one or more ideal impulse functions. This situation could never actually exist in practice, but there are situations in which the assumption of ideal impulse functions is convenient. In such a case, the spectrum will contain a portion having no roll-off at all; that is, the spectrum would theoretically be infinite in bandwidth.

If the signal has finite discontinuities or jumps, the spectrum has a -6 dB/octave roll-off rate. The ideal square wave fits this case, as will be illustrated in Example 9-11.

If the signal is continuous (i.e., no jumps), but the first derivative or slope changes abruptly at one or more points, the spectrum has a -12 dB/octave roll-off rate. The triangular wave is a good example of this type (see Problems 9-13b and 9-14b).

If the signal and its first derivative are both continuous, but the second derivative has a finite discontinuity, the spectrum has a -18 dB/octave roll-off rate. It is nearly impossible to detect this condition visually, but a knowledge of the fact may occur as a result of other information.

The table could be continued, but the range shown covers most of the normal requirements for common waveforms. In general, if a function and its first $k-1$ derivatives are continuous, but its kth derivative has a finite discontinuity, the spectral roll-off rate will be $-6(k+1)/$dB/octave.

Example 9-9

Derive the Fourier transform of the first derivative of a time signal [i.e., pair (O-2) of Table 9-5].

Solution The theorem is best derived by considering the inverse transform of $\bar{X}(f)$.

$$x(t) = \int_{-\infty}^{\infty} \bar{X}(f)\epsilon^{j\omega t}\, df \qquad (9\text{-}71)$$

Both sides of this equation are now differentiated with respect to time. This yields

$$\frac{dx(t)}{dt} = \int_{-\infty}^{\infty} j\omega \bar{X}(f)\epsilon^{j\omega t}\, df \tag{9-72}$$

By comparing Eq. (9-72) with Eq. (9-71), the quantity $j\omega \bar{X}(f)$ is seen to represent the Fourier transform of the derivative on the left. By induction, this result is readily extended to the case of the nth derivative.

Example 9-10

Derive operation pair (O-4) of Table 9-5.

   Solution   Application of the definition of the Fourier transform to $x(t - \tau)$ yields

$$\mathscr{F}[x(t - \tau)] = \int_{-\infty}^{\infty} x(t - \tau)\epsilon^{-j\omega t}\, dt \tag{9-73}$$

A change in variables will now be made. Let $u = t - \tau$, which results in $du = dt$. Substitution of these values yields

$$\begin{aligned}
\mathscr{F}[x(t - \tau)] &= \int_{-\infty}^{\infty} x(u)\epsilon^{-j\omega u}\epsilon^{-j\omega \tau}\, du \\
&= \epsilon^{-j\omega T} \int_{-\infty}^{\infty} x(u)\epsilon^{-j\omega u}\, du = \epsilon^{-j\omega \tau}\bar{X}(f) \\
&= \epsilon^{-j2\pi f\tau}\bar{X}(f)
\end{aligned} \tag{9-74}$$

Example 9-11

Using Table 9-6, determine the spectral roll-off rates for the functions corresponding to the example problems indicated below. Verify by comparing with the results of (a) Example 9-7, (b) Example 9-8, and (c) Example 9-2.

   Solution   (a) The function of Example 9-7 has a finite discontinuity at $t = 0$. Hence the spectrum should display a roll-off rate of $1/f$ or -6 dB/octave. From Eq. (9-65), the high-frequency asymptotic behavior of $X(f)$ approaches

$$X(f) \approx \frac{A}{2\pi f} \qquad \text{for } f \gg \frac{\alpha}{2\pi} \tag{9-75}$$

which is a $1/f$ form, as predicted.

   (b) The function of Example 9-8 is an impulse, which should display no roll-off at all. Indeed, this prediction is readily verified by Eq. (9-69).

   (c) The function of Example 9-2 has two finite discontinuities in each cycle, a positive jump and a negative jump. The function is periodic, so it should display a roll-off rate of $1/n$. Observe from Eq. (9-40) that this property is true, so the spectrum has a -6 dB/octave roll-off. The absence of the even harmonics in the spectrum should not confuse the issue, since the roll-off rate is correct for those components present.

Example 9-12

Derive the Fourier transform of the two-sided exponential function shown in Fig. 9-25a having a damping factor α. Sketch the form of the spectrum.

Solution   This function may be expressed as

$$x(t) = A\epsilon^{-\alpha|t|} \tag{9-76}$$

or

$$x(t) = \begin{cases} A\epsilon^{\alpha t} & \text{for } t < 0 \\ A\epsilon^{-\alpha t} & \text{for } t > 0 \end{cases} \tag{9-77}$$

where $\alpha > 0$. Since the function is even, the transform may be evaluated by integrating only over positive time, and this operation is

$$\bar{X}(f) = 2 \int_0^\infty A\epsilon^{-\alpha t} \cos \omega t \, dt \tag{9-78}$$

This integral may be evaluated by parts or the result may be found in a standard integral table. The result is

$$\bar{X}(f) = \frac{2A\epsilon^{-\alpha t}(-\alpha \cos \omega t + \omega \sin \omega t)}{\alpha^2 + \omega^2} \bigg]_0^\infty$$

$$= \frac{2\alpha A}{\alpha^2 + \omega^2} = \frac{2\alpha A}{\alpha^2 + 4\pi^2 f^2} \tag{9-79}$$

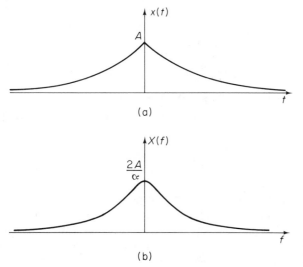

(a)

(b)

Figure 9-25 Exponential function of Example 9-12 and its transform.

It is interesting to point out that the spectrum is real and is an even function of frequency as expected. Furthermore, the function itself is continuous, but its first derivative has a finite discontinuity at $t = 0$. This property indicates that the spectrum should have a roll-off rate of $1/f^2$, which is readily observed in Eq. (9-79). A sketch of the spectrum is shown in Fig. 9-25b. This two-sided exponential function and its spectrum can be compared with the one-sided exponential function and its spectrum as considered in Example 9-7. In the present example, the spectrum has a roll-off rate of $1/f^2$ (or -12 dB/octave); in the earlier example, the discontinuity at $t = 0$ resulted in a $1/f$ (or -6 dB/octave) roll-off rate.

GENERAL PROBLEMS

9-1. A certain periodic signal as viewed on an oscilloscope has a period of 2 ms. The net positive area in one cycle appears to be greater than the net negative area. List the five lowest frequencies on a one-sided basis that would probably appear in the spectrum.

9-2. A certain periodic signal as viewed on an oscilloscope has a period of 5 μs. The net negative area in one cycle appears to be greater than the net positive area. List the five lowest frequencies on a one-sided basis that would probably appear in the spectrum.

9-3. A certain periodic bandlimited signal has the following frequencies in the Fourier series: dc, 20 Hz, and 40 Hz. The signal can be expressed in sine–cosine form as

$$x(t) = 5 + 6 \cos 40\pi t + 8 \sin 40\pi t - 5 \cos 80\pi t + 12 \sin 80\pi t$$

Express the signal in **(a)** amplitude-phase form and **(b)** complex exponential form.

9-4. A certain periodic bandlimited signal has the following frequencies in the Fourier series: dc, 500 Hz, and 1 kHz. The signal can be expressed in sine–cosine form as

$$x(t) = 12 + 20 \cos 1000\pi t - 20 \sin 1000\pi t + 12 \cos 2000\pi t - 6 \sin 2000\pi t$$

Express the signal in **(a)** amplitude-phase form and **(b)** complex exponential form.

9-5. A certain periodic bandlimited signal has the following Fourier series amplitude-phase representation:

$$x(t) = 6 + 10 \sin (100\pi t + 30°) + 8 \sin (200\pi t - 120°)$$

Express the signal in **(a)** sine–cosine form and **(b)** complex exponential form.

9-6. A certain periodic bandlimited signal has the following Fourier series amplitude-phase representation:

$$x(t) = 20 + 12 \cos (2000\pi t - 60°) + 6 \cos (4000\pi t + 150°)$$

Express the signal in **(a)** sine–cosine form and **(b)** complex exponential form.

9-7. A certain periodic voltage has the form of the square wave shown in Table 9-2. The amplitude is $A = 10$ V and the period is $T = 1$ ms. Prepare a table listing all frequencies below 10 kHz and the corresponding amplitude of the components on a one-sided basis.

9-8. A certain periodic voltage has the form of the triangular wave shown in Table 9-2. The amplitude is $A = 20$ V and the period is $T = 2$ μs. Prepare a table listing all frequencies below 5 MHz and the corresponding amplitudes on a one-sided basis.

9-9. For the voltage of Problem 9-7, plot the **(a)** one-sided and **(b)** two-sided spectra for the frequency range -10 to 10 kHz.

9-10. For the voltage of Problem 9-8, plot the **(a)** one-sided and **(b)** two-sided spectra for the frequency range -5 to 5 MHz.

9-11. A certain rectangular pulse has the form shown in Table 9-4. The amplitude is $A = 10$ V, and the pulse width is $\tau = 1$ ms.
   (a) Write an equation for the Fourier transform.
   (b) Sketch the form of the amplitude spectrum between dc and 2 kHz.

9-12. A certain triangular pulse has the form shown in Table 9-4. The amplitude is $A = 20$ V and the pulse width is $2\tau = 1$ ms.
   (a) Write an equation for the Fourier transform.
   (b) Sketch the form of the amplitude spectrum between dc and 4 kHz.

9-13. By inspection, determine the spectral roll-off rates for the periodic signals of Table 9-2.
   (a) Square wave
   (b) Triangular wave
   (c) Sawtooth wave
   (d) Half-wave rectified wave
   (e) Full-wave rectified wave
   (f) Pulse train
   Express the results both as $1/n^k$ and in decibels per octave. Check conclusions with the tabulated results.

9-14. By inspection, determine the spectral roll-off rates for the nonperiodic signals of Table 9-4.
   (a) Rectangular pulse
   (b) Triangular pulse
   (c) Sawtooth pulse
   (d) Half-cycle cosine pulse
   Express the results both as $1/f^k$ and in decibels per octave. Check your conclusions with the tabulated results.

DERIVATION PROBLEMS

9-15. Starting with the expression for $\bar{X}_n$ as given by Eq. (9-16), apply Euler's formula and establish the formula of Eq. (9-18).

9-16. Starting with the exponential form of the Fourier series in Eq. (9-15), apply Eqs. (9-17) and (9-18) and show that the exponential series is equivalent to the sine–cosine series as given by Eq. (9-2).

9-17. Derive the even-function symmetry condition for a Fourier series; that is, show that if $x(-t) = x(t)$, integration need be performed only over half a cycle, and the result is doubled. (*Hint:* This theorem is easier to derive by performing the basic integration from $-T/2$ to $T/2$.)

9-18. Derive the odd-function symmetry condition for a Fourier series; that is, show that if $x(-t) = -x(t)$, integration need be performed only over half a cycle, and the result is doubled (see the hint in Problem 9-17).

9-19. Without using any symmetry conditions, derive the sine–cosine form of the Fourier series of the square wave of Table 9-2. Note that the amplitude-phase form is the same in this case.

9-20. Without using any symmetry conditions, derive the sine–cosine form of the Fourier series of the triangular wave of Table 9-2. Note that the amplitude-phase form is the same in this case.

9-21. Repeat the derivation of Problem 9-19 using a symmetry condition to simplify the analysis.

9-22. Repeat the derivation of Problem 9-20 using a symmetry condition to simplify the analysis.

9-23. Determine the complex exponential form of the series of Problems 9-19 and 9-21.

9-24. Determine the complex exponential form of the series of Problems 9-20 and 9-22.

9-25. Without using any symmetry condition, derive the Fourier transform of the triangular pulse of Table 9-4.

9-26. Repeat the derivation of Problem 9-25 using a symmetry condition to simplify the analysis.

9-27. Derive the modulation theorem as given by (O-5) of Table 9-5.

9-28. Derive the time-scaling transform pair as given by (O-6) of Table 9-5.

COMPUTER PROBLEMS

9-29. Write a computer program that will perform a *sine–cosine* Fourier series summation of a finite number N of the terms in the range from t_1 to t_2 in steps of Δt. You will thus be working with a summation of the form

$$x(t) = A_0 + \sum_{n=1}^{N} (A_n \cos n\omega_1 t + B_n \sin n\omega_1 t)$$

Input data: N, $A_n(n = 0 \text{ to } n = N)$, $B_n(n = 1 \text{ to } n = N)$, $f_1(\text{or } \omega_1)$, t_1, t_2, Δt

Output data: t, $x(t)$

Note: Check the sine and cosine functions in the software library of your system to ensure compatibility with radian input since $n\omega_1 t$ is measured in radians.

9-30. Write a computer program similar to the one discussed in Problem 9-29 except that the amplitude-phase form of the Fourier series will be used. The summation will then be

$$x(t) = C_0 + \sum_{n=1}^{N} C_n \cos (n\omega_1 t + \phi_n)$$

Input data: N, $C_n(n = 0 \text{ to } n = N)$, $\phi_n(n = 0 \text{ to } n = N)$, $f_1 \text{ (or } \omega_1)$, t_1, t_2, Δt

Output data: t, $x(t)$

See note at end of Problem 9-29. The angles ϕ_n must also be specified in radians.

10

INTRODUCTION
TO DISCRETE-TIME SYSTEMS†

OVERVIEW

The use of digital technology for performing many functions previously associated with analog equipment has increased tremendously in recent years. This includes the use of dedicated microprocessors and microcomputers as well as general-purpose computers. Prominent among the areas in which digital technology has been most significant is that of *digital signal processing*. This includes both *digital filtering* and *digital spectral analysis*.

A fundamental process in the development and understanding of digital signal processing is the concept of the *discrete-time system*. The theory of discrete-time systems includes the concept of the *sampled-data signal*, the *difference equation*, and the *discrete transfer function*. The latter concept is established with the aid of an operational technique called the *z-transform*. The preceding concepts are developed in this chapter as a basis for further study in the rapidly developing application areas of digital processing.

OBJECTIVES

After completing this chapter, the reader should be able to

1. Define the terms *analog signal*, *continuous-time signal*, *discrete-time signal*, *sampled-data signal*, and *digital signal*.

2. Define the terms *quantization*, *quantized variable*, and *quantization noise*.

3. Draw the block diagram of a typical digital signal processing system, and discuss the functions of the different blocks.

---

†Portions of this chapter have been adapted from the author's text *Digital Signal Processing*, 2nd ed. (with G. Dougherty and R. Dougherty), Reston Publishing Co., Reston, Va., 1984.

4. Write the equation of a sampled-data signal and sketch its form.
5. Write the equation of the spectrum of a sampled-data signal and sketch its form.
6. State Shannon's sampling theorem and discuss its significance.
7. Define *aliasing* and discuss its effect.
8. Define *folding frequency*.
9. Write the equation of an ideal impulse sampled signal and sketch its form.
10. Write the equation of the spectrum of an ideal impulse sampled signal and sketch its form.
11. Discuss the operation of a *holding circuit*.
12. Define the z-transform of a discrete-time signal.
13. Show how the z-transform is related to the Laplace transform.
14. Determine the z-transforms of common discrete-time functions, (e.g., step function, exponential function, etc.).
15. Discuss the use of various z-transform operation pairs and their effects on signals.
16. State the form of a linear difference equation for a discrete-time, linear, time-invariant system.
17. Determine the solution of a linear difference equation by an algorithm involving arithmetic operations.
18. Define the form of the *transfer function* of a discrete-time system, and show how it is related to the difference equation.
19. Determine the response of a discrete-time system from an arbitrary input function using the transfer function concept.
20. Determine the impulse response of a discrete-time system and discuss its significance.
21. Determine the inverse z-transform of a function using partial fraction expansion.
22. Determine the inverse z-transform of a function using power series expansion.
23. Define stability as it relates to a discrete-time system.
24. Show how relative stability can be determined from a knowledge of the poles in the z-plane.
25. State the form of the convolution summation of a discrete-time system, and show how it relates to the transfer function.

10-1 GENERAL DISCUSSION

At the beginning of this chapter, it is appropriate to discuss a few of the common terms that will be used and some of the assumptions that will be made. Wherever possible, the definitions and terminology will be established in accordance with the recommendations of the IEEE.

An *analog* signal is a function that is defined over a continuous range of time and in which the amplitude may assume a continuous range of values. Common examples are the sinusoidal function, the step function, the output of a microphone, and so on. The term "analog" apparently originated from the field of analog computation, in which voltages and currents are used to represent physical variables, but it has been extended in usage.

A *continuous-time* signal is a function that is defined over a continuous range of time, but in which the amplitude may either have a continuous range of values or a finite number of possible values. In this context, an analog signal could be considered as a special case of a continuous-time signal. In practice, however, the terms "analog" and "continuous time" are interchanged casually in usage and are often used to mean the same thing. Because of the association of the term "analog" with physical analogies, preference has been established for the term "continuous time," and this practice will be followed for the most part in this chapter.

The term *quantization* describes the process of representing a variable by a set of distinct values. A *quantized variable* is one that may assume only distinct values.

A *discrete-time* signal is a function that is defined only at a particular set of values of time. This means that the independent variable, time, is quantized. If the amplitude of a discrete-time signal is permitted to assume a continuous range of values, the function is said to be a *sampled-data* signal. A sampled-data signal could arise from sampling an analog signal at discrete values of time.

A *digital* signal is a function in which both time and amplitude are quantized. A digital signal may always be represented by a sequence of numbers in which each number has a finite number of digits.

The terms "discrete-time" and "digital" are often interchanged in practice and are often used to mean the same thing. A great deal of the theory underlying discrete-time signals is applicable to purely digital signals, so it is not always necessary to make rigid distinctions. The term "discrete-time" often is used in pursuing theoretical developments, and the term "digital" is often used in describing hardware or software realizations.

A system can be described by any of the preceding terms according to the type of hardware or software employed and the type of signals present. Thus reference can be made to "analog systems," "continuous-time systems," "discrete-time systems," "digital systems," and so on.

A *linear* system is one in which the parameters of the system are not dependent on the nature or the level of the input excitation. This statement is equivalent to the statement that the principle of superposition applies. A linear system can be described by linear differential or difference equations. A *time-invariant* linear system is one in which the parameters are fixed and do not vary with time. A *lumped* system is one that is composed of finite nonzero elements satisfying ordinary differential or difference equation relationships (as opposed to a distributed system, satisfying partial differential equation relationships). These latter definitions were considered in Chapter 7 for continuous-time systems.

The standard form for numerical processing of a digital signal is the binary number system. The binary number system makes use only of the values 0 and 1 to represent all possible numbers. The number of levels m that can be represented by a

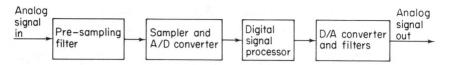

Figure 10-1   Block diagram of a possible digital processing system.

number having *n* *bi*nary dig*its* (bits) is given by

$$m = 2^n \tag{10-1}$$

Conversely, if *m* is the number of possible levels required, the number of bits required is the smallest integer greater than or equal to $\log_2 m$.

The process by which digital signal processing is achieved will be illustrated by a simplified system in which the signal is assumed to vary from 0 to 7 volts and in which eight possible levels (at 1-V increments) are used for the binary numbers. A block diagram is shown in Fig. 10-1, and some waveforms of interest are shown in Fig. 10-2. The signal is first passed through a continuous-time presampling filter whose function is discussed later. The signal is then read at intervals of *T* seconds by a sampler. These samples must then be quantized to one of the standard levels. Although there are different strategies employed in the quantization process, one common approach, which is assumed here, is that a sample is *rounded* to the *nearest* level. Thus a sample of value 4.2 V would be quantized to 4 V, and a sample of value 4.6 V would be quantized to 5 V.

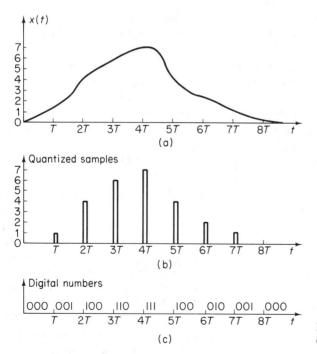

Figure 10-2   Sampling and digital conversion process.

This process for the signal given is illustrated in Fig. 10-2a and b. The pulses representing the signal have been made very narrow to illustrate the fact that other signals may be inserted, or *multiplexed*, in the empty space. These pulses may then be represented as binary numbers as illustrated in (c). In order that these numbers could be seen on the figure, each has been shown over much of the space in a given interval. In practice, if other signals are to be inserted, the pulses representing the bits of the binary numbers could be made very short. A given binary number could then be read in a very short interval at the beginning of a sampling period, thus leaving most of the time available for other signals.

The process by which an analog sample is quantized and converted to a binary number is called *analog-to-digital* (*A/D*) *conversion*. In general, the dynamic range of the signal must be compatible with that of the A/D converter employed, and the number of bits employed must be sufficient for the required accuracy.

The signal can now be processed by the type of unit appropriate for the application intended. This unit may be a general-purpose computer or microcomputer, or it may be a special unit designed specifically for this purpose. At any rate, it is composed of some combination of standard digital circuits capable of performing the various arithmetic functions of addition, subtraction, multiplication, etc. In addition, it has logic and storage capability.

At the output of the processor, the digital signal can be converted to analog form again. This is achieved by the process of *digital-to-analog* (*D/A*) *conversion*. In this step, the binary numbers are first successively converted back to continuous-time pulses. The "gaps" between the pulses are then filled in by a *reconstruction filter*. This filter may consist of a holding circuit, which is a special circuit designed to hold the value of a pulse between successive sample values. In some cases, the holding circuit may be designed to extrapolate the output signal between successive points according to some prescribed curve-fitting strategy. In addition to a holding circuit, a basic continuous-time filter may be employed to provide additional smoothing between points.

A fundamental question that may arise is whether or not some information has been lost in the process. After all, the signal has been sampled only at discrete intervals of time; is there something that might be missed in the intervening time intervals? Furthermore, in the process of quantization, the actual amplitude is replaced by the nearest standard level, which means that there is a possible error in amplitude.

In regard to the sampling question, it will be shown in Section 10-2 that if the signal is bandlimited, and if the sampling rate is greater than or equal to twice the highest frequency, the signal can theoretically be recovered from its discrete samples. This corresponds to a minimum of two samples per cycle at the highest frequency. In practice, this sampling rate is usually chosen to be somewhat higher than the minimum rate (say, three or four times the highest frequency) in order to ensure practical implementation. For example, if the highest frequency of the analog signal is 5 kHz, the theoretical minimum sampling rate is 10,000 samples per second, and a practical system would employ a rate somewhat higher. The input continuous-time signal is often passed through a low-pass analog *presampling filter* to ensure that the highest frequency is within the bounds for which the signal can be recovered.

If a signal is not sampled at a sufficiently high rate, a phenomenon known as *aliasing* results. This concept results in a frequency's being mistaken for an entirely different frequency upon recovery. For example, suppose a signal with frequencies ranging from dc to 5 kHz is sampled at a rate of 6 kHz, which is clearly too low to ensure recovery. If recovery is attempted, a component of the original signal at 5 kHz now appears to be at 1 kHz, resulting in an erroneous signal. A common example of this phenomenon is one we will call the "wagon wheel effect," probably noticed by the reader in western movies as the phenomenon in which the wheels appear to be rotating backwards. Since each individual frame of a film is equivalent to a discrete sampling operation, if the rate of spokes passing a given angle is too large for a given movie frame rate, the wheels appear to be turning either backwards or at a very slow speed. The effect of a presampling filter removes the possibility that a spurious signal whose frequency is too high for the system will be mistaken for one in the proper frequency range.

With respect to the quantization error, it can be seen that the error can be made as small as one chooses if the number of bits can be made arbitrarily large. Of course, there is a practical maximum limit, so it is necessary to tolerate some error from this phenomenon. Even in continuous-time systems, there may be noise present which would introduce uncertainty in the actual magnitude. In fact, the uncertainty present in the digital sampling process is called *quantization noise*.

10-2 SAMPLED-DATA SIGNALS

At this point, we will introduce an important class of signals whose properties serve as a link between continuous-time signals and discrete-time or digital signals. A *sampled-data signal* can be considered as arising from sampling a continuous-time signal at periodic intervals of time T as illustrated in Fig. 10-3. The *sampling rate* or *sampling frequency* is $f_s = 1/T$.

Initially, we will assume that each sample has a width τ, so that the resulting signal consists of a series of relatively narrow pulses whose amplitudes are modulated by the original continuous-time signal. This particular form of a sampled-data signal is designated in communications systems as a *pulse amplitude modulated* (PAM) signal.

Let $x^*(t)$ represent the sampled-data signal, and let $x(t)$ represent the original continuous-time signal. We may consider $x^*(t)$ as the product of $x(t)$ and a hypothetical pulse train $p(t)$ as illustrated in Fig. 10-3. Thus,

$$x^*(t) = x(t)p(t) \tag{10-2}$$

An important property of the sampled-data signal is its spectrum $X^*(f)$. This can be derived by first expressing $p(t)$ in the Fourier series form

$$p(t) = \sum_{-\infty}^{\infty} \bar{P}_m \epsilon^{jm\omega_s t} \tag{10-3}$$

where $\omega_s = 2\pi f_s = 2\pi/T$. The coefficients $\bar{P}_m$ in Eq. (10-3) converge as $1/f$. However, it is not necessary at this point to actually introduce the values for $\bar{P}_m$ into Eq. (10-3) as long as we understand their general behavior.

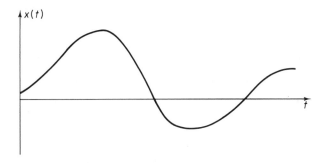

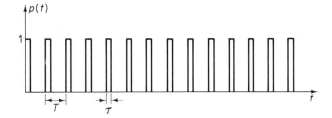

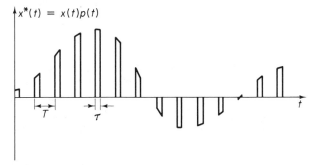

Figure 10-3 Development of sampled-data signal using nonzero width pulse sampling.

Substitution of Eq. (10-3) in Eq. (10-2) results in the expression

$$x^*(t) = \sum_{-\infty}^{\infty} \bar{P}_m x(t) \epsilon^{jm\omega_s t} \tag{10-4}$$

The spectrum may now be determined by taking the Fourier transforms of both sides of Eq. (10-4). Each term of the series on the right may be transformed with the help of operation (O-5) of Table 9-5. The result is

$$\bar{X}^*(f) = \sum_{-\infty}^{\infty} \bar{P}_m \bar{X}(f - mf_s) \tag{10-5}$$

Typical sketches of $|\bar{X}(f)|$ and $|\bar{X}^*(f)|$ are shown in Fig. 10-4. Due to lack of space, only a small section of the negative frequency range of $|\bar{X}^*(f)|$ is shown, but since it is an even function of frequency, its behavior in the negative frequency range is readily understood.

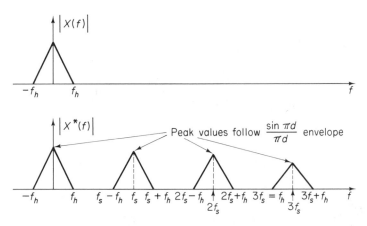

Figure 10-4   Spectrum of sampled-data signal using nonzero width pulse sampling.

It can be observed that the spectrum of a sampled-data signal consists of the original spectrum plus an infinite number of translated versions of the original spectrum. These various translated functions are shifted in frequency by amounts equal to the sampling frequency and its harmonics. The magnitudes are multiplied by the $\bar{P}_m$ coefficients, so that they diminish with frequency. However, for a very short duty cycle ($\tau \ll T$), the components drop off very slowly, and the spectrum would be extremely wide in this case.

Assume that the spectrum of $x(t)$ is bandlimited to $0 \leq f \leq f_h$ in the positive frequency sense as illustrated in Fig. 10-4 where f_h is the highest possible frequency. In order to be able to eventually recover the original signal from the sampled-data form, it is necessary that none of the shifted spectral components overlap each other. If portions of any of the shifted functions overlap, certain frequencies appear to be different from their actual values, and it becomes impossible to separate or recover these particular components. This process of spectral overlap is called *aliasing*, and it can occur if either of the following conditions exist: (a) the signal is not bandlimited to a finite range, or (b) the sampling rate is too low.

Theoretically, if the signal is not bandlimited, there is no way of avoiding the aliasing problem with the basic sampling scheme employed. However, the spectra of most real-life signals are such that they may be assumed to be bandlimited. Furthermore, a common practice employed in many sampled-data systems is to filter the continuous-time signal before sampling to ensure that it does meet the bandlimited criterion closely enough for all practical purposes.

Let us now turn to the concept of the sampling rate. In order to avoid aliasing in Fig. 10-4 it is necessary that $f_s - f_h \geq f_h$. This leads to the important inequality

$$f_s \geq 2f_h \tag{10-6}$$

Equation (10-6) is a statement of *Shannon's sampling theorem*, which states that a signal must be sampled at a rate at least as high as twice the highest frequency in the spectrum. In practice, the sampling rate must be chosen to be somewhat greater than $2f_h$ to ensure recovery with practical hardware limitations.

If no aliasing occurs, the original signal can be recovered by passing the sampled-data signal through a low-pass filter having a cutoff frequency somewhere between f_h and $f_s - f_h$. It is impossible to build filters having an infinite sharpness of cutoff, so that a *guard band* between f_h and $f_s - f_h$ is desired. This illustrates the need for a sampling rate somewhat greater than the theoretical minimum.

A convenient definition that is useful in sampling analysis is the *folding frequency* f_0. It is given by

$$f_0 = \frac{f_s}{2} = \frac{1}{2T} \tag{10-7}$$

The folding frequency is simply the highest frequency that can be processed by a given discrete-time system with sampling rate f_s. Any frequency greater than f_0 will be "folded" and cannot be recovered. In addition, it will obscure data within the correct frequency range; so it is important to clearly limit the frequency content of a signal before sampling.

A word about terminology should be mentioned here. The highest frequency f_h in the signal is called the *Nyquist frequency*, and the minimum sampling rate $2f_h$ at which the signal could theoretically be recovered is called the *Nyquist rate*.

A point of ambiguity is that the frequency $f_0 = f_s/2$ is also referred to as the Nyquist frequency in some references. To avoid confusion in terminology, we will use the term *folding frequency* in reference to $f_0 = f_s/2$, as discussed previously.

10-3 IDEAL IMPULSE SAMPLING

The sampled-data signal of the last section was derived on the assumption that each of the samples had a nonzero width τ. We now wish to consider the limiting case that results when the width τ is assumed to approach zero. In this case, the samples will be represented as a sequence of impulse functions.

While the analog samples of any real sampled-data signal derived directly from a continuous-time signal could never reach the extreme limit of zero width, the limiting concept serves two important functions: (a) If the widths of the actual pulse samples are quite small compared with the various time constants of the system under consideration, the impulse function assumption is a good approximation, and it leads to simplified analysis; (b) When a signal is sampled, converted from analog to digital form, and subsequently processed with digital circuitry, it may be considered simply as a number occurring at a specific instant of time. A very convenient way of modeling a digital signal of this form is through the impulse sampling representation.

The form of the ideal impulse sampled-data signal is illustrated in Fig. 10-5. As in Section 10-2, $x^*(t)$ will be used to represent the sampled-data signal, and $x(t)$ will represent the original continuous-time signal. The pulse function is designated as $p_\delta(t)$ and is assumed to be a train of impulse functions of the form

$$p_\delta(t) = \sum_{-\infty}^{\infty} \delta(t - nT) \tag{10-8}$$

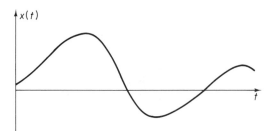

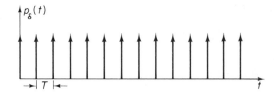

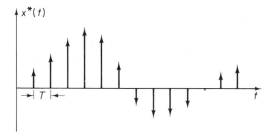

Figure 10-5 Development of sampled-data signal using ideal impulse sampling.

The sampled-data signal $x^*(t)$ can be expressed as

$$x^*(t) = x(t)p_\delta(t) = x(t) \sum_{-\infty}^{\infty} \delta(t - nT) \tag{10-9}$$

The only values of $x(t)$ having significance in Eq. (10-9) are those at $t = nT$. Hence an alternative form for the sampled-data signal is

$$x^*(t) = \sum_{-\infty}^{\infty} x(nT)\delta(t - nT) \tag{10-10}$$

Both the forms (10-9) and (10-10) will be used in subsequent work. The first expression is useful in deriving spectral relationships, due to the product form given. The second form provides the interpretation that the sampled-data signal is composed of a series of equally spaced impulses whose weights represent the values of the original signal at sampling instants.

The spectrum of the ideal impulse train can be derived by utilizing the integration property of an ideal impulse function, which states

$$\int_{-\infty}^{\infty} \delta(t - a) \, dt = 1 \tag{10-11}$$

Since the impulse train is periodic, it may be expanded in a Fourier series. The coefficients $\bar{P}_m$ may be determined from the work of Chapter 9 with the help of Eq.

(10-11). The result is

$$\bar{P}_m = \frac{1}{T} \qquad (10\text{-}12)$$

This result indicates that all of the spectral components have equal weights, and there is no convergence at all for the spectral components! The function $p_\delta(t)$ may then be written as

$$p_\delta(t) = \sum_{-\infty}^{\infty} \frac{1}{T} \epsilon^{jm\,\omega_s t} \qquad (10\text{-}13)$$

Substitution of Eq. (10-13) in Eq. (10-9) yields

$$x^*(t) = \frac{1}{T} \sum_{-\infty}^{\infty} x(t)\epsilon^{jm\,\omega_s t} \qquad (10\text{-}14)$$

The Fourier transform may now be applied to both sides of Eq. (10-14), and transform operation (O-5) can be applied to each of the terms on the right. The result is

$$\bar{X}^*(f) = \frac{1}{T} \sum_{-\infty}^{\infty} \bar{X}(f - mf_s) \qquad (10\text{-}15)$$

The form of the spectrum of the ideal impulse sampled-data signal is shown in Fig. 10-6. The general form is similar to that of the sampled-data signal derived from the nonzero sampling process shown in Fig. 10-4, and the basic sampling requirements developed in Section 10-2 apply here. Comparison of Figs. 10-4 and 10-6 and Eqs. (10-5) and (10-15) indicates that the major difference is the behavior of the levels of the spectral components. The spectral components derived with nonzero pulse widths gradually diminish with frequency. However, the spectral components derived from ideal impulse sampling are all of equal magnitude and do not diminish with frequency.

An important deduction from this discussion is that *the spectrum of an impulse sampled-data signal is a periodic function of frequency.* The period in the frequency domain is equal to the sampling frequency f_s. The sampling process in the time domain leads to a periodic function in the frequency domain.

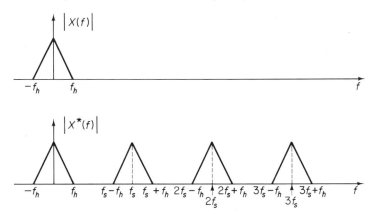

Figure 10-6 Spectrum of sampled-data signal using ideal impulse sampling.

10-4 HOLDING CIRCUIT

It was mentioned in Section 10-2 that a continuous-time signal could be recovered from its sampled-data form by passing the sampled-data signal through a low-pass filter having a cutoff somewhere between f_h and $f_s - f_h$. This process of reconstruction can be aided by the use of a *holding circuit*, which actually performs a portion of the filtering required, thus permitting the use of a less complex filter for the final smoothing. Although a number of holding circuits of varying complexity have been devised, we will restrict the consideration here to the *zero-order* holding circuit.

The zero-order holding circuit is best explained by first assuming that we are dealing with samples having sufficiently small widths that the variation in the peak is insignificant during the interval τ. Hence, a given pulse may be assumed as rectangular. The holding circuit simply accepts the value of the pulse at the beginning of a sampling interval and holds it to the beginning of the next interval, at which time it changes to the new value. This process is illustrated in Fig. 10-7. The resulting function is, of course, not normally the same as the original signal before sampling, but it is now in the form of a continuous-time function, and it will be easier to perform subsequent processing on it in this form.

For analytical purposes, the implementation of a zero-order holding circuit shown in Fig. 10-8 will be considered. The delay block represents an ideal analog

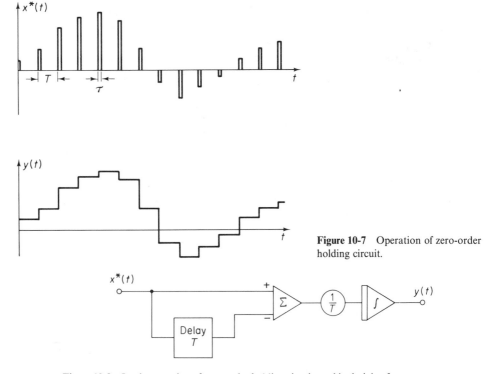

Figure 10-7 Operation of zero-order holding circuit.

Figure 10-8 Implementation of zero-order holding circuit used in deriving frequency response.

delay line having a delay of T seconds. The delayed signal is subtracted from the direct signal, and the net difference is integrated over the sampling interval to yield the output.

The sequence of events for a sampled-data signal begins with the appearance of a very short pulse at the beginning of an interval. Since the delayed signal will not initially appear at the input of the difference circuit, the integrator reaches a value proportional to the area of the pulse in a time τ. This value is held until the delayed signal reaches the inverted input after T seconds. By superposition, the effect of the delayed and inverted pulse is to cancel out the output of the integrator previously established. Hence the next output value of the integrator will be a function of the next input pulse only.

A continuous transfer function may be derived for the zero-order holding circuit. This is best achieved by assuming now that a given input pulse may be approximated by an impulse. For convenience, assume that the given impulse occurs at $t = 0$. The output $y(t)$ produced by this impulse is illustrated in Fig. 10-9. It can be expressed as

$$y(t) = \frac{1}{T}[u(t) - u(t - T)] \tag{10-16}$$

We may now take the Laplace transforms of both sides of Eq. (10-16). Furthermore, since $X(s) = 1$, then $Y(s)$ is the same as the transfer function $G(s)$, and we have

$$G(s) = \frac{1}{sT}(1 - \epsilon^{-sT}) \tag{10-17}$$

The steady-state frequency response $G(j\omega)$ is obtained by setting $s = j\omega$ in Eq. (10-17). We will leave as an exercise for the reader to show that the amplitude response $A(\omega)$ and the phase response $\beta(\omega)$ of this circuit may be expressed as

$$A(\omega) = \frac{\sin \pi f T}{\pi f T} \tag{10-18}$$

and

$$\beta(\omega) = \pi f T \tag{10-19}$$

(As in the case of some previous functions, we are permitting $A(\omega)$ to assume negative values here.) The amplitude response is shown in Fig. 10-10. We see that

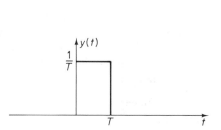

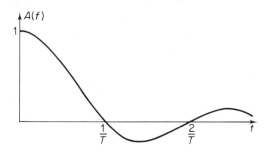

Figure 10-9 Impulse response of zero-order holding circuit.

Figure 10-10 Amplitude response of zero-order holding circuit.

the circuit does function as a type of low-pass filter, although it is not particularly outstanding in this capacity. Normally, additional filtering of the signal will be required to effectively remove components of the sampled-data signal about the sampling frequency and its harmonics, but the presence of the holding circuit eases the requirements.

10-5 DISCRETE-TIME SIGNALS

In preceding sections, the concept of a sampled-data signal was established. This was achieved by expressing the sampled signal as the product of a reference continuous-time or analog signal and a pulse train consisting of narrow rectangular pulses. If the pulses are assumed to become very narrow, the pulse train can be conveniently represented mathematically as an impulse train.

To enhance the process of steady-state Fourier analysis, the sampled-data signals in earlier sections were permitted to extend over both the negative and positive time regions. On the other hand, future developments are best achieved by assuming that the signal exists only for positive time. Thus, we will begin by representing any sampled-data signal of interest in the form

$$x^*(t) = \sum_{n=0}^{\infty} x(nT)\delta(t - nT) \tag{10-20}$$

The Laplace transform of Eq. (10-20) is given by

$$X^*(s) = \sum_{n=0}^{\infty} x(nT)\epsilon^{-nTs} \tag{10-21}$$

The interpretation of Eq. (10-20), along with its Laplace transform in Eq. (10-21), is a very important one, and we will return to it frequently in developing various discrete-time system results. In fact, this result serves as a link in relating some of the purely continuous-time system results to those of discrete-time systems.

Consider now the case of a general discrete-time signal that is defined only at integer multiples of a basic interval T. This signal differs from the sampled-data signal $x^*(t)$ only in the sense that it may not necessarily have arisen from sampling a continuous-time signal. Instead, it may have arisen from some purely discrete or digital process. Nevertheless, we can still interpret the signal in the form of Eq. (10-20) whenever desirable.

Except where it is desirable to use the sampled-data interpretation, the most straightforward notation for a discrete-time signal is simply $x(n)$, where n is an integer defined over some range $n_1 \le n \le n_2$. The integer n defines the particular location in the sequence corresponding to a given sample. If the discrete-time signal is derived from sampling a continuous-time signal $x_1(t)$, the signals are related by

$$x(n) = \begin{cases} x_1(nT) & \text{for } n \text{ an integer} \\ 0 & \text{otherwise} \end{cases} \tag{10-22}$$

In effect, Eq. (10-22) states that the discrete-time signal is equal to the continuous-time signal at sample points and is zero elsewhere.

10-6 z-TRANSFORM

The z-transform is an operational function that may be applied to discrete-time systems in the same manner as the Laplace transform is applied to continuous-time systems. We will develop this concept through the use of the *one-sided z*-transform, which is most conveniently related to the concepts of continuous-time systems as discussed earlier in the book. The z-transform of a discrete-time signal $x(n)$ is denoted by $X(z)$. The symbolic forms for the z-transform and inverse z-transform are given by

$$X(z) = \mathbf{Z}\left[x(n)\right] \tag{10-23}$$

and

$$x(n) = \mathbf{Z}^{-1}[X(z)] \tag{10-24}$$

The actual definition of the one-sided z-transform is

$$X(z) = \sum_{n=0}^{\infty} x(n)z^{-n} \tag{10-25}$$

The function $X(z)$ is an infinite series, but it can often be expressed in closed form.

A comparison of Eqs. (10-21) and (10-25), with recognition of the fact that T does not appear in Eq. (10-25), yields some useful relationships between the s-plane and the z-plane. We note that

$$X(z) = \left[X^*(s)\right]_{z=\epsilon^{sT}} \tag{10-26}$$

The s and z variables are related by

$$z = \epsilon^{sT} \tag{10-27}$$

and

$$s = \frac{1}{T}\ln z \tag{10-28}$$

It is of interest to note the effect of the transformation of Eqs. (10-27) and (10-28) as shown in Fig. 10-11. The left-hand half of the s-plane maps to the interior of the unit circle in the z-plane, and the right-hand half of the s-plane maps to the exterior of the unit circle in the z-plane. The $j\omega$-axis in the s-plane maps to the boundary of the unit circle in the z-plane. The transformation from z to s is a multivalued transformation, as can be seen from Eq. (10-28), with recognition of the properties of the complex logarithm. In fact, there are an infinite number of values in the s-plane corresponding to a given point in the z-plane. This property is closely related to the concept of the spectrum of a sampled signal developed in the previous chapter.

The boundaries of major interest in this transformation are the $j\omega$ axis in the s-plane and the unit circle in the z-plane. This situation results from letting $s = j\omega$, which is equivalent to

$$z = \epsilon^{j\omega T} \tag{10-29}$$

As the cyclic frequency f varies over the range $-1/2T \leq f \leq 1/2T$, the argument of Eq. (10-29) varies from $-\pi$ to π. This is equivalent to a complete rotation around

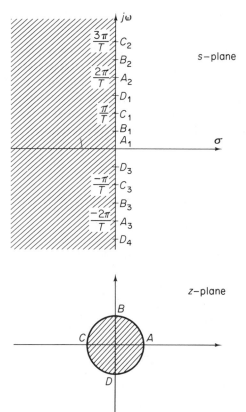

Figure 10-11 Complex mapping relationship between s-plane and z-plane.

the unit circle in the z-plane. As the frequency increases beyond $1/2T$, the locus in the z-plane continues to rotate around the same path again, with a complete rotation for each increase of $\omega T = 2\pi$. Once again, the concept of the sampling theorem is evident.

We consider next the process of actually calculating the z-transform of a given discrete signal. In general, this may be accomplished by either (a) use of the definition as given by Eq. (10-25), or (b) application of a contour integral to the Laplace transform of a corresponding continuous-time signal (if such can be found). Only the first method will be considered in this text.

Application of the basic definition of Eq. (10-25) results in a series in which the value of a particular sample in the sequence is readily observed by the weight of the corresponding z coefficient. For a finite-length signal, the basic power series may be the most ideal form in which to express the transform, particularly if the series is fairly short in duration. On the other hand, for either long or infinite series, it is often desirable, if possible, to represent the transform as a closed-form expression. This is rarely possible with a "real-life" random signal. However, in the same spirit that continuous-time systems are continually analyzed with simple inputs such as the impulse, step, sinusoid, etc., so it is with discrete-time systems. The response of a discrete-time system to such standard waveforms serves to define clear impres-

TABLE 10-1   z-TRANSFORM FUNCTION PAIRS

$x(n)$	$X(z)$	$X(s)$	
$\delta(n)$	1	1	(ZT-1)
1 or $u(n)$	$\dfrac{z}{z-1}$	$\dfrac{1}{s}$	(ZT-2)
nT	$\dfrac{Tz}{(z-1)^2}$	$\dfrac{1}{s^2}$	(ZT-3)
ϵ^{-naT}	$\dfrac{z}{z-\epsilon^{-aT}}$	$\dfrac{1}{s+a}$	(ZT-4)
a^n	$\dfrac{z}{z-a}$	$\dfrac{1}{s-\ln a/T}$	(ZT-5)
$\sin naT$	$\dfrac{z\sin aT}{z^2-2z\cos aT+1}$	$\dfrac{a}{s^2+a^2}$	(ZT-6)
$\cos naT$	$\dfrac{z^2-z\cos aT}{z^2-2z\cos aT+1}$	$\dfrac{s}{s^2+a^2}$	(ZT-7)

sions and boundaries of what the response due to any waveform would be like, and thus it is very useful to give these signals great attention.

A summary of some of the most common function pairs is given in Table 10-1, and a summary of certain operations is given in Table 10-2. As a point of convenience, the corresponding Laplace transforms of the unsampled functions are given in the last column of Table 10-1. These are not intended as comprehensive tables, but they provide adequate information to deal with most of the standard waveforms of interest in digital signal processing. The examples that follow and the problems at the end of the chapter illustrate the derivation of some of these entries.

Example 10-1

Derive the z-transform of the discrete unit step function $u(n)$.

Solution   The discrete unit step function $u(n)$ is defined by

$$u(n) = \begin{cases} 1 & \text{for } n \geq 0 \\ 0 & \text{for } n < 0 \end{cases} \tag{10-30}$$

Utilization of the definition of the z-transform yields

$$X(z) = \sum_{0}^{\infty} x(n)z^{-n} = \sum_{0}^{\infty} (1)z^{-n} \tag{10-31}$$

The infinite summation given by Eq. (10-31) can be expressed in closed form for $|z| > 1$ as

$$X(z) = \frac{1}{1-z^{-1}} = \frac{z}{z-1} \tag{10-32}$$

TABLE 10-2 z-TRANSFORM OPERATION PAIRS

$x(n)$	$X(z) = \mathbf{Z}[x(n)]$	
$ax_1(n) + bx_2(n)$	$aX_1(z) + bX_2(z)$	(ZO-1)
$^a x(n - m)$	$z^{-m}X(z)$	(ZO-2)
$\epsilon^{-naT}x(n)$	$X(\epsilon^{aT}z)$	(ZO-3)
$a^{-n}x(n)$	$X(az)$	(ZO-4)
$n^l x(n)$	$\left(-z\dfrac{d}{dz}\right)^l X(z)$	(ZO-5)
$x(0)$	$\displaystyle\lim_{z \to \infty} X(z)$	(ZO-6)
$x(\infty)$	$^b\displaystyle\lim_{z \to 1} \dfrac{z-1}{z} X(z)$	(ZO-7)
$x(n)h(n)$	$\dfrac{1}{2\pi j}\displaystyle\int_c \dfrac{X(\bar{z})H(z/\bar{z})\,d\bar{z}}{\bar{z}}$	(ZO-8)
$\displaystyle\sum_{m=0}^{n} x(m)h(n - m)$	$X(z)H(z)$	(ZO-9)

^{a} It is assumed that $x(n - m) = 0$ for $n < m$. Otherwise, initial condition terms are required.

^{b} This theorem is valid only if all the poles of $(z - 1)/z\, X(z)$ lie inside the unit circle.

Example 10-2

Derive operation pair (ZO-2).

   Solution   Considering that $x(n - m) = 0$ for $n < m$, application of the definition of the z-transform yields

$$\mathbf{Z}[x(n - m)] = \sum_{n=m}^{\infty} x(n - m)z^{-n} \tag{10-33}$$

Let $n - m = k$. Substitution of this quantity yields

$$\mathbf{Z}[x(n - m)] = \sum_{k=0}^{\infty} x(k)z^{-k}z^{-m} \tag{10-34}$$

$$= z^{-m}\sum_{k=0}^{\infty} x(k)z^{-k} = z^{-m}X(z)$$

10-7 TRANSFER FUNCTION

Let us now consider a discrete-time, linear, time-invariant system consisting of a single input $x(n)$ and a single output $y(n)$. Such a system can be described by a linear difference equation with constant coefficients of the form

$$y(n) + b_1 y(n - 1) + b_2 y(n - 2 + \cdots + b_k y(n - k)$$
$$= a_0 x(n) + a_1 x(n - 1) + a_2 x(n - 2) + \cdots + a_k x(n - k) \tag{10-35}$$

This equation describes an ordinary difference equation of order k with constant coefficients. For convenience in notation, the difference order k has been chosen to be the same on both sides of Eq. (10-35). In the event they are different, one need only specify that certain coefficients are zero. It should be observed that this equation has certain features similar to the differential equation input-output relationship of a continuous-time system, as described in Chapter 7.

If Eq. (10-35) is solved for $y(n)$, the following algorithm is obtained:

$$y(n) = \sum_{i=0}^{k} a_i x(n-i) - \sum_{i=1}^{k} b_i y(n-i) \tag{10-36}$$

A very interesting feature of Eq. (10-36) is that it can be completely solved by the basic arithmetic operations of multiplication, addition, and subtraction. All that is required to start a solution is to specify the input function $x(n)$ and the first k values of the output $y(n)$. The algorithm of Eq. (10-36) is then applied step by step. As each successive value of $y(n)$ is calculated, the integer n is advanced one step, and a computation of the next value is made. The solution of a difference equation is seen to be considerably simpler in concept than that of the corresponding differential equation.

Let us now consider a *relaxed* system, that is, one with no initial values stored in the system. If we take the z-transforms of both sides of Eq. (10-35) and employ operation (ZO-2) of Table 10-2, we obtain, after factoring,

$$(1 + b_1 z^{-1} + b_2 z^{-2} + \cdots + b_k z^{-k}) Y(z)$$
$$= (a_0 + a_1 z^{-1} + a_2 z^{-2} + \cdots + a_k z^{-k}) X(z) \tag{10-37}$$

Solving for $Y(z)$, we obtain

$$Y(z) = \frac{(a_0 + a_1 z^{-1} + a_2 z^{-2} + \cdots + a_k z^{-k})}{(1 + b_1 z^{-1} + b_2 z^{-2} + \cdots + b_k z^{-k})} X(z) \tag{10-38}$$

We now may define the *transfer function* $H(z)$ of the discrete-time system as

$$H(z) = \frac{N(z)}{D(z)} = \frac{a_0 + a_1 z^{-1} + a_2 z^{-2} + \cdots + a_k z^{-k}}{1 + b_1 z^{-1} + b_2 z^{-2} + \cdots + b_k z^{-k}} \tag{10-39}$$

where $N(z)$ is the numerator polynomial and $D(z)$ is the denominator polynomial.

The expression of Eq. (10-39) is arranged in *negative* powers of z, which is usually the most natural form in which the function occurs. On the other hand, it is frequently desirable to express $H(z)$ in *positive* powers of z, particularly when we wish to factor the polynomials or to perform a partial fraction expansion. This is done by multiplying numerator and denominator by z^k, and the result is

$$H(z) = \frac{a_0 z^k + a_1 z^{k-1} + a_2 z^{k-2} + \cdots + a_k}{z^k + b_1 z^{k-1} + b_2 z^{k-2} + \cdots + b_k} \tag{10-40}$$

Using the transfer function concept, the input–output relationship becomes

$$Y(z) = H(z)X(z) \tag{10-41}$$

Thus, a discrete-time system can be represented by the same type of transfer function relationship as for a continuous-time system. In this case, however, the transfer function and the transformed variables are functions of the discrete variable z. To distinguish this transfer function from that of a continuous-time system, we will refer to $H(z)$ simply as a *discrete transfer function*.

In the same fashion as for a continuous transfer function, we may determine poles and zeros for the discrete transfer function and represent them, for this case, in the z-plane. Various geometrical techniques have been developed for analyzing system performance in terms of relative pole and zero locations. Let $z_1, z_2, \ldots, z_k$ represent the k zeros, and let $p_1, p_2, \ldots, p_k$ represent the k poles. The transfer function may then be expressed in positive powers of z as

$$H(z) = \frac{a_0(z - z_1)(z - z_2) \cdots (z - z_k)}{(z - p_1)(z - p_2) \cdots (z - p_k)} \tag{10-42}$$

If desired, this result may also be expressed in negative powers of z as

$$H(z) = \frac{a_0(1 - z_1 z^{-1})(1 - z_2 z^{-1}) \cdots (1 - z_k z^{-1})}{(1 - p_1 z^{-1})(1 - p_2 z^{-1}) \cdots (1 - p_k z^{-1})} \tag{10-43}$$

As in the case of a continuous-time system, we may define the *impulse response* by assuming that the input is simply $x(n) = \delta(n)$. In the case of a continuous-time system, the impulse response is often somewhat difficult to implement physically. However, for a discrete-time system, the "impulse function" is simply a number (usually unity) applied at a single sampling instant, which is readily implemented in an actual system. Since $\mathbf{Z}[\delta(n)] = 1$, the impulse response $h(n)$ is seen to be

$$h(n) = \mathbf{Z}^{-1}[H(z)] \tag{10-44}$$

In general, when a discrete-time system is excited by an arbitrary input $x(n)$, the transform $X(z)$ is determined and multiplied by $H(z)$ to obtain the output transform $Y(z)$. The discrete-time output signal is then determined by inverting $Y(z)$ to give $y(n)$. Of course, it may actually be faster to program the original difference equation directly on a digital computer or programmable calculator and obtain a solution point by point. However, the major advantage of the z-transform approach is the powerful conceptual and operational basis that it provides in studying discrete-time system behavior. For that reason, the next section will be devoted to various procedures for inverting z-transforms.

10-8 INVERSE z-TRANSFORM

The next problem to be considered is that of finding the discrete-time signal $y(n)$ corresponding to a given transform $Y(z)$, which is the process of inverse transformation. In general, the inverse z-transform may be determined by at least three separate procedures: (a) partial fraction expansion and use of transform pair tables, (b) power series expansion, and (c) inversion integral method. The first two methods will be considered in this text.

Partial Fraction Expansion

The method most widely used in routine problems is probably the partial fraction expansion technique. This approach is very similar to the one employed with Laplace transforms in continuous-time system analysis. One must be careful to make sure that all the terms in the expansion fit forms that may be easily recognized from a combination of the table of pairs and the table of operations. As in the case of all partial fraction expansion methods, there are "tricks" that one acquires with experience. In the end, the expansion can always be recombined as a check on its validity if there is any doubt.

Assume that we are given a z-transform $Y(z)$ for which we wish to determine the inverse $y(n)$. Any transform of interest in most digital signal processing systems can usually be described by a rational function of z (ratio of polynomials). Since the transfer function $H(z)$ is also a rational function, as can be deduced from Eqs. (10-39) and (10-40), the resultant $Y(z)$ is also a rational function of z and will have the same general form as the transfer function, except that it will normally have more poles and zeros due to the excitation $X(z)$. We will assume that $Y(z)$ can be represented in positive powers of z as

$$Y(z) = \frac{c_0 z^l + c_1 z^{l-1} + \cdots + c_l}{z^l + d_1 z^{l-1} + \cdots + d_l} \tag{10-45}$$

Assume now that the poles of $Y(z)$ are known. The function may then be expressed in the form

$$Y(z) = \frac{c_0 z^l + c_1 z^{l-1} + \cdots + c_l}{\displaystyle\prod_{m=1}^{l} (z - p_m)} \tag{10-46}$$

The form of the partial fraction expansion and the subsequent inverse transform will depend on the nature of the given poles. The simplest, and by far the most common, case is where all the poles are of simple order; that is, there are no repeated roots in the denominator polynomial. In general, poles may be either real or complex. In the case of complex poles, a pole will always be accompanied by its complex conjugate in the case where all the polynomial coefficients are real.

Probably the most straightforward procedure for the case where all the poles are of simple order is to first divide both sides of Eq. (10-46) by z and then expand $Y(z)/z$ in a partial fraction expansion. Such an expansion will usually be of the form

$$\frac{Y(z)}{z} = \frac{A_1}{z - p_1} + \frac{A_2}{z - p_2} + \cdots + \frac{A_l}{z - p_l} \tag{10-47}$$

A given coefficient A_m may be determined by multiplying both sides of Eq. (10-47) by $z - p_m$ and setting $z = p_m$. This results in zero for all the resulting terms on the right except the A_m term, in which the multiplicative factor has been canceled by the denominator. The result is then

$$A_m = (z - p_m) \frac{Y(z)}{z} \bigg]_{z = p_m} \tag{10-48}$$

This expression is valid only for poles of simple order.

If the coefficients of both the numerator and denominator polynomials of $Y(z)$ are real, then it can be shown that complex poles of $Y(z)$ always occur in conjugate pairs. Furthermore, in this case the corresponding coefficients are also complex conjugates. This means that such a pair can be manipulated into the product of an exponential function and a sinusoidal function. To illustrate this process, assume that a given first-order complex pole $p_r = |p_r|\underline{/\theta_r}$ and its conjugate $\tilde{p}_r = |p_r|\underline{/-\theta_r}$ are present. Assume that the corresponding coefficients are $A_r = |A_r|\underline{/\phi_r}$ and $\tilde{A}_r = |A_r|\underline{/-\phi_r}$. The reader is invited to show that this combination can be expressed as

$$A_r(p_r)^n + \tilde{A}_r(\tilde{p}_r)^n = 2|A_r|(|p_r|)^n \cos(n\theta_r + \phi_r) \tag{10-49}$$

Example 10-3

By partial fraction expansion, obtain the inverse z-transform of

$$Y(z) = \frac{1}{(1 - z^{-1})(1 - 0.5z^{-1})} \tag{10-50}$$

Solution   As a first step, we eliminate the negative powers of z by multiplying numerator and denominator by z^2.

$$Y(z) = \frac{z^2}{(z - 1)(z - 0.5)} \tag{10-51}$$

We now form $Y(z)/z$ and express it in partial fraction form as

$$\frac{Y(z)}{z} = \frac{z}{(z - 1)(z - 0.5)} = \frac{A_1}{z - 1} + \frac{A_2}{z - 0.5} \tag{10-52}$$

Application of Eq. (10-48) yields $A_1 = 2$ and $A_2 = -1$. Multiplication of both sides of Eq. (10-52) by z and inversion yields

$$y(n) = 2 - (0.5)^n \tag{10-53}$$

Several values are tabulated as follows:

n	0	1	2	3	4	5	6	∞
$y(n)$	1	1.5	1.75	1.875	1.9375	1.96875	1.984375	2

Example 10-4

The difference equation describing the input-output relationship of a certain initially relaxed discrete-time system is given by

$$y(n) - y(n - 1) + 0.5y(n - 2) = x(n) + x(n - 1) \tag{10-54}$$

Find (a) the transfer function $H(z)$, (b) the impulse response $h(n)$, and (c) the output response when a unit step function is applied at $n = 0$.

Solution   (a) Taking the z-transform of both sides of Eq. (10-54) and arranging in the form of Eq. (10-39) yields

$$H(z) = \frac{1 + z^{-1}}{1 - z^{-1} + 0.5z^{-2}} \qquad (10\text{-}55)$$

The transfer function can be arranged in positive powers of z by multiplying numerator and denominator by z^2. This yields

$$H(z) = \frac{z(z + 1)}{z^2 - z + 0.5} \qquad (10\text{-}56)$$

The zeros are $z_1 = 0$ and $z_2 = -1$, and the poles are $p_1, p_2 = 0.5 \pm j0.5 = 0.707107 \underline{/\pm 45°}$. The factored form of $H(z)$ is then

$$H(z) = \frac{z(z + 1)}{(z - 0.5 - j0.5)(z - 0.5 + j0.5)} \qquad (10\text{-}57)$$

(b) To obtain the impulse response, we expand $H(z)$ in a partial fraction expansion according to the procedure previously discussed. The result is

$$H(z) = \frac{Az}{z - 0.5 - j0.5} + \frac{\tilde{A}z}{z - 0.5 + j0.5} \qquad (10\text{-}58)$$

$A = 1.581139 \underline{/-71.5651°}$ and $\tilde{A} = 1.581139 \underline{/71.5651°}$. Inversion and use of Eq. (10-49) yield

$$h(n) = 3.162278(0.707107)^n \cos(45n° - 71.5651°) \qquad (10\text{-}59)$$

where the argument of the cosine function is expressed in degrees for convenience. An alternative approach would be to force the given transform into the approximate forms of pairs (ZT-6) and (ZT-7) with modification by use of operation (ZO-4). However, this approach is probably more cumbersome.

(c) In order to solve the response due to a step excitation, we multiply the transfer function by the transform of the step function in accordance with Eq. (10-41). The result is

$$Y(z) = \frac{z^2(z + 1)}{(z - 1)(z^2 - z + 0.5)} \qquad (10\text{-}60)$$

Expansion yields

$$Y(z) = \frac{A_1 z}{z - 1} + \frac{A_2 z}{z - 0.5 - j0.5} + \frac{\tilde{A}_2 z}{z - 0.5 + j0.5} \qquad (10\text{-}61)$$

where $A_1 = 4$, $A_2 = 1.581139 \underline{/-161.5651°}$, and $\tilde{A}_2 = 1.581139 \underline{/161.5651°}$. Inversion and subsequent simplification result in

$$y(n) = 4 + 3.162278(0.707107)^n \cos(45n° - 161.5651°) \qquad (10\text{-}62)$$

Power Series Expansion

The second method that we will consider for inverting z-transforms is the power series method. This method is particularly useful when the inverse transform has no simple closed-form solution or when it is desired to represent the signal as a sequence of numbers defined at sample points. The key to this approach is the basic definition of the z-transform as given by Eq. (10-25). A given transform is manipulated to yield a power series of the appropriate form. The values of the signal at sample points are then read directly from the coefficients of the terms in the power series. The pertinent power series may be obtained by dividing the numerator polynomial by the denominator polynomial both arranged in descending powers of z. The process will be illustrated with an example.

Example 10-5

Using a power series expansion, determine several terms in the inverse transform of the function of Example 10-3.

Solution   The two binomials in the denominator of Eq. (10-51) are first multiplied together to yield a single polynomial. The function $Y(z)$ is then expressed as

$$Y(z) = \frac{z^2}{z^2 - 1.5z + 0.5} \tag{10-63}$$

The power series is obtained by a division process as follows:

$$
\begin{array}{r}
1 + 1.5z^{-1} + 1.75z^{-2} + 1.875z^{-3} + \cdots \\
z^2 - 1.5z + 0.5 \overline{\smash)z^2 } \\
\underline{z^2 - 1.5z + 0.5} \\
1.5z - 0.5 \\
\underline{1.5z - 2.25 + 0.75z^{-1}} \\
1.75 - 0.75z^{-1} \\
\underline{1.75 - 2.625z^{-1} + 0.875z^{-2}} \\
1.875z^{-1} - 0.875z^{-2}
\end{array}
\tag{10-64}
$$

The first few terms of the series may then be written as

$$Y(z) = 1 + 1.5z^{-1} + 1.75z^{-2} + 1.875z^3 + \cdots \tag{10-65}$$

By inspection, we obtain the following values:

n	0	1	2	3
$y(n)$	1	1.5	1.75	1.875

These results are in agreement with those of Example 10-3.

Example 10-6

Determine the inverse transform of

$$Y(z) = 1 + 5z^{-1} - 3z^{-2} + 2z^{-4} \qquad (10\text{-}66)$$

Solution   This transform is already in the form of a power series. Note that this corresponds to a signal of finite length having only a few specified points. The values are tabulated below.

n	0	1	2	3	4
$y(n)$	1	5	-3	0	2

At all other sample points, $y(n) = 0$.

10-9 RESPONSE FORMS AND STABILITY

We will not investigate the different forms associated with the response terms of discrete-time systems. It will be seen that this development closely parallels the corresponding situation for a continuous-time system as given in Section 7-6.

It has previously been shown that the transfer function of a discrete-time system can be expressed in the form

$$H(z) = \frac{a_0(z - z_1)(z - z_2) \cdots (z - z_k)}{(z - p_1)(z - p_2) \cdots (z - p_k)} \qquad (10\text{-}67)$$

When the system is excited by an arbitrary signal $x(n)$, the transform of the output signal $y(n)$ is obtained by multiplying $H(z)$ by $X(z)$. The poles contained in $Y(z)$ may result from two sources: poles due to the transfer function $H(z)$, and poles due to the input $X(z)$.

The *natural response* is defined as that portion of the response due to the poles of $H(z)$. The *forced response* is defined as that portion of the response due to the poles of $X(z)$. If the natural response vanishes after a sufficiently long time, it is called a *transient response*. In this case only the forced response remains, and it is called a steady-state response. In order for this latter condition to exist, the system must be *stable*.

As in the case of a continuous-time system, a discrete-time system is said to be stable if every finite input produces a finite output. The stability concept may be readily expressed by conditions relating to the impulse response $h(n)$. These conditions are:

1. *Stable system.* A discrete-time system is stable if $h(n)$ vanishes after a sufficiently long time.
2. *Unstable system.* A discrete-time system is unstable if $h(n)$ grows without bound after a sufficiently long time.

3. *Marginally stable system.* A discrete-time system is marginally stable if $h(n)$ approaches a constant nonzero value or a bounded oscillation after a sufficiently long time.

Stability may be determined directly from the transfer function if the poles are given. A complex pole of the form $p_m = |p_m|\underline{/\phi_m}$ can be thought of as producing one or more time response terms of the form

$$y_m(n) = An^r p_m^n \tag{10-68}$$

It can be shown that the only condition required for $y_m(n)$ to eventually vanish is that $|p_m| < 1$. Similarly, if $|p_m| > 1$, then $y_m(n)$ will grow without bound. If $|p_m| = 1$ and $r = 0$ (corresponding to a first-order pole), $y_m(n)$ will be either a constant ($p_m = 1$) or a constant amplitude oscillation. However, if $|p_m| = 1$ and $r > 0$ (corresponding to a multiple-order pole), $y_m(n)$ will grow without bound.

A summary of the preceding points and a few other inferences follow:

1. Poles of a discrete transfer function inside the unit circle represent stable terms regardless of their order.
2. Poles of a discrete transfer function outside the unit circle represent unstable terms regardless of their order.
3. First-order poles on the unit circle represent marginally stable terms, but multiple-order poles on the unit circle represent unstable terms.
4. A discrete system is only as stable as its least stable part. Thus all poles of a perfectly stable system must lie *inside* the unit circle.
5. In general, zeros are permitted to lie anywhere in the z-plane.

Example 10-7

A system is described by the difference equation

$$y(n) + 0.1y(n - 1) - 0.2y(n - 2) = x(n) + x(n - 1) \tag{10-69}$$

(a) Determine the transfer function $H(z)$ and discuss its stability.
(b) Determine the impulse response $h(n)$.
(c) Determine the response due to a unit step function excitation if the system is initially relaxed.

    Solution (a) Taking the z-transforms of both sides of Eq. (10-69) and solving for $H(z)$, we obtain

$$H(z) = \frac{Y(z)}{X(z)} = \frac{1 + z^{-1}}{1 + 0.1z^{-1} - 0.2z^{-2}} \tag{10-70}$$

The poles and zeros are best obtained by momentarily arranging numerator and denominator polynomials in positive powers of z.

$$H(z) = \frac{z^2 + z}{z^2 + 0.1z - 0.2} = \frac{z(z + 1)}{(z - 0.4)(z + 0.5)} \tag{10-71}$$

The poles are located at $+0.4$ and -0.5, which are inside the unit circle. Thus, the system is stable.

(b) The impulse response may be obtained by expanding $H(z)$ in a partial fraction expansion according to the procedure of the preceding section. This yields

$$H(z) = \frac{1.555556z}{z - 0.4} - \frac{0.555556z}{z + 0.5} \tag{10-72}$$

Inversion of Eq. (10-72) yields

$$h(n) = 1.555556(0.4)^n - 0.555556(-0.5)^n \tag{10-73}$$

It can be readily seen that the impulse response $h(n)$ vanishes after a sufficiently long time, as expected, since this is a stable transfer function.

(c) To obtain the response due to $x(n) = 1$, we multiply $X(z)$ by $H(z)$ and obtain

$$Y(z) = \frac{z^2(z + 1)}{(z - 1)(z - 0.4)(z + 0.5)} \tag{10-74}$$

Partial fraction expansion yields

$$Y(z) = \frac{2.222222z}{z - 1} - \frac{1.037037z}{z - 0.4} - \frac{0.185185z}{z + 0.5} \tag{10-75}$$

The inverse transform is

$$y(n) = 2.222222 - 1.037037(0.4)^n - 0.185185(-0.5)^n \tag{10-76}$$

10-10 DISCRETE-TIME CONVOLUTION

An alternative approach for relating the input and output of a discrete-time system is through the convolution concept. We will choose to develop this concept through a rather intuitive approach, which provides some insight into the process itself. Assume that a given discrete-time system has an impulse $h(n)$. This means that an impulse (unit sample) occurring at $n = 0$ will produce a response $h(n)$. A delayed impulse $\delta(n - m)$ occurring at $n = m$ will produce a delayed response $h(n - m)$. The discrete-time input signal can be thought of as an impulse train in which each successive impulse has a weight equal to that particular sample value. The forms for the various impulses and the responses they produce can be outlined as follows:

$$
\begin{aligned}
x(0)\delta(n) &\longrightarrow x(0)h(n) \\
x(1)\delta(n - 1) &\longrightarrow x(1)h(n - 1) \\
x(2)\delta(n - 2) &\longrightarrow x(2)h(n - 2) \\
&\vdots \\
x(m)\delta(n - m) &\longrightarrow x(m)h(n - m)
\end{aligned}
\tag{10-77}
$$

In general, the response at any arbitrary value of n is obtained by summing all the components that have occurred up to that point, that is,

$$y(n) = \sum_{m=0}^{n} x(m)h(n - m) \qquad (10\text{-}78)$$

The convolution operation can be shown to be commutative, which means that Eq. (10-78) can be expressed as

$$y(n) = \sum_{m=0}^{n} h(m)x(n - m) \qquad (10\text{-}79)$$

From earlier work in this chapter, it is known that

$$Y(z) = H(z)X(z) \qquad (10\text{-}80)$$

Performing the z-transformation on both sides of Eq. (10-78) and comparing with Eq. (10-80), operation (ZO-9) of Table 10-2 is readily obtained. In some discrete-time system developments, it is desirable to replace the upper limit on the summations of Eqs. (10-78) and (10-79) with ∞. This change in notation does not affect the value of the summation since $h(n - m) = 0$ for $n < m$.

The convolution approach represents an alternative technique for analyzing a discrete system or for signal processing as compared with the direct difference equation approach. With the convolution approach, the values of $h(n)$ may be stored in the system memory. As the samples of the input signal enter the system, the operation of Eq. (10-78) or (10-79) is performed to yield successive output samples.

The direct convolution approach to signal processing, as discussed in this section, is used primarily when the impulse response is relatively short in duration. Otherwise, the number of operations required to compute each new value of $y(n)$ will become excessive.

GENERAL PROBLEMS

10-1. A certain digital signal processing system utilizes 8-bit words. Determine the number of possible levels than can be encoded.

10-2. A certain digital signal processing system utilizes 16-bit words. Determine the number of possible levels that can be encoded.

10-3. In a certain digital signal processing system, it is desired to encode no less than 2000 levels. Determine the minimum number of bits required.

10-4. In a certain digital signal processing system, it is desired to encode no less than 10,000 levels. Determine the minimum number of bits required.

10-5. A certain signal $x(t)$ is sampled by a narrow pulse train as shown in Fig. 10-3. The signal is periodic and contains only two frequencies: 100 Hz and 300 Hz. The pulse train has a frequency of 1 kHz. List all positive frequencies in the spectrum of $x^*(t)$ between dc and 4.5 kHz.

10-6. A certain signal $x(t)$ is sampled by an impulse train as shown in Fig. 10-5. The signal is periodic and contains three frequencies: 1 kHz, 2 kHz, and 4 kHz. The impulse train has a frequency of 12 kHz. List all positive frequencies in the spectrum of $x^*(t)$ between dc and 42 kHz.

10-7. A certain signal is bandlimited from dc to 4 kHz. It is to be sampled and converted to digital form. If the sampling rate is chosen to be 25% greater than the theoretical minimum, determine the **(a)** sampling rate f_s and **(b)** interval T between successive samples.

10-8. A certain signal is bandlimited from dc to 800 Hz. It is to be sampled and converted to digital form. If the sampling rate is chosen to be 50% larger than the theoretical minimum, determine the **(a)** sampling rate f_s and **(b)** interval T between successive samples.

10-9. A certain algorithm is described by the difference equation

$$y(n) = 0.8y(n + 1) + 0.5x(n) - 0.7x(n - 1)$$

Determine the transfer function.

10-10. A certain algorithm is described by the difference equation

$$y(n) = 0.4y(n - 1) + 0.2y(n - 2) + 0.5x(n) - 0.8x(n - 2)$$

Determine the transfer function

10-11. A certain transfer function is given by

$$H(z) = \frac{Y(z)}{X(z)} = \frac{2 + 3z^{-1}}{1 - 0.5z^{-1}}$$

Write the algorithm for generating $y(n)$.

10-12. A certain transfer function is given by

$$H(z) = \frac{Y(z)}{X(z)} = \frac{z^2 + 2z - 3}{z^2 - 0.5z + 0.8}$$

Write the algorithm for generating $y(n)$.

10-13. Determine the inverse transform of

$$X(z) = 5 + 2z^{-1} + 4z^{-3}$$

10-14. Determine the inverse transform of

$$X(z) = 2 + 4z^{-1} + 5z^{-2} + 6z^{-4}$$

10-15. By the simplest procedure, determine the first three terms of the inverse transform of

$$Y(z) = \frac{6z^2 + 13z + 4}{2z^2 + z + 4}$$

10-16. By the simplest procedure, determine the first three terms of the inverse transform of

$$Y(z) = \frac{3z^3 + 2z^2 + 2z + 5}{z^3 + 4z^2 + 3z + 2}$$

10-17. Using partial fraction expansion, determine the inverse transform of the function

$$Y(z) = \frac{10}{(1 - 0.2z^{-1})(1 - 0.4z^{-1})}$$

10-18. Using partial fraction expansion, determine the inverse transform of the function

$$Y(z) = \frac{20z^2}{(z - 0.4)(z - 1)}$$

10-19. Using partial fraction expansion, determine the inverse transform of the function

$$Y(z) = \frac{10}{(1 + z^{-1})(1 + 0.5z^{-2})}$$

10-20. Using partial fraction expansion, determine the inverse transform of the function

$$Y(z) = \frac{40z^2}{z^2 + 0.6z + 0.08}$$

10-21. A system is described by the difference equation

$$y(n) - 0.5y(n - 1) = x(n) + 0.5x(n - 1)$$

Determine the **(a)** transfer function and **(b)** impulse response. **(c)** Determine if the system is stable.

10-22. A system is described by the difference equation

$$y(n) + 0.5y(n - 1) = x(n) - 0.5x(n - 1)$$

Determine the **(a)** transfer function and **(b)** impulse response. **(c)** Determine if the system is stable.

10-23. The system of Problem 10-21 is excited by the step function $x(n) = 10u(n)$. Determine the response $y(n)$.

10-24. The system of Problem 10-22 is excited by the step function $x(n) = 10u(n)$. Determine the response $y(n)$.

10-25. A zero-order integrator is described by the algorithm

$$y(n) = y(n - 1) + Tx(n - 1)$$

(a) Determine the transfer function.
(b) Show that the system is marginally stable.

10-26. A first-order trapezoidal integrator is described by the algorithm

$$y(n) = \frac{T}{2}[x(n) + x(n - 1)] + y(n - 1)$$

(a) Determine the transfer function.
(b) Show that the system is marginally stable.

10-27. The integrator of Problem 10-25 is excited by the exponential input

$$x(n) = \epsilon^{-n\alpha T}$$

(a) Determine the response $y(n)$.
(b) Perform the definite integral of $\epsilon^{-\alpha t}$ from 0 to t and compare with the result of (a) for $t = nT$.

10-28. Repeat Problem 10-27 as applied to the integrator of Problem 10-26.

10-29. One type of first-order, low-pass, discrete-time (or digital) filter is achieved with the algorithm

$$y(n) = Tx(n) + Ky(n - 1)$$

where K is a positive constant in the range $0 < K < 1$.
(a) Determine the transfer function $H(z)$.
(b) Show that the transfer function is stable.

10-30. One type of first-order, low-pass, discrete-time (or digital) filter is achieved with the algorithm

$$Y(n) = C[x(n) + x(n-1)] + KY(n-1)$$

where C and K are both positive constants in the range $0 < C < 1$ and $0 < K < 1$.
(a) Determine the transfer function $H(z)$.
(b) Show that the transfer function is stable.

10-31. Determine the degree of stability of the transfer function

$$H(z) = \frac{3z + 6}{8z^2 - 2z - 3}$$

10-32. Determine the degree of stability of the transfer function

$$H(z) = \frac{4(1 - z^{-1} + z^{-2})}{2 + 5z^{-1} + 2z^{-2}}$$

10-33. Determine the degree of stability of the transfer function

$$H(z) = \frac{4z^2 + 8}{2z^2 + z - 1}$$

10-34. Determine the degree of stability of the transfer function

$$H(z) = \frac{2(1 + z^{-1})}{1 - z^{-1} + z^{-2}}$$

DERIVATION PROBLEMS

10-35. Derive the z-transform of the exponential function as given by (ZT-4).

10-36. Derive the z-transform of the alternate form of the exponential function as given by (ZT-5).

10-37. Starting with the z-transform of the step function as given by (ZT-2), apply operation pair (ZO-5) to derive the z-transform of nT (ZT-3).

10-38. Derive transform pairs (ZT-6) and (ZT-7) from (ZT-4) with the help of Euler's equation.

10-39. Derive operation pair (ZO-3).

10-40. Derive operation pair (ZO-4).

10-41. Derive $x(n)$ from $X(z)$ for transfer pair (ZT-4) from a power series expansion.

10-42. Derive $x(n)$ from $X(z)$ for transform pair (ZT-5) from a power series expansion.

COMPUTER PROBLEMS

10-43. Assume a first-order difference equation in which the algorithm for generating $y(n)$ is of the form

$$y(n) = a_0 x(n) + a_1 x(n-1) - b_1 y(n-1)$$

Write a computer program for generating $y(n)$ in the range $0 \leq n \leq N$.

Input data: $a_0, a_1, b_1, N, X(n)$
Output data: $n, y(n)$

10-44. Assume a second-order difference equation in which the algorithm for generating $y(n)$ is of the form

$$y(n) = a_0x(n) + a_1x(n-1) + a_2x(n-2) - b_1y(n-1) - b_2y(n-2)$$

Write a computer program for generating $y(n)$ in the range $0 \leq n \leq N$.

Input data: $a_0, a_1, a_2, b_1, b_2, N, x(n)$
Output data: $n, y(n)$

10-45. Write a program to perform discrete-time convolution as given by Eq. (10-78). The program should generate $y(n)$ in the range $0 \leq n < 2N$

Input data: $N, h(n),$ and $x(n)$ for $0 \leq n \leq N$
Output data: n and $y(n)$ for $0 \leq n \leq 2N$

Appendix A

COMPLEX
ALGEBRA

OVERVIEW

The use of complex algebra greatly facilitates the solution of many circuit problems involving both the steady-state sinusoidal approach and the general Laplace transform approach. The reader who is acquainted with steady-state ac circuit theory will undoubtedly be familiar with this subject, and this treatment will serve as a review when necessary. The reader not familiar with ac circuit theory should find this treatment adequate to deal with the complex number operations involved in transform analysis.

A-1 COMPLEX NUMBERS

The concept of a complex number system is derived from a two-dimensional representation as shown in Fig. A-1. In this representation we define a complex number

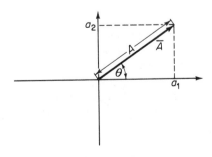

Figure A-1 Representation of the complex number system.

$\bar{A}$ as

$$\bar{A} = a_1 + ja_2 \qquad (A\text{-}1)$$

where $j = \sqrt{-1}$.

The quantity a_1 is called the *real part* of $\bar{A}$, and the quantity a_2 is called the *imaginary part* of $\bar{A}$. Of course, both a_1 and a_2 are themselves real numbers. The real part a_1 is read along the x-axis, and the x-axis is called the *real axis*. The imaginary part a_2 is read along the y-axis, and the y-axis is called the *imaginary axis*. Complex numbers are often called *phasors* when applied to ac circuit theory. The bar above $\bar{A}$ emphasizes that the quantity is complex.

Thus, we may represent a complex number by specifying its real and imaginary parts, two quantities being required to completely specify the number. In the event that the imaginary part is zero, the locus of all complex numbers is the real axis. Thus the real axis represents the domain of all real numbers in the complex-number system. The form of a complex number in which the real and imaginary parts are specified is called the *rectangular* or *cartesian* form of a complex number.

Referring back to Fig. A-1, we observe that if the complex number is interpreted as being a phasor or vector from the origin to the point (a_1, a_2), we can also describe the complex number by specifying its *magnitude A* and its angle (also called the *argument*) with respect to the x-axis. (Positive angles are interpreted in a counterclockwise rotational direction.) From the geometry, it can be seen that

$$a_1 = A \cos \theta \qquad (A\text{-}2)$$

$$a_2 = A \sin \theta \qquad (A\text{-}3)$$

Substitution of Eqs. (A-2) and (A-3) into Eq. (A-1) yields

$$\begin{aligned} \bar{A} &= A \cos \theta + jA \sin \theta \\ &= A(\cos \theta + j \sin \theta) \end{aligned} \qquad (A\text{-}4)$$

The quantity in parentheses above can be related to a compact expression by means of *Euler's formula*. The proof of this formula is given in most basic calculus books. Euler's formula states that

$$\epsilon^{j\theta} = \cos \theta + j \sin \theta \qquad (A\text{-}5)$$

Substitution of Eq. (A-5) into Eq. (A-4) yields

$$\bar{A} = A\epsilon^{j\theta} \qquad (A\text{-}6)$$

This form of a complex number is called the *polar form*, in which the magnitude A and the angle θ are directly specified. The beauty of the exponential form is that it obeys all the algebraic laws of exponential functions, and as we shall see shortly, this form is ideal for multiplying and dividing complex numbers.

Now let us consider a shorthand way of writing the polar form of complex numbers that is universally employed in electrical theory. Referring back to Eq. (A-6), we define the following notation:

$$\bar{A} = A\epsilon^{j\theta} \stackrel{\text{def}}{=} A \underline{/\theta} \qquad (A\text{-}7)$$

In subsequent work, we will use the form defined by Eq. (A-7) extensively.

To summarize, a complex number may be expressed in either rectangular or polar form

$$\bar{A} = a_1 + ja_2 \tag{A-8}$$

or

$$\bar{A} = A\epsilon^{j\theta} \stackrel{\text{def}}{=} A \underline{/\theta} \tag{A-9}$$

Conversion from rectangular to polar form is achieved by the relationships

$$A = \sqrt{a_1^2 + a_2^2} \tag{A-10}$$

$$\theta = \tan^{-1}\frac{a_2}{a_1} \tag{A-11}$$

Conversion from polar to rectangular form is achieved by the relationships

$$a_1 = A\cos\theta \tag{A-12}$$

$$a_2 = A\sin\theta \tag{A-13}$$

Example A-1

Convert each of the following numbers from rectangular to polar form
    (a) $1 + j\sqrt{3}$ (b) $-1 + j\sqrt{3}$ (c) $-1 - j\sqrt{3}$ (d) $1 - j\sqrt{3}$
    Solution   The basic angle associated with a base of unity and an altitude $\sqrt{3}$ is $\pi/3$ or $60°$. Strictly speaking, the angle should be expressed in radians when placed in the exponential polar form. However, it is a common practice to express the angle in degrees, and we will maintain this practice. Inspection of the signs of the real and imaginary parts of the four numbers results in the answers that follow. Note that (c) and (d) are expressed both in positive and negative angles, since the use of the negative angles results in the smallest possible angle.
    (a) $1 + j\sqrt{3} = 2\underline{/60°}$
    (b) $-1 + j\sqrt{3} = 2\underline{/120°}$
    (c) $-1 - j\sqrt{3} = 2\underline{/240°} = 2\underline{/-120°}$
    (d) $1 - j\sqrt{3} = 2\underline{/300°} = 2\underline{/-60°}$
The reader may find it instructive to convert the polar forms obtained back to rectangular forms.

A-2 OPERATIONS WITH COMPLEX NUMBERS

Let us now consider the basic arithmetic operations applied to complex numbers. For reference in subsequent operations, let us define two complex numbers $\bar{A}$ and

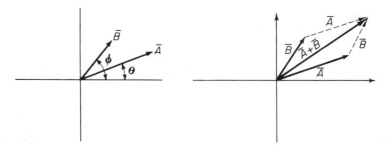

Figure A-2 Two complex numbers or phasors.

Figure A-3 Addition of complex numbers.

$\bar{B}$, where

$$\bar{A} = a_1 + ja_2 = A\,\underline{/\theta} \tag{A-14}$$

$$\bar{B} = b_1 + jb_2 = B\,\underline{/\phi} \tag{A-15}$$

as shown in Fig. A-2.

Addition

The sum of two complex numbers is given by

$$\bar{A} + \bar{B} = a_1 + b_1 + j(a_2 + b_2) \tag{A-16}$$

According to Eq. (A-16), to add complex numbers, we simply add the real parts and the imaginary parts separately. Geometrically, this operation may be represented as shown in Fig. A-3, in which one of the phasors is "picked up" and placed on the end of the other phasor. The sum is the resultant phasor between the origin and the terminal point of the second phasor. Thus, complex numbers may be added analytically by Eq. (A-16) or by the graphical procedure just discussed, if desired.

Subtraction

The difference between two complex numbers is given by

$$\bar{A} - \bar{B} = a_1 - b_1 + j(a_2 - b_2) \tag{A-17}$$

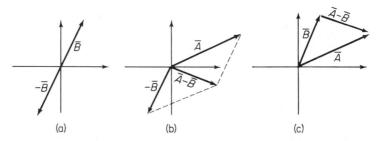

(a)   (b)   (c)

Figure A-4 Subtraction of complex numbers.

Since, in an algebraic sense, subtraction is considered as the addition of a negative number, a geometric interpretation is obtained by first forming the quantity $-\bar{B}$ as shown in Fig. A-4a. This is equivalent to rotating the phasor by 180°. Now we add $\bar{A}$ and $-\bar{B}$ as shown in Fig. A-4b. An alternative geometric procedure is to keep the original orientation as shown in Fig. A-4c and to construct the difference phasor between the terminal points of $\bar{A}$ and $\bar{B}$. This phasor is displaced from the origin, but its magnitude and angle are preserved.

Multiplication

Multiplication of two complex numbers is best achieved by means of the polar forms. We have

$$\bar{A} \times \bar{B} = (A\underline{/\theta})(B\underline{/\phi})$$
$$= AB\underline{/\theta + \phi} \tag{A-18}$$

Thus to form the product of two complex numbers, we *multiply the magnitudes and add the angles.* The geometric interpretation is shown in Fig. A-5.

We may also multiply two complex numbers in rectangular form by expanding the products and utilizing the fact that $j^2 = -1$. We thus have

$$\bar{A} \times \bar{B} = (a_1 + ja_2)(b_1 + jb_2) = a_1 b_1 - a_2 b_2 + j(a_1 b_2 + a_2 b_1) \tag{A-19}$$

The reader should not try to remember the result expressed by Eq. (A-19) as this can be easily worked out in each case when numbers are given.

Division

Division of two complex numbers is also best achieved by means of the polar forms. We have

$$\frac{\bar{A}}{\bar{B}} = \frac{A\underline{/\theta}}{B\underline{/\phi}} = \frac{A}{B}\underline{/\theta} \tag{A-20}$$

Thus, to divide two complex numbers, we *divide the magnitudes and subtract the angles.* The geometric interpretation is shown in Fig. A-6.

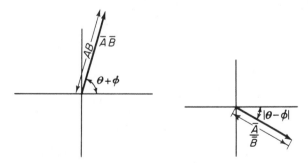

Figure A-5 Multiplication of complex numbers.

Figure A-6 Division of complex numbers.

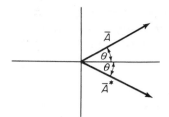

Figure A-7 Illustration of the complex conjugate.

Division may also be accomplished in rectangular form by use of the complex conjugate concept. The complex conjugate of a vector $\bar{A}$ is designated by $\tilde{\bar{A}}$. If $\bar{A}$ is given by

$$\bar{A} = a_1 + ja_2 = A\underline{/\theta} \tag{A-21}$$

then

$$\tilde{\bar{A}} = a_1 - ja_2 = A\underline{/-\theta} \tag{A-22}$$

as shown in Fig. A-7.

The product of $\bar{A}$ and $\tilde{\bar{A}}$ is given by

$$\bar{A}\tilde{\bar{A}} = A^2 \tag{A-23}$$

Thus the product of a complex number and its conjugate is the square of the magnitude of the number.

In dividing with rectangular forms, an expression of the following form appears:

$$\frac{\bar{A}}{\bar{B}} = \frac{a_1 + ja_2}{b_1 + jb_2} \tag{A-24}$$

If we now multiply both numerator and denominator by the complex conjugate of the denominator, we reduce the denominator to a real number. Hence

$$\frac{\bar{A}}{\bar{B}} = \frac{(a_1 + ja_2)}{(b_1 + jb_2)} \times \frac{(b_1 - jb_2)}{(b_1 - jb_2)} = \frac{a_1 b_1 + a_2 b_2 + j(a_2 b_1 - a_1 b_2)}{b_1^2 + b_2^2}$$

$$= \frac{a_1 b_1 + a_2 b_2 + j(a_2 b_1 - a_1 b_2)}{B^2} \tag{A-25}$$

Example A-2

Given:

$$\bar{A} = 3 + j4 = 5\underline{/53.13°}$$

$$\bar{B} = -2 + j2 = 2.828\underline{/135°}$$

$$\bar{C} = 2 - j1 = 2.236\underline{/-26.57°}$$

Determine (a) $\bar{D} = \bar{A} + \bar{B} - \bar{C}$, and (b) $\bar{E} = \dfrac{\bar{A} \times \bar{C}}{\bar{B}}$ (two ways)

Solution (a) Application of the rules for addition and subtraction yields

$$\bar{D} = \bar{A} + \bar{B} - \bar{C} = (3 + j4) + (-2 + j2) - (2 - j1) = -1 + j7 \quad \text{(A-26)}$$

(b) Application of the rules for multiplication and division in polar form yields

$$\bar{E} = \frac{\bar{A} \times \bar{C}}{\bar{B}} = \frac{(5\underline{/53.13^\circ})(2.236\underline{/-26.57^\circ})}{(2.828\underline{/135^\circ})} \quad \text{(A-27)}$$

$$= 3.953\underline{/-108.44^\circ}$$

To perform these operations in rectangular form, we proceed as follows:

$$\bar{E} = \frac{(3 + j4)(2 - j1)}{(-2 + j2)} = \frac{10 + j5}{-2 + j2}$$

$$= \frac{(10 + j5)(-2 - j2)}{(-2 + j2)(-2 - j2)} = \frac{-10 - j30}{8} = -1.25 - j3.75 \quad \text{(A-28)}$$

Notice that multiplication and division in polar form resulted in $\bar{E}$ being expressed in polar form, whereas the same operation performed in rectangular form resulted in $\bar{E}$ being expressed in rectangular form. It can be readily shown that the two results are identical.

Appendix B

NORMALIZATION

OVERVIEW

The solution of network problems involving realistic practical component values can often be a tedious chore if the component values are very large or very small. However, due to the principle of linearity, the element values can often be scaled or normalized to yield a simpler combination of component values, thus facilitating the numerical operations involved in solving the circuit. Of course, when the desired solution is obtained, it must be rescaled or denormalized to obtain the correct solution.

For very simple problems, there is very little to be gained by normalizing the element values. The problem must require a reasonable degree of numerical computation to warrant the extra effort required to normalize and denormalize the quantities. Also, normalization is, to some extent, a personal matter in the sense that some people are quite enthusiastic about it, whereas others do not seem to use it very often. For that reason, we have chosen to include it here in an appendix rather than in the main text.

There are three types of normalization that may be performed in a circuit: (a) source-amplitude-level normalization, (b) resistance or impedance-level normalization, and (c) frequency or time scale normalization. Let us consider each of these techniques separately.

B-1 SOURCE-AMPLITUDE-LEVEL NORMALIZATION

Source-amplitude-level normalization is the simplest and most intuitive of the three types of normalization and is a direct consequence of the linearity principle. If, for some reason, we desire to scale the amplitudes of all the sources, the amplitudes of all the responses (voltage and currents) are scaled by the same factor as the sources. Thus if we multiply the values of all sources by a constant K_s, the responses in the normalized network are K_s times the actual responses. To obtain the actual responses, we must multiply the responses obtained from the normalized circuit by $1/K_s$.

The only cautious point to make here is that *all sources should be multiplied by the same constant*. Source-level normalization is often used in conjunction with other types of normalization, and we will postpone consideration of an example until we consider resistance-level normalization.

B-2 IMPEDANCE- OR RESISTANCE-LEVEL NORMALIZATION

Quite often, the impedance level of a practical circuit is such that an analysis involves rather small or rather large numbers. In such a case, it is frequently desirable to scale the impedance level to a hypothetical level in which the impedances have values, perhaps, in the neighborhood of a few ohms.

The impedance of a resistor is simply the value of the resistance, the impedance of an inductor is directly proportional to the inductance, and the impedance of a capacitor is inversely proportional to the capacitance. These properties lead to the following set of rules for impedance or resistance level normalization:

To multiply the impedance or resistance level of a network by a constant K_r, element values are modified as follows:

1. Multiply resistances by K_r.
2. Multiply inductances by K_r.
3. Divide capacitances by K_r.

The normalized circuit obtained from this process has the property that all impedances defined in the network are K_r times the corresponding impedances in the original network. On the other hand, the frequency and time scales are not affected by this process. *Note that it is necessary to modify all element values in the circuit by the proper factor*.

The effect of this normalization on voltages and currents can be summarized as follows:

1. If *all* sources are *voltage sources*, the currents in the normalized network are $1/K_r$ times the actual currents in the real network. Therefore, currents solved from the scaled network must be multiplied by K_r to yield the actual currents. However, since voltages in the scaled network divide in the same proportion as in the actual network, the values obtained from the normalized circuit are the same as in the actual circuit.
2. If *all* sources are *current sources*, the currents in the normalized network divide in the same proportion as in the actual network; therefore, the values of currents obtained are the same as in the actual network. On the other hand, the voltages in the normalized network are K_r times the voltages in the actual network. Thus voltages obtained from the scaled network must be multiplied by $1/K_r$ to yield the actual voltages.

3. If the sources are mixed between voltage and current sources, usually they may all be converted to the same type of source before normalizing, in which case, item 1 or 2 will be applicable, depending on the type of sources obtained.

Example B-1

The resistive circuit shown in Fig. B-1a is given. Normalize both the source level and the resistance level so that the current source is 2 A and the 100-kΩ resistor is 1 Ω.

  Solution The source level is multiplied by

$$K_s = 100 \tag{B-1}$$

The resistance level is multiplied by

$$K_r = 10^{-5} \tag{B-2}$$

The normalized circuit is shown in Fig. B-1b.

  Suppose, for example, that we are interested in the voltage and current associated with the original 200-kΩ resistor. The normalized quantities are shown in Fig. B-1b with the primes emphasizing that these are normalized quantities. An analysis yields

$$i'_0 = 0.4 \text{ A} \tag{B-3}$$

$$v'_0 = 0.8 \text{ V} \tag{B-4}$$

  To obtain the actual values, we reason as follows: Since the single source is a current source, the resistance normalization did not affect the current. However, the source-level normalization did multiply the current by 100. Thus

$$i_0 = \frac{i'_0}{100} = 0.004 \text{ A} = 4 \text{ mA} \tag{B-5}$$

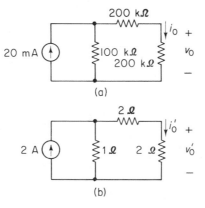

(a)

(b)

Figure B-1 Circuits of Example B-1.

On the other hand, the voltage was affected by both scale factors. Resistance normalization made the normalized voltage lower by a factor 10^{-5}, and source normalization made the normalized voltage higher by a factor 10^2. Thus

$$v_0 = \frac{10^5}{10^2} \times v_0' = 10^3 v_0' = 800 \text{ V} \tag{B-6}$$

The reader might note that Ohm's law is certainly satisfied with v_0, i_0, and the original 200-kΩ resistor. This particular problem was chosen primarily for illustration, as very little was gained from normalization in this case.

Example B-2

The filter circuit shown in Fig. B-2a is given. Normalize the impedance level of the circuit so that the right-hand resistor becomes 1 Ω.

Solution   The normalization factor is

$$K_r = \frac{1}{500} = 0.002 \tag{B-7}$$

The resistances and inductance are multiplied by 0.002, whereas the capacitances are divided by 0.002 (or multiplied by 500). The normalized network is shown in Fig. B-2b.

The resistance values now have "nice" values, but the reactive components do not. This problem will be continued after a discussion of frequency normalization in the next section. At that time we will also normalize the frequency scale to obtain simple values for the reactive components.

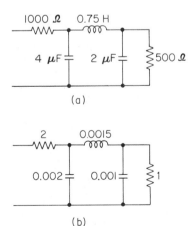

(a)

(b)

Figure B-2   Circuits of Example B-2.

B-3 FREQUENCY OR TIME-SCALE NORMALIZATION

The last type of normalization to consider is that of scaling the frequency or the time scale for purposes of convenience. In this sense, we are interpreting the term frequency in a general sense to mean the rate of variation of the time function. In a special case, this may apply to a steady-state sinusoidal repetition frequency, or it may apply to the "frequency" of a more general time-varying response.

Let us illustrate the general concept of a frequency or time scale normalization by means of the exponential function shown in Fig. B-3a. This function is

$$v(t) = A\epsilon^{-1000t} = A\epsilon^{-t/0.001} \tag{B-8}$$

in which the time constant is $\tau = 0.001$ s. Suppose that we wish to change the time scale by an appropriate factor such that the time constant of the normalized response is 1 s. This is equivalent to a normalized function

$$v'(t') = A\epsilon^{-t'} \tag{B-9}$$

where v' represents the normalized voltage and t' represents the normalized time scale. This function is shown in Fig. B-3b and is, of course, identical to the unnormalized function except for the time scale. The new time scale is related to the original time scale by

$$t' = 1000t \tag{B-10}$$

or

$$t = \frac{t'}{1000} \tag{B-11}$$

If considerable computation is necessary, the primes may be dropped from the normalized time scale, so long as we keep in mind exactly what we are doing. In this sense, we can interpret this normalization as *replacing t* by $t/1000$ to accomplish

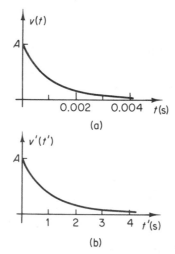

(a)

(b)

Figure B-3 Time scale of normalization.

the desired purpose. We shall use this latter interpretation in the remainder of this section.

Note that in this example we have, in a sense, multiplied the time scale by 1000 since it takes the normalized function 1000 times as long to reach a given value as before. On the other hand, we have, in a sense, multiplied the "frequency" scale of this function by 1/1000 since the normalized function varies only 1/1000 as rapidly as the original function.

Thus, the concept of time scale and frequency scale normalization are essentially identical but are inversely related. Therefore, if we define a time scale normalization factor by K_t and the corresponding frequency scale normalization factor by K_f, it follows that

$$K_f = \frac{1}{K_t} \qquad (B-12)$$

Therefore, it is not really necessary to consider them as separate procedures, but as different ways of looking at the same procedure. However, for the sake of simplicity, we will tabulate the procedures separately later in this section.

Before tabulating these procedures, let us observe the effect of the normalization on the network response. Referring again to the exponential function of Fig. B-3, we observe that the behavior of a network to the normalized exponential function may be entirely different than the response to the original excitation. The desired result from normalization is that the response on the new time scale must have the identical shape in terms of the new time scale as the original function on the original scale.

We first deduce that the voltage–current relationship for resistors is independent of the frequency or time scale of the waveforms. Thus, *resistors are unchanged by a frequency or time scale normalization.*

On the other hand, due to the derivative and integral relationships for inductors and capacitors, it is necessary to modify the original inductance and capacitance values, to preserve the same relative magnitude of voltages and currents in the normalized circuit. The proofs of the operations that follow can be verified if we choose appropriate examples.

Time Scale Normalization

We define multiplication of the time scale of a given problem as follows:

$$t(\text{normalized}) = K_t \times t(\text{actual}) \qquad (B-13)$$

Sources and element values are modified as follows:

1. The quantity t appearing in all sources is replaced by t/K_t.

2. Capacitances and inductances are multiplied by K_t.

3. Resistances are unchanged.

If any voltage or current time response is obtained from the normalized network, we may obtain the actual response of the original network by replacing t in the normalized response by $K_t t$. On the other hand, if any *transform* or *frequency* response is obtained from the normalized network, we may obtain the actual response by replacing s by s/K_t or ω by ω/K_t.

Frequency Scale Normalization

We define multiplication of the frequency scale of a given problem as follows:

$$\omega(\text{normalized}) = K_f \times \omega(\text{actual}) \tag{B-14}$$

Again we emphasize that this is identical to time scale normalization with

$$K_f = \frac{1}{K_t} \tag{B-15}$$

Sources and element values are modified as follows:

1. The quantity t appearing in all sources is replaced by $K_f t$.
2. Inductances and capacitances are divided by K_f.
3. Resistances are unchanged.

If any voltage or current time response is obtained from the normalized network, we may obtain the actual response of the original network by replacing t in the normalized response by t/K_f. One the other hand, if any *transform* or *frequency* response is obtained from the normalized network, we may obtain the actual response by replacing s by $K_f s$ or ω by $K_f \omega$.

Before looking at some examples, we might consider one point that often leads to error in frequency scale normalization. This point is connected with the concept of angular frequency ω and repetition frequency f. The problem is caused by the fact that normalization frequently is enhanced by the use of some convenient value of ω, say $\omega = 1$ rad/s, as a reference normalization frequency, while the original frequency corresponding to $\omega = 1$ is specified as a repetition frequency f in Hz. In this case, it is usually desirable to convert f to rad/s by multiplying by 2π so that the two frequencies are both angular frequencies. The normalization factor then follows directly.

Example B-3

Referring back to Example B-2 and assuming that the resistance level has been normalized as previously stated, normalize the frequency scale so that an *angular frequency* of 1000 rad/s in the actual network corresponds to an *angular frequency* of 1 rad/s in the normalized network.

Solution   The frequency normalization scale factor K_f is

$$K_f = \frac{1}{1000} \tag{B-16}$$

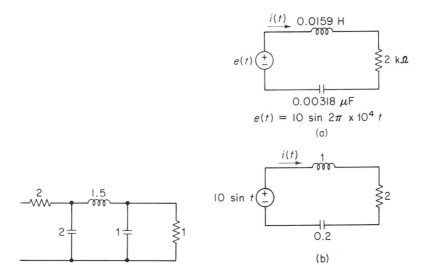

Figure B-4   Circuits of Example B-3.   **Figure B-5**   Circuits of Example B-4.

Inductances and capacitances are divided by K_f, resulting in the normalized network shown in Fig. B-4. The element values are now "nice" values for analysis purposes. This problem has illustrated how a combination of impedance and frequency normalization frequently results in a simple set of component values.

Example B-4

The *RLC* circuit shown in Fig. B-5a is excited at $t = 0$ by a sinusoidal voltage with a magnitude of 10 V and a repetition frequency of 10 kHz. Normalize the frequency or time scale so that the normalized excitation has an angular frequency of 1 rad/s, and normalize the impedance level so that the resistor becomes a 2-Ω resistor.

Solution   The actual *angular frequency* of the source is $\omega = 2\pi \times 10^4$ rad/s which is to be normalized to $\omega = 1$ rad/s. This corresponds to a frequency normalization factor of

$$K_f = \frac{1}{2\pi \times 10^4} \tag{B-17}$$

The resistance level is to be multiplied by

$$K_r = \frac{1}{10^3} \tag{B-18}$$

As a result of the two processes, we multiply the resistance by K_r, the inductance by K_r/K_f, and the capacitance by $1/(K_r K_f)$. The normalized circuit is shown in Fig. B-5b. The reader might wish to show that the solution of the normalized circuit is

$$i(t) = 2.5\epsilon^{-t} \sin(2t - 126.87°) + 2.236 \sin(t + 63.44°) \tag{B-19}$$

To obtain the current response of the actual circuit, we replace t by t/K_f and multiply the normalized current by 10^{-3}. The latter step amounts to expressing the current in mA. Thus the *actual* current in mA is

$$i(t) = 2.5\epsilon^{-2\pi \times 10^4 t} \sin(4\pi \times 10^4 t - 126.87°)$$
$$+ 2.236 \sin(2\pi \times 10^4 t + 63.44°) \tag{B-20}$$

Appendix C

TRANSFORM DERIVATIONS

C-1 LAPLACE TRANSFORM OF A DERIVATIVE

Theorem

$$\mathscr{L}[f'(t)] = sF(s) - f(0) \tag{C-1}$$

Proof. By definition,

$$\mathscr{L}[f'(t)] = \int_0^\infty f'(t)\epsilon^{-st}\,dt \tag{C-2}$$

Integrating by parts, we have

$$u = \epsilon^{-st} \qquad dv = f'(t)\,dt$$
$$du = -s\epsilon^{-st} \qquad v = f(t)$$
$$\mathscr{L}[f'(t)] = f(t)\epsilon^{-st}\Big]_0^\infty + s\int_0^\infty f(t)\epsilon^{-st}\,dt$$
$$= sF(s) - f(0) \tag{C-3}$$

C-2 LAPLACE TRANSFORM OF AN INTEGRAL

Theorem

$$\mathscr{L}\left[\int_0^t f(t)\,dt\right] = \frac{F(s)}{s} \tag{C-4}$$

Proof. By definition,

$$\mathscr{L}\left[\int_0^t f(t)\,dt\right] = \int_0^\infty \left[\int_0^t f(t)\,dt\right]\epsilon^{-st}\,dt \tag{C-5}$$

Integrating by parts, we have

$$u = \int_0^t f(t)\,dt \qquad dv = \epsilon^{-st}\,dt$$

$$du = f(t) \qquad\qquad v = \frac{-\epsilon^{-st}}{s} \tag{C-6}$$

$$\mathscr{L}\left[\int_0^t f(t)\,dt\right] = \frac{-\epsilon^{-st}\int_0^t f(t)\,dt}{s}\Bigg]_0^\infty + \frac{1}{s}\int_0^\infty f(t)\epsilon^{-st}\,dt$$

$$= \frac{F(s)}{s}$$

C-3 LAPLACE TRANSFORM OF A SHIFTED FUNCTION

Theorem

$$\mathscr{L}[f(t-a)u(t-a)] = \epsilon^{-as}F(s) \tag{C-7}$$

Proof. By definition,

$$\mathscr{L}[f(t-a)u(t-a)] = \int_a^\infty f(t-a)\epsilon^{-st}\,dt \tag{C-8}$$

Letting $t - a = u$, Eq. (C-8) becomes

$$\mathscr{L}[f(t-a)u(t-a)] = \int_0^\infty f(u)\epsilon^{-st}\epsilon^{-as}\,du = \epsilon^{-as}\int_0^\infty f(u)\epsilon^{-st}\,du$$

$$= \epsilon^{-as}F(s) \tag{C-9}$$

C-4 LAPLACE TRANSFORM OF $\epsilon^{-\alpha t}f(t)$

Theorem

$$\mathscr{L}[\epsilon^{-\alpha t}f(t)] = F(s+\alpha) \tag{C-10}$$

Proof. By definition,

$$\mathscr{L}[\epsilon^{-\alpha t}f(t)] = \int_0^\infty \epsilon^{-\alpha t}f(t)\epsilon^{-st}\,dt = \int_0^\infty f(t)\epsilon^{-(s+\alpha)t}\,dt \tag{C-11}$$

$$= F(s+\alpha)$$

C-5 INVERSION FORMULA FOR COMPLEX POLES

Theorem.   Assume that $F(s)$ contains a simple quadratic factor of the form $(s + \alpha)^2 + \omega^2$ so that

$$F(s) = \frac{Q(s)}{(s + \alpha)^2 + \omega^2} \tag{C-12}$$

Let $f_1(t)$ represent the time response due to the poles at $-\alpha \pm j\omega$. Then

$$f_1(t) = \frac{M}{\omega} \epsilon^{-\alpha t} \sin (\omega t + \theta) \tag{C-13}$$

where

$$M \underline{/\theta} = Q(-\alpha + j\omega) \tag{C-14}$$

Proof. Partial fraction expansion of $F(s)$ yields

$$F(s) = \frac{K_1}{s + \alpha - j\omega} + \frac{K_1^*}{s + \alpha + j\omega} + R(s) \tag{C-15}$$

where $R(s)$ is the remainder function. Furthermore,

$$K_1 = \frac{Q(-\alpha + j\omega)}{2j\omega} = \frac{M\epsilon^{j\theta}}{2j\omega} \tag{C-16}$$

$$K_1^* = \frac{Q(-\alpha - j\omega)}{-2j\omega} = \frac{M\epsilon^{-j\theta}}{-2j\omega} \tag{C-17}$$

according to the notation of Eq. (C-14). Let $F_1(s)$ represent the portion of the expansion of interest.

$$F_1(s) = \frac{M\epsilon^{j\theta}}{2j\omega(s + \alpha - j\omega)} + \frac{M\epsilon^{-j\theta}}{-2j\omega(s + \alpha + j\omega)} \tag{C-18}$$

The inverse transform of $F_1(s)$ is $f_1(t)$:

$$f_1(t) = \left[\frac{M\epsilon^{j(\omega t + \theta)} - M\epsilon^{-j(\omega t + \theta)}}{2j\omega} \right] \epsilon^{-\alpha t}$$

$$= \frac{M}{\omega} \epsilon^{-\alpha t} \sin(\omega t + \theta) \tag{C-19}$$

Appendix D

TABLE OF
LAPLACE TRANSFORMS

D-1 LAPLACE TRANSFORMS

$f(t)$	$F(s) = \mathscr{L}[f(t)]$	
$\delta(t)$	1	(T-1)
1 or $u(t)$	$\dfrac{1}{s}$	(T-2)
t	$\dfrac{1}{s^2}$	(T-3)
$\epsilon^{-\alpha t}$	$\dfrac{1}{s + \alpha}$	(T-4)
$\sin \omega t$	$\dfrac{\omega}{s^2 + \omega^2}$	(T-5)
$\cos \omega t$	$\dfrac{s}{s^2 + \omega^2}$	(T-6)
$\epsilon^{-\alpha t} \sin \omega t$	$\dfrac{\omega}{(s + \alpha)^2 + \omega^2}$ ª	(T-7)
$\epsilon^{-\alpha t} \cos \omega t$	$\dfrac{s + \alpha}{(s + \alpha)^2 + \omega^2}$ ª	(T-8)
t^n	$\dfrac{n!}{s^{n+1}}$	(T-9)
$\epsilon^{-\alpha t} t^n$	$\dfrac{n!}{(s + \alpha)^{n+1}}$	(T-10)

ªComplex roots.

D-2 TRANSFORM OPERATIONS

$f(t)$	$F(s)$	
$f'(t)$	$sF(s) - f(0)$	(O-1)
$\int_0^t f(t)\,dt$	$\dfrac{F(s)}{s}$	(O-2)
$f(t-a)u(t-a)$	$\epsilon^{-as}F(s)$	(O-3)
$\epsilon^{-\alpha t}f(t)$	$F(s+\alpha)$	(O-4)
$tf(t)$	$\dfrac{-dF(s)}{ds}$	(O-5)
$f(at)$	$\dfrac{1}{a}F\left(\dfrac{s}{a}\right)$	(O-6)
$\lim\limits_{t\to 0} f(t)$ (initial value)	$\lim\limits_{s\to\infty} sF(s)$	(O-7)
$\lim\limits_{t\to\infty} f(t)$ (final value)	$^a\lim\limits_{s\to 0} sF(s)$	(O-8)

[a]If poles of $sF(s)$ are located in left-hand half-plane.

ANSWERS TO SELECTED ODD-NUMBERED PROBLEMS

Chapter 2

2-1. $30u(t)$

2-3. $12u(t - 5)$

2-5. $60u(t + 8)$

2-7. $(5t + 8t^2)u(t)$

2-9. $(2t + 6)u(t)$

2-11. $8tu(t)$

2-13. -0.5×10^6 V/s

2-15. $(6 + 8t)u(t)$

2-17. $(4t + 3t^2)u(t)$

2-19. (a) 10^6, 1 μs (b) 0.2, 5 s
(c) 0.02, 50 s (d) 25, 0.04 s

2-21. (a) $80\epsilon^{-500t}$
(b) 48.52 V, 29.43 V, 10.83 V,
0.54 V

2-23. (a) 50 V (b) 2000 rad/s
(c) 318.3 Hz (d) 3.142 ms

2-25. $50 \sin 1000\pi t$

2-27. $12 \cos 3tu(t)$

2-29. (a) $6 \cos (\omega t - 40°)$
(b) $80 \cos (\omega t - 125°)$
(c) $2 \sin (\omega t + 125°)$
(d) $9 \cos (\omega t + 70°)$
(e) $8 \cos (\omega t + 65°)$
(f) $4 \cos (\omega t - 130°)$

(g) $2 \cos (\omega t - 120°)$
(h) $7 \cos (\omega t + 80°)$

2-31. (a) $6 \sin (\omega t + 140°)$
(b) $80 \sin (\omega t + 55°)$
(c) $2 \sin (\omega t - 55°)$
(d) $9 \sin (\omega t - 110°)$
(e) $8 \sin (\omega t - 115°)$
(f) $4 \sin (\omega t + 50°)$
(g) $2 \sin (\omega t - 150°)$
(h) $7 \sin (\omega t + 80°)$

2-33. (a) e_2 leads e_1 by 90°
(b) e_2 leads e_1 by 120°
(c) i_1 leads i_2 by 50°
(d) i_2 leads i_1 by 110°

2-35. $5 \sin (100t + 53.13°)$

2-37. $9.4849 \sin (\omega t + 138.09°)$

2-39. (a) Envelope has a time constant
of 0.5 s. Oscillation has a frequency of 1 Hz.
(b) Envelope has a time constant
of 1 s. Oscillation has a frequency of 100 Hz.

2-41. $5(t - 4)u(t - 4)$

2-43. Ramp starting at $t = 2$ s with a
slope of 8 V/s.

2-45. $160(t - 0.05)u(t - 0.05)$

2-47. $2 \sin [100\pi(t - 0.02)]u(t - 0.02)$

2-49. $200(t - 0.01)u(t - 0.01)$

2-51. Pulse starting at $t = 3$ s with a width of 4 s and a height of 2.

2-53. Current ramps upward from 0 at $t = 0$ to 5 A at $t = 1$ s, remains at 5 A until $t = 2$ s, ramps upward to 10 A at $t = 3$ s, and remains at 10 A thereafter.

2-55. $20u(t - 0.01) - 20u(t - 0.016)$

2-57. $5u(t) + 5u(t - 2) + 5u(t - 4) - 15u(t - 6)$

2-59. $\dfrac{Et}{T} u(t) - E \sum\limits_{n=1}^{\infty} u(t - nT)$

2-61. $2 \sin 200\pi t u(t) + 2 \sin 200\pi(t - 0.005)u(t - 0.005)$

2-63. $0.01\delta(t)$

2-65. (a) Pulse of height 2 V and width 1 s.

(b) Pulse of height 20 V and width 0.1 s.

(c) $2\delta(t)$

2-67. $50\delta(t)$

2-69. $0.01u(t)$

2-71. $20\delta(t) - 20\delta(t - 0.016)$

2-73. $2u(t) - 2u(t - 1) - 2u(t - 3) + 2u(t - 4)$

2-75. $20(t - 0.01)u(t - 0.01) - 20(t - 0.016)u(t - 0.016)$

2-77. $5tu(t) + 5(t - 2)u(t - 2) + 5(t - 4)u(t - 4) - 15(t - 6)u(t - 6)$

2-79. (a) 1.6 V, 6.928 V   (b) 9.6 W

2-87. $\dfrac{\tau}{T} E, \sqrt{\dfrac{\tau}{T}} E$

2-89. $\dfrac{E}{\pi}, \dfrac{E}{2}$

Chapter 3

3-1. $i_1(t) = 2.30769u(t) + 0.923077tu(t)$
$i_2(t) = 0.769231u(t) + 2.30769tu(t)$

3-3. $v_1(t) = 6.85714u(t) - 2.85714tu(t)$
$v_2(t) = 2.28571u(t) - 14.2857tu(t)$

3-5. $4 \sin \omega t$

3-7. Voltage source of value $8u(t) + 6tu(t)$ in series with 7 Ω.

3-9. 24 mA

3-11. 16 V

3-13. $8 \times 10^{-6}\epsilon^{-10t}$

3-15. $20t - 6\epsilon^{-4t} + 6$

3-17. $20t - 6\epsilon^{-4t} + 31$

3-19. Positive pulse of 6 A from $t = 0$ to $t = 2$ s followed by negative pulse of -3 A from $t = 2$ s to $t = 6$ s.

3-21. $6u(t) - 9u(t - 2) + 3u(t - 6)$

3-23. Voltage ramps upward from 0 to 60 V at $t = 2$ s, changes slope and ramps upward to 90 V at $t = 4$ s, ramps downward to 30 V at $t = 8$ s, and remains at 30 V for $t > 8$ s.

3-25. $30tu(t) - 15(t - 2)u(t - 2) - 30(t - 4)u(t - 4) + 15(t - 8)u(t - 8)$

3-27. 6-V step voltage source ($+$ terminal up) in series with 3 F.

3-29. Impulse current source of weight 18 (directed upward) in parallel with 3 F.

3-31. 12 V

3-33. 3 A

3-35. $24 \cos 3t$

3-37. $0.01 \sin 4000t$

3-39. Positive pulse of 4 V from $t = 5$ ms to $t = 10$ ms followed by negative pulse of -4 V from $t = 10$ ms to $t = 15$ ms. Impulse at $t = 5$ ms with area 0.02 V·s and impulse at $t = 15$ ms with area -0.02 V·s

3-41. $0.02\delta(t - 5 \times 10^{-3}) + 4u(t - 5 \times 10^{-3}) - 8u(t - 10 \times 10^{-3}) + 4u(t - 15 \times 10^{-3}) - 0.02\delta(t - 15 \times 10^{-3})$

3-43. 2 A step current source (directed downward) in parallel with 4 H.

3-45. Impulse voltage source of weight 8 (+terminal down) in series with 4 H.

3-47. 16 cos 20t, 24 cos 20t

3-49. 28 cos 20t, 56 cos 20t

3-51. 0.75

3-53. $i_1 = 4 \sin \omega t$, $i_2 = 2 \sin \omega t$, $v_1 = 48 \sin \omega t$, $v_2 = 96 \sin \omega t$

3-59. Values of reflected components are 0.1 μF, 200 μH, and 150 Ω

3-61. $2tu(t) + 12u(t)$

3-63. $5 \sin (2t + 53.13°)$

3-65. $(4 + 2t^2)u(t)$

Chapter 4

4-1. $L\dfrac{d^2i}{dt^2} + R\dfrac{di}{dt} + \dfrac{i}{C} = \omega E \cos \omega t$,

2nd order

4-3. $R_1 LC\dfrac{d^2 i_2}{dt^2} + (L + R_1 R_2 C)\dfrac{di_2}{dt} +$

$(R_1 + R_2)i_2 = E \sin \omega t$, 2nd order

4-5. A sin ωt

4-7. A

4-9. A cos ωt

4-11. $i_{C1} = 0$, $i_{C2} = 0$, $v_L = 0$, $i_{R1} = i_L = i_{R2} = 1.5$ A, $v_{R1} = 12$ V, $v_{C1} = v_{C2} = v_{R2} = 18$ V

4-13. $i_C = i_{L2} = i_{R3} = 0$, $v_{L1} = 0$, $v_{L2} = 0$, $v_{R3} = 0$, $i_{R1} = i_{L1} = i_{R2} = 1.6$ A, $v_{R1} = 24$ V, $v_{R2} = v_C = 56$ V

4-15. $v_{L1} = 0$, $v_{L2} = 0$, $i_C = 0$, $i_{L1} = i_{R2} = i_{R3} = 1$ A, $i_{L2} = i_{R1} = 2$ A, $v_{R1} = 40$ V, $v_{R2} = 15$ V, $v_{R3} = 25$ V, $v_C = 15$ V

4-17. (a) $i_C = i_R = 2$ A, $v_C = 0$, $v_R = 20$ V
(b) $i_C = i_R = 0$, $v_R = 0$, $v_C = 20$ V

4-19. (a) $i_C = i_R = 5$ A, $v_C = -20$ V, $v_R = 50$ V
(b) $i_C = i_R = 0$, $v_R = 0$, $v_C = 30$ V

4-21. (a) $i_L = i_C = i_R = 2$ A, $v_C = 6$ V, $v_R = 8$ V, $v_L = -14$ V
(b) All voltages and currents are zero.

4-23. (a) $v_C = 12$ V
(b) $v_C = 12$ V, $i_R = i_C = -5$ A, $v_R = -20$ V
(c) $i_R = i_C = 0$, $v_R = 0$, $v_C = -8$ V

4-25. (a) $i_L = 2$ A, $v_C = 20$ V
(b) $i_L = 2$ A, $v_C = 20$ V, $v_{R1} = 60$ V, $i_{R1} = 12$ A, $v_{R2} = 20$ V, $i_{R2} = 2$ A, $i_C = 0$, $v_L = 40$ V
(c) $v_{R1} = v_{R2} = v_C = 60$ V, $v_L = 0$, $i_{R1} = 12$ A, $i_C = 0$, $i_L = i_{R2} = 6$ A

4-27. (a) $i_L = 3$ A, $v_C = 27$ V
(b) $i_{R1} = i_L = 3$ A, $v_{C1} = v_{R2} = v_{R3} = 27$ V, $v_{R4} = 0$, $i_{R4} = 0$, $v_{R1} = 9$ V, $v_L = 0$, $i_{R2} = 3$ A, $i_{R3} = i_{C2} = 4.5$ A, $i_{C1} = -4.5$ A
(c) $v_L = 0$, $i_{C1} = 0$, $i_{C2} = 0$, $i_{R1} = i_L = 4$ A, $v_{R1} = 12$ V, $v_{C1} = v_{R3} = 24$ V, $i_{R2} = 2.667$ A, $i_{R3} = i_{R4} = 1.333$ A, $v_{R3} = 8$ V, $v_{R4} = 16$ V

4-29. $i(t) = 5\epsilon^{-5t}$, $v_C(t) = 10(1 - \epsilon^{-5t})$

4-31. $i(t) = 7\epsilon^{-5t}$, $v_C(t) = 10 - 14\epsilon^{-5t}$

4-33. $i(t) = 3\epsilon^{-2t}$, $v_C(t) = 12 - 18\epsilon^{-2t}$

4-35. $80 - 32\epsilon^{-2t}$

4-37. (b) $2\dfrac{di}{dt} + 3i + 4\displaystyle\int_0^t i\,dt = e(t)$
(c) $i(0^+) = 0$

4-39. (b) $6i_1 + 2\displaystyle\int_0^t i_1\,dt - 2\displaystyle\int_0^t i_2\,dt = 20 \cos 5t - 8 - 2\displaystyle\int_0^t i_1\,dt +$

$2\displaystyle\int_0^t i_2\,dt + 2\dfrac{di_2}{dt} + 4i_2 = 8$
(c) $i_1(0^+) = 2$ A, $i_2(0^+) = -3$ A

4-41. (b) $\dfrac{v}{4} + 2\dfrac{dv}{dt} + 3\displaystyle\int_0^t v\,dt = i(t) - 2$
(c) $v(0^+) = 5$ V

Chapter 5

5-1. $\dfrac{8}{s}$

5-3. $\dfrac{5}{s^2}$

5-5. $\dfrac{60}{s^2 + 9}$

5-7. $\dfrac{12}{s^2 + 4s + 13}$

5-9. $\dfrac{5.196s + 9}{s^2 + 9}$

5-11. $\dfrac{5.196s + 19.392}{s^2 + 4s + 13}$

5-13. $\dfrac{72}{(s + 5)^4}$

5-15. $\dfrac{5}{s^2}\, \epsilon^{-2s}$

5-17. $\dfrac{5(1 - \epsilon^{-2s} - \epsilon^{-4s} + \epsilon^{-6s})}{s^2}$

5-19. (a) -4; real, first order; $2(s + 4)$
(b) $-1, -4$; real, first order; $4(s + 1)(s + 4)$
(c) $-2 \pm j5$; complex, first order; $(s + 2 - j5)(s + 2 + j5)$
(d) $-3, -3$; real, second order; $(s + 3)^2$
(e) $\pm j5$; imaginary, first order; $(s - j5)(s + j5)$

5-21. $8\epsilon^{-3t}$

5-23. $4 \sin 8t$

5-25. $6 \cos 3t + 5 \sin 3t$

5-27. $4\epsilon^{-2t} \cos 5t - 6\epsilon^{-2t} \sin 5t$

5-29. $\delta(t) + \epsilon^{-t}$

5-31. $5\epsilon^{-2t} + 4\epsilon^{-3t}$

5-33. $4 + 2\epsilon^{-t} - 3\epsilon^{-2t}$

5-35. $2\epsilon^{-2t} + 3\epsilon^{-4t}$

5-37. $0.2\epsilon^{-t} + 0.6325\epsilon^{-2t} \sin (3t - 18.43°)$

5-39. $-0.5172\epsilon^{-2t} + 2.841 \sin (5t + 10.49°)$

5-41. $40 - 100\epsilon^{-t} + 63.246\epsilon^{-t} \sin (t + 108.43°)$

5-43. $0.15\epsilon^{-t} - 0.058824\epsilon^{-2t} + 0.10931\epsilon^{-3t} \sin (4t - 123.48°)$

5-45. $5t^3\epsilon^{-2t}$

5-47. $20\epsilon^{-t} - (20t + 20)\epsilon^{-2t}$

Chapter 6

6-1. $Z(s) = \dfrac{5}{s}$, $Y(s) = 0.2s$

6-3. $Z(s) = 4s$, $Y(s) = \dfrac{0.25}{s}$

6-5. Voltage source $30/s$ (+ at top) in series with $4/s$. Current source 7.5 (directed upward) in parallel with $4/s$.

6-7. Current source $4/s$ (directed downward) in parallel with $2s$. Voltage source 8 (+ at bottom) in series with $2s$.

6-9. $(5 + 2s)I_1(s) - 2sI_2(s) = \dfrac{10s}{s^2 + 4} + 6 - 2sI_1(s) + \left(8s + 4 + \dfrac{10}{s}\right) I_2(s) = -18 + \dfrac{6}{s} - \dfrac{60}{s^2 + 9}$

6-11. (a) Voltage source $40/[s(s + 0.6667)]$ in series with impedance $8/(s + 0.6667)$.
(b) $3.333(1 - \epsilon^{-2t})$

6-13. $\dfrac{2(s^2 + s + 1)}{s + 1}$

6-15. $\dfrac{5(s + 1)}{s(s^2 + s + 1)}$

6-17. (a) $v(t) = A_1 + A_2\epsilon^{-t} + A_2\epsilon^{-3t}$
(b) First term is steady-state. Remainder is transient.

6-19. (a) $v(t) = B_1\epsilon^{-3t} \sin (4t + \theta_1) + B_2 \sin (2t + \theta_2)$
(b) First term is transient. Second term is steady-state.

6-21. $i(t) = 5\epsilon^{-5t}$, which is transient. Steady-state is zero. $v_C(t) = 10 - 10\epsilon^{-5t}$, First term is steady-state. Second term is transient.

6-23. $i(t) = 2.5\epsilon^{-5t} + 3.536 \sin(5t + 135°)$
$v_C(t) = -5\epsilon^{-5t} + 7.071 \sin(5t + 45°)$. First term of each function is transient, and second term is steady-state.

6-25. $i(t) = 5 - 5\epsilon^{-3t}$
First term is steady-state, and second term is transient. $v_L(t) = 30\epsilon^{-3t}$, which is transient. Steady-state is zero.

6-27. $i(t) = -1.8\epsilon^{-3t} + 3 \sin(4t + 36.87°)$
$v_L(t) = 10.8\epsilon^{-3t} + 24 \sin(4t + 126.87°)$
First term of each function is transient, and second term is steady-state.

6-29. $16.706 + 4.851 \sin(4t - 75.96°)$

6-31. underdamped; 3, 5 rad/s, 4 rad/s

6-33. overdamped; 1000, 2000

6-35. $5\epsilon^{-t} \sin 2t$, $20 + 22.36\epsilon^{-t} \sin(2t - 116.57°)$

6-37. $2.236 \sin(5t - 63.43°) + 2.5\epsilon^{-t} \sin(2t + 126.87°)$

6-39. overdamped; 417, 9583

6-41. underdamped; 25×10^3, 10^5 rad/s, 96.825×10^3 rad/s

6-43. $20\epsilon^{-t} \sin 2t$

6-45. 4th order

6-47. 5th order

6-49. $10 + 14.14 \sin(0.7071t - 135°)$

6-51. $10 + 11.547\epsilon^{-0.5t} \sin(0.8660t - 120°)$

6.53. $6 + 8\epsilon^{-t} - 2\epsilon^{-2t}$

6-55. $0.39528 \sin(t + 161.57°) + 1.7129\epsilon^{-2t} \sin(3t - 4.1849°)$

Chapter 7

7-1. $\dfrac{2}{s + 2}$, $2\epsilon^{-2t}$

7-3. $\dfrac{2}{s + 2}$, $2\epsilon^{-2t}$

7-5. $\dfrac{5}{s^2 + 2s + 5}$, $2.5\epsilon^{-t} \sin 2t$

7-7. $\dfrac{1.5858}{s^2 + 1.4142s + 1}$

7-9. $-10\epsilon^{-4t} + 14.14 \sin(4t + 45°)$

7-15. $\dfrac{12.5(s + 2)}{s^2 + 4s + 13}$

7-17. $\dfrac{d^2v_2}{dt^2} + 3\dfrac{dv_2}{dt} + 4v_2 = 2\dfrac{d^2v_1}{dt^2} + 5v_1$

7-19. $1 - \epsilon^{-2t}$

7-21. $22.36\epsilon^{-t} \sin(2t + 86.565°)$

7-23. (a) $\dfrac{s + 1}{s^2 + s + 1}$ (b) $\dfrac{s^2 + s + 1}{s + 1}$

7-25. $\dfrac{5}{s^2 + 2s + 5}$

7-27. Zeros: $-1, -2, \infty$; Poles: $0, -3, -4$

7-29. Zeros: $0, 1 \pm j2, \infty$; Poles: $\pm j2, -2 \pm j3$

7-31. Zeros: $0, -2, 2, \infty(5)$; Poles: $\pm j$ (2 each), $\pm j\sqrt{2}, \pm j\sqrt{3}$

7-33. $G(s) = \dfrac{51(s + 2)^2(s^2 + 1)}{(s + 3)(s + 4)(s^2 + 2s + 17)}$

7-35. $G(s) = \dfrac{10s(s + 2)}{(s + 3)(s^2 + 4s + 13)}$

7-37. Stable

7-39. Unstable

7-41. $\dfrac{G_1(s)G_2(s)}{1 + G_1(s)G_2(s)G_3(s)}$

7-43. $1 - \epsilon^{-5t}$

7-47. $\dfrac{5(s^2 + 2s + 5)}{s(s + 2.5)}$

7-49. $\dfrac{3(s + 2)(s + 4)}{(s + 1)(s + 3)}$

Chapter 8

8-1. $i(t) = 2 \sin (5t + 53.13°)$,
$v_R(t) = 12 \sin (5t + 53.13°)$
$v_L(t) = 24 \sin (5t + 143.13°)$,
$v_C(t) = 40 \sin (5t - 36.87°)$

8-3. $i_R(t) = 0.25 \sin 200t$,
$i_C(t) = 0.5 \sin (200t + 90°)$,
$i_L(t) = 0.125 \sin (200t - 90°)$,
$i_0(t) = 0.45069 \sin (200t + 56.310°)$

8-5. (a) $R = 60 \, \Omega$, $X = 25 \, \Omega$
(c) $G = 0.014201$ S,
$B = -0.0059172$ S

8-7. (a) $R(\omega) = \dfrac{4}{1 + \omega^2}$,
$X(\omega) = \dfrac{2\omega(\omega^2 - 1)}{1 + \omega^2}$

8-9. Voltage source $30/180°$ in series with impedance $-j24$.

8-11. (a) $-4.8\epsilon^{-4t} + 6 \sin (3t + 53.13°)$
$14.4\epsilon^{-4t} + 24 \sin (3t - 36.87°)$
(b) $6 \sin (3t + 53.13°)$
$24 \sin (3t - 36.87°)$

8-13. $G(j\omega) = \dfrac{2}{2 + j\omega}$, $A(\omega) = \dfrac{2}{\sqrt{4 + \omega^2}}$,
$A_{db}(\omega) = 20 \log_{10} \left(\dfrac{2}{\sqrt{4 + \omega^2}} \right)$,
$\beta(\omega) = -\tan^{-1} \dfrac{\omega}{2}$

8-15. Flat level of 20 dB for $\omega <$ 5 rad/s. Slope of -6 dB/octave for $\omega > 5$ rad/s.

8-17. Flat level of -20 dB for $\omega <$ 2 rad/s. Slope of 6 dB/octave for 2 rad/s $< \omega < 20$ rad/s. Flat level of 0 dB for $\omega > 20$ rad/s

8-19. Flat level of 34 dB for $\omega <$ 10 rad/s. Slope of -6 dB/octave for 10 rad/s $< \omega < 100$ rad/s. (Drops to a level of 14 dB at $\omega = 100$ rad/s.) Slope of -12 dB/octave for $\omega > 100$ rad/s.

Chapter 9

9-1. dc, 500, 1000, 1500, 2000 Hz

9-3. (a) $5 + 10 \sin (40\pi t + 36.87°) +$
$13 \sin (80\pi t - 22.62°)$
(b) $5 + (3 - j4)\epsilon^{j40\pi t} +$
$(3 + j4)\epsilon^{-j40\pi t} +$
$(-2.5 - j6)\epsilon^{j80\pi t} +$
$(-2.5 + j6)\epsilon^{-j80\pi t}$

9-5. (a) $6 + 5 \cos 100\pi t +$
$8.660 \sin 100\pi t -$
$6.928 \cos 200\pi t - 4 \sin 200\pi t$
(b) $6 + (2.5 - j4.330)\epsilon^{j100\pi t} +$
$(2.5 + j4.330)\epsilon^{-j100\pi t} +$
$(-3.464 + j2)\epsilon^{j200\pi t} +$
$(-3.464 - j2)\epsilon^{-j200\pi t}$

9-7.

f, kHz	1	3
Amplitude, V	12.732	-4.244

5	7	9
2.546	-1.819	1.415

9-11. (a) $0.01 \dfrac{\sin 10^{-3} \pi f}{10^{-3} \pi f}$

9-13. (a) $1/n$ (b) $1/n^2$ (c) $1/n$
(d) $1/n^2$ (e) $1/n^2$ (f) $1/n$

Chapter 10

10-1. 256

10-3. 11

10-5. 100, 300, 700, 900, 1100, 1300 Hz; 1.7, 1.9, 2.1, 2.3, 2.7, 2.9, 3.1, 3.3, 3.7, 3.9, 4.1, 4.3 kHz

10-7. (a) 10 kHz (b) 100 μs

10-9. $\dfrac{0.5z - 0.7}{z - 0.8}$

10-11. $y(n) = 2x(n) + 3x(n - 1) + 0.5y(n - 1)$

10-13. $x(0) = 5$, $x(1) = 2$, $x(3) = 4$, $x(n) = 0$ for all other values of n.

10-15. $y(0) = 3$, $y(1) = 5$, $y(2) = -6.5$

10-17. $y(n) = -10(0.2)^n + 20(0.4)^n$

10-19. $y(n) = 13.3333(-1)^n - 5(-0.5)^n + 1.6667(0.5)^n$

10-21. (a) $\dfrac{z + 0.5}{z - 0.5}$

(b) $h(n) = -\delta(n) + 2(0.5)^n$

(c) stable

10-23. $y(n) = -20(0.5)^n + 30$

10-25. (a) $\dfrac{T}{z - 1}$

10-27. (a) $y(n) = \dfrac{T}{1 - \epsilon^{-\alpha T}}(1 - \epsilon^{-n\alpha T})$

(b) $\dfrac{1}{\alpha}(1 - \epsilon^{-\alpha t})$

10-29. (a) $\dfrac{Tz}{z - K}$

10-31. Stable

10-33. Marginally stable

INDEX